DESIGN, CONTROL, AND APPLICATION OF MODULAR MULTILEVEL CONVERTERS FOR HVDC TRANSMISSION SYSTEMS

DESIGN, CONTROL, AND APPLICATION OF MODULAR MULTILEVEL CONVERTERS FOR HVDC TRANSMISSION SYSTEMS

Kamran Sharifabadi

Research & Technology, Statoil ASA, Norway

Lennart Harnefors

ABB Corporate Research, Sweden

Hans-Peter Nee

School of Electrical Engineering, KTH Royal Institute of Technology, Sweden

Staffan Norrga

School of Electrical Engineering, KTH Royal Institute of Technology, Sweden

Remus Teodorescu

Department of Energy Technology, Aalborg University, Denmark

This edition first published 2016
© 2016 John Wiley & Sons

Registered office
John Wiley & Sons Ltd, The Atrium, Southern Gate, Chichester, West Sussex, PO19 8SQ, United Kingdom

For details of our global editorial offices, for customer services and for information about how to apply for permission to reuse the copyright material in this book please see our website at www.wiley.com.

The right of the author to be identified as the author of this work has been asserted in accordance with the Copyright, Designs and Patents Act 1988.

Library of Congress Cataloging-in-Publication Data

Names: Sharifabadi, Kamran, 1963- author. | Harnefors, Lennart, 1968- author. | Nee,
    Hans-Peter, 1963- author. | Norrga, Staffan, 1968- author. | Teodorescu,
    Remus, author.
Title: Design, control, and application of modular multilevel converters for
    HVDC transmission systems / Kamran Sharifabadi, Research & Technology,
    Statoil ASA, Norway, Lennart Harnefors, ABB Corporate Research, Sweden, Hans-Peter Nee,
    KTH, Sweden, Staffan Norrga, KTH, Sweden, Remus
    Teodorescu, Aalborg University, Denmark.
Description: Chichester, West Sussex, United Kingdom : John Wiley & Sons,
    Inc., 2016. | Includes bibliographical references and index.
Identifiers: LCCN 2016011610 (print) | LCCN 2016013688 (ebook) | ISBN
    9781118851562 (cloth) | ISBN 9781118851524 (pdf) | ISBN 9781118851548
    (epub)
Subjects: LCSH: Electric current converters–Design and construction. |
    Electric current converters–Automatic control. | Electric power
    transmission–Direct current–Equipment and supplies.
Classification: LCC TK2796 .S53 2016 (print) | LCC TK2796 (ebook) | DDC
    621.31/7–dc23
LC record available at http://lccn.loc.gov/2016011610

A catalogue record for this book is available from the British Library.

Typeset in 10/12pt TimesLTStd by SPi Global, Chennai, India

1   2016

Contents

Preface

State of art power electronics plays a major role in our society and our daily life. Voltage-source converter (VSC) technologies are utilized in applications such as electric vehicles, variable-speed drives, high-voltage direct current (HVDC) transmission and flexible alternating-current transmission systems (FACTS). Nowadays, converter technologies are an integral part of all renewable power generation technologies such as wind turbine generators and photovoltaic generation units. VSC technologies are key elements for achieving enhanced system yield in power generation and power transmission.

During the past decade, multilevel VSC technology has dominated the market, but the new modular multilevel converter (MMC) technology adopted by the industry in recent years has demonstrated clear advantages with inherent properties such as built-in redundancy, higher efficiency, and lower harmonic content. State-of-the-art MMCs are core components in FACTS equipment and VSC-HVDC schemes.

The fast speed of MMC technology development has left a knowledge gap between the technology specialists and VSC-HVDC project developers and key personnel involved in those projects. There is a need for field engineers, project developers, and those involved in the development and management of these projects and graduate students to gain insights into the design, control, and application of this new MMC technology. In this respect, what is needed is a reference book to fill the gap between field engineers and HVDC specialists.

This book is a result of collaboration between experts from industry and academia. It provides theoretical insights into the design and control of MMC technology and investigate practical aspects of the project planning, design, manufacture, implementation, and commissioning of MMC-HVDC and multi-terminal HVDC transmission technologies.

Separated into three distinct parts, the first part of the book offers an overview of MMC technology, including information on converter component sizing, control and communication, protection and fault management, and generic modeling and simulation. The second part covers the applications of MMC-HVDC transmission technology in offshore wind power plants including planning, technical, and economic requirements and optimization options, fault management, and dynamic and transient stability. Finally, the third part explores the applications of MMC in HVDC transmission and multi-terminal configurations, including supergrids.

The authors hope the reader will find the book useful and stimulating and wish the reader an interesting journey into the world of MMCs.

Acknowledgements

The authors are deeply indebted to Dr. Kalle Ilves, ABB AB, Sweden. He generally contributed to chapters 1, and 5 and brought many valuable viewpoints on the text and figures. He also, more specifically, contributed material concerning the predictive submodule sorting methods for Chapter 5. Furthermore, the authors would also like to express their gratitude to Mr. Arman Hassanpoor, ABB AB, Sweden who provided valuable material to the section concerning tolerance-band submodule balancing methods in Chapter 5.

The authors are also acknowledging Massimo Bongiorno, Professor, Chalmers University, for general support to Chapter 4 and Laszlo Mathe, Associate Professor, Aalborg University, for general support to the development of the simulation models.

About the Companion Website

Don't forget to visit the companion website for this book:

www.wiley.com/go/Sharifabadi/ModularConverters

The companion website material consists of a collection of self-explanatory simulation models in Matlab/Simulink and PLECS for the MMC following closely the theory of control, modulation and modelling described in the chapters 3,4,5 and 6.

Scan this QR code to visit the companion website

Nomenclature

A list of the important symbols that used in this book can be found in Tables 1–4. A list of acronyms can be found in Tables 5–7.

Table 1 Superscripts, subscripts, circumflexes, and prefixes.

$*$	Complex conjugate
$\star$	Reference
f,F	Filtered value
0	Nominal value
$+$	Positive sequence
$-$	Negative sequence
u,l	Upper, lower arm
0	Mean value, zero-sequence component
a,b,c	Phases a, b, c
α,β	Components of the stationary $\alpha\beta$ reference frame
d,q	Components of the synchronous dq reference frame
Σ	Sum
Δ	Difference
$\overline{\cdot}$	Mean value
$\hat{\cdot}$	Peak value, estimated value
$\tilde{\cdot}$	Difference
Δ	Ripple quantity, parasitic quantity, difference, increment

Table 2   Variables.

h	Signed multiple of the fundamental frequency
i	Submodule index
k	Phase number
m	Carrier index
n	Sample index, sideband index
$s = d/dt$	Differential operator (or, where appropriate, complex Laplace variable)
t	Time
z	Forward-shift operator (or, where appropriate, complex z-transform variable)
$\delta = z - 1$	Delta operator
$i_{u,l}$	Arm current
$i_s = i_u - i_l$	Output current
$i_c = (i_u + i_l)/2$	Circulating current
i_d	DC-bus current
$e = i_s^\star - i_s$	Output-current control error
e'	Modified control error
$v_{cu,l}^i$	Capacitor voltage in submodule i
$v_{cu,l}^\Sigma = \sum_{i=1}^N v_{cu,l}^i$	Sum capacitor voltage per arm
$v_{cu}^\Sigma = v_{cu}^\Sigma + v_{cl}^\Sigma$	Sum capacitor voltage per phase
$v_c^\Delta = v_{cu}^\Sigma - v_{cl}^\Sigma$	Imbalance sum capacitor voltage
$n_{u,l}^i$	Submodule insertion index
$n_{u,l} = \left(\sum_{i=1}^N n_{u,l}^i \right) /N$	Insertion index per arm
$v_{u,l} = \sum_{i=1}^N n_{u,l}^i v_{cu,l}^i$	Inserted arm voltage
$v_s = (-v_u + v_l)/2$	Output voltage
$v_c = (v_u + v_l)/2$	Internal voltage
v_a	AC-bus voltage
v_g	Grid voltage
v_{PCC}	PCC voltage
$v_{du,l}$	Pole-to-ground dc-bus voltage
$v_d = v_{du} + v_{dl}$	Pole-to-pole dc-bus voltage
$v_d^\Delta = v_{du} - v_{dl}$	Imbalance dc-bus voltage
v_R	R-part output
$W_{u,l} = C(v_{cu,l}^\Sigma)^2/(2N)$	Stored energy per arm
$W_\Sigma = W_u + W_l$	Stored energy per phase
$W_\Delta = W_u - W_l$	Imbalance stored energy
$W_d = C_d' v_d^2/2$	Effective stored dc-bus energy
P	Active output power
P_d	DC-side input power
Q	Reactive output power
ω	Instantaneous angular frequency of the control-system dq frame
$\theta = \int \omega \, dt$	Angle of the control-system dq frame

Table 3 Parameters and functions.

f_1	Fundamental frequency
$\omega_1 = 2\pi f_1$	Fundamental angular frequency
f_s	Sampling frequency
f_{sw}	Switching frequency
$T_s = 1/f_s$	Sampling period
T_c	Computational time delay
$T_d = T_c + 0.5T_s$	Total time delay
K	Space-vector scaling constant
M	Number of phases
N	Number of submodules per arm
C	Submodule capacitance
C_d	Installed dc-bus capacitance
$C_d' = C_d + 2MC/N$	Effective dc-bus capacitance
L	Arm inductance
R	Parasitic arm resistance
R_I	Insertion resistance
$\widehat{V}_s$	Peak value, fundamental component of v_s
$\widehat{I}_s$	Peak value, fundamental component of i_s
$\widehat{V}_{max}$	Maximum allowed $\widehat{V}_s$
$\widehat{I}_{max}$	Maximum allowed $\widehat{I}_s$
φ	Phase angle (lagging) of current relative voltage
φ_h	Phase angle of order-h symmetric component
δ_a	AC-bus-voltage phase angle
δ_g	Grid-voltage phase angle
θ_1	Voltage-reference phase shift
ω_c	Carrier angular frequency
θ_c	Carrier phase shift
m_f	Frequency ratio
m_a	Modulation index
C_h	Complex Fourier series coefficient
C_{mn}	Double complex Fourier series coefficient
J_n	Bessel function of order n
L	Sorted list of submodules
Re	Real part
Im	Imaginary part
sat	Saturation function
satv	Vectorial saturation function

Table 4   Controller parameters and transfer functions.

α_b	PLL low-pass-filter bandwidth
α_c	Output-current control-loop bandwidth
α_d	DC-bus-voltage control-loop bandwidth
α_f	Voltage-feedforward-filter bandwidth
α_h	R-part bandwidth
α_{id}	DC-bus-voltage integrator bandwidth
α_{ip}	PLL integrator bandwidth
α_l	Power-synchronization control low-pass-filter bandwidth
α_p	PLL bandwidth
α_s	Power-synchronization control-loop bandwidth
K_h	R-part gain
K_i	I-part gain
K_p	P-part gain
K_s	Power-synchronization-control gain
K_v	Voltage droop gain
R_a	"Active resistance" for circulating-current control
R_s	"Active resistance" for power-synchronization control
ϕ_h	Compensation angle for resonant filter
F_h	R part
G_c	Closed-loop system
G_k	Open-loop system
H_h	Resonant filter
H_p	PLL low-pass filter

Table 5 Acronyms A—G.

Ag	Silver
ALA	Arm-Level Averaged
ALA-BLK	Arm-Level Averaged with Blocking capabilities
Al	Aluminum
AlN	Aluminum Nitride
AlSiC	Aluminum Silicon Carbide
APOD	Alternative Phase Opposite Disposition
APS	Auxiliary Power Supply
ARCP	Auxiliary Resonant-Commutated Pole
ATB	Average-voltage Tolerance Band
BCA	Bilateral Connection Agreement
BFOM	Baliga's Figure Of Merit
BJT	Bipolar Junction Transistor
BLK	Blocking
BPF	Band-Pass Filter
CCC	Capacitor-Commutated Converter
CCU	Central Control Unit
CTB	Cell-voltage Tolerance Band
CTBoptimized	optimized Cell-voltage Tolerance Band
CTE	Coefficient of Thermal Expansion
CUSC	Connection and Use of System Code
D	Derivative
DBC	Direct-Bonded Copper
DBS	Dynamic Braking System
DCB	DC Circuit Breaker
DCU	Distributed Control Unit
DDSRF	Decoupled Double Synchronous Reference Frame
DECC	Department of Energy & Climate Change
DFT	Discrete Fourier Transform
DPS	Dynamic Performance Study
DPWM	Digital Pulse-Width Modulation
DSOGI	Double Second-Order Generalized Integrator
DSP	Digital Signal Processor
DTC	Direct Torque Control
EMC	Electromagnetic Compatibility
EMF	Electromotive Force
EMI	Electromagnetic Interference
ENTSO-E	European Network of Transmission System Operators for Electricity
EOA	Emergency Overnight Accommodation
ESL	Equivalent Series inductance
ESR	Equivalent Series Resistance
FACTS	Flexible Alternating-Current Transmission System
FAT	Factory Acceptance Test
FEED	Front-End Engineering Design
FB	Full Bridge
FIT	Failures In Time
FPGA	Field-Programmable Gate Array
FPNSC	Flexible Positive/Negative-Sequence Control
FRT	Fault Ride Through
GC	Grid Code
GDU	Gate-Drive Unit
GPS	Global Positioning System
GTO	Gate Turn-Off

Table 6   Acronyms H—R.

HB	Half Bridge
HMI	Human—Machine Interface
HVDC	High-Voltage Direct Current
HW	Hardware
I	Integral
IEA	International Energy Agency
IEC	International Electrotechnical Commission
IGBT	Insulated-Gate Bipolar Transistor
IGCT	Integrated Gate-Commutated Thyristor
ISOP	Input-Series Output-Parallel
IPT	Instantaneous Power Theory
JFET	Junction Field-Effect Transistor
LCC	Line-Commutated Converter
LLA	Leg-Level Averaged
LPF	Low-Pass Filter
LVRT	Low-Voltage Ride Through
MMC	Modular Multilevel Converter
Mo	Molybdenum
MOSFET	Metal-Oxide-Semiconductor Field-Effect Transistor
MOV	Metal-Oxide Varistor
MSI	Mixed-Sequence Injection
NLC	Nearest-Level Control
NSCOGI	North Sea Countries Offshore Grid Initiative
NSI	Negative-Sequence Injection
OEM	Original Equipment Manufacturer
OfGEM	Office of Gas and Electricity Markets
OFTO	Offshore Transmission Owner
PA	PolyAmide
PBT	PolyButylene Terephthalate
PCC	Point of Common Coupling
PD	Phase Disposition
PEEC	Partial-Element Equivalent Circuit
PET	PolyEthylene Terephthalate
PFC	Power-Flow Controller
PI	Proportional—Integral
PID	Proportional—Integral—Derivative
PLL	Phase-Locked Loop
PNSE	Positive/Negative Sequence Extraction
PNSRG	Positive/Negative Sequence Reference Generator
POD	Phase Opposite Disposition
PPA	PolyPhthalAmide
PPS	PolyPhenylene Sulphide
P	Proportional
PR	Proportional—Resonant
PSC	Phase-Shifted Carrier
PSI	Positive-Sequence Injection
PWM	Pulse-Width Modulation
QSG	Quadrature Signal Generation
R	Resonant
RES	Renewable Energy Source
RTS	Real-Time Simulator

Table 7   Acronyms S—Z.

SAT	Site Acceptance Test
SC	Short Circuit
SCFM	Short-Circuit Failure Mode
SCR	Short-Circuit Ratio
Si	Silicon
SiC	Silicon Carbide
SLA	Submodule-Level Averaged
SLS	Submodule-Level Switched
SM	Submodule
SOA	Safe Operating Area
SOGI	Second-Order Generalized Integrator
SSOA	Safe Switching Operator Area
STATCOM	Static synchronous Compensator
SVC	Static VAr Compensator
TC	Thermal Conductivity
THD	Total Harmonic Distortion
TIB	Tapped-Inductor Buck
TSO	Transmission System Operator
UPS	Uninterruptible Power Supply
VMM	Voltage-Margin Method
VSC	Voltage-Source Converter
VT	Voltage Transformer
WPP	Wind-Power Plant
WT	Wavelet Transform
WTHD	Weighted Total Harmonic Distortion
XLPE	eXtruded cross-bound PolyEthylene
ZVS	Zero-Voltage Switching

Introduction

The global demand for clean energy and electricity is rising fast. The International Energy Agency (IEA) predicts that, by 2030, global electricity consumption will be close to 30,000 TWh a year, which will be almost twice the amount of energy consumed in 2010. Thermal power plants burning fossil fuels are still the main source of power generation worldwide. According to the organization Renewable Energy Policy Network for the 21st Century (REN21), an estimated 77.2% of global electricity was generated from fossil fuels during 2014 [1].

Modern civilization cannot rely solely on fossil-fuel-based power generation in the future, even though today it is the dominant technology for electric power generation. As planet earth is facing some serious climatological problems related to the continuous increasing level of greenhouse-gas emissions, there are major international attempts to curb these emissions and prevent a global climate crisis. The following facts may support the motivation for a paradigm shift from fossil-fuel-based power generation to a future with higher levels of renewable energy sources (RES) in combination with other energy sources with less emission.

- Approximately 30% of globally extracted fossil fuels are used for electric power generation annually and approximately 40% of all energy-related CO_2 emissions originate from the fossil fuel used to generate electricity [2].
- Approximately 80% of the energy is lost in the chain from harvesting the oil and gas resources to end-use electricity consumers. It is estimated that approximately 9% is lost in power transmission and distribution [2].
- Approximately 40% of energy consumption is for heating and cooling in buildings and industry [3].

Consequently, the power sector could play a major role in reducing greenhouse-gas emissions by incorporating new technologies to improve power generation and power transmission efficiency and by integrating renewables [4].

According to the IEA, a combination of energy-efficiency measures and renewable power generation could deliver 75% of the required emissions reduction over the next two decades.

Design, Control, and Application of Modular Multilevel Converters for HVDC Transmission Systems, First Edition.
Kamran Sharifabadi, Lennart Harnefors, Hans-Peter Nee, Staffan Norrga, and Remus Teodorescu.
© 2016 John Wiley & Sons, Ltd. Published 2016 by John Wiley & Sons, Ltd.
Companion Website URL: www.wiley.com/go/Sharifabadi/ModularConverters

Improving the efficiency of this energy chain remains the most cost-effective way of lowering consumption and emissions.

Third-world countries and emerging economies will increase their level of industrialization and the world's megacities will consume most of the world's energy, and thereby account for the highest level of greenhouse-gas emissions. On average 500 g of CO_2 emission is produced per 1 kWh of electricity generation with fossil fuels in the industrial countries. This value is much higher in developing countries, with extensive use of brown-coal-fired power plants.

Transmission and distribution losses and the low energy efficiency of end-user applications are other elements contributing to the reduced exploitation of generated electricity. There are major efforts and research programs to respond to the aforementioned issues. Some of the solutions which are under investigation and implementation are:

- integrating large amounts of renewable energy sources and distributed power generation to the power systems;
- increasing energy efficiency in power generation;
- reducing electrical losses in transmission and distribution systems;
- increasing the energy efficiency of end-user applications and consumers;
- using innovative demand-side control (with demand-side management, loads are influenced to respond to power imbalances by reducing their power demand).

The development and integration of renewable energy sources such as wind, solar and fuel cells have gained significant momentum during recent years, providing an estimated 19.1% of global power generation in 2013 [1]. The power systems are managed by generation capacities following demand patterns, but in the future, as more renewable and intermittent generation is integrated into the power system, intelligent demand control will play a major role in balancing the power system. This will require the active involvement of power generators, utilities, and consumers.

Power electronics is identified as a key enabling technology for the efficient integration of RES into the grid. Power electronics and new converter technologies may support some of the sustainability goals by enabling the cost-efficient and reliable integration of RES into the power system.

Modern power electronics and converter technologies constitute the core of the modern voltage-source-converter high-voltage direct current (VSC-HVDC) transmission technology. VSC-HVDC transmission technology is ideal for linking remote offshore wind-power plants to mainland networks as it overcomes the limitations of long distance ac transmission, while ensuring robust performance and minimal electrical losses. VSC-HVDC is seen as the backbone of future multi-terminal dc grids and supergrids [5]. The supergrid is defined as an overlay wide-area transmission network that resides on top of the existing grid and makes it possible to trade high volumes of electricity across great distances.

Modular multilevel converter (MMC) technology, with its inherent advantages—such as built-in redundancy, higher efficiency, and lower harmonics—is gaining momentum in different applications such as VSC-HVDC schemes and utility grade flexible alternating current transmission systems (FACTS).

The increase of the system yield will resolve some of the energy challenges associated with reaching the goals of sustainable power generation and transmission. Some developed countries have allocated significant resources to the integration of renewable energy sources into the grid and the development of intelligent demand control in order to comply with the international sustainability goals.

Increasing the efficiency of renewable power generation as well as power transmission and distribution without the loss of system security is among the major challenges in the near future.

Challenges with Integrating Renewable Energy Sources into the Power System

Wind and solar energy are seen as the major renewable contributors for the near future. High penetration levels for RES to the power system will require a mindset change for the stakeholders of the electricity sector and will depend on a substantial willingness to support the required financial, political, and technical changes. The existing ac network infrastructures were designed for the transmission of electricity from large centralized power plants located near large cities or industrial centers and consumers. In contrast, RES such as offshore wind power plants are located away from the existing grid infrastructures and demand centers. The integration of RES into the power system requires the development and reinforcement of distribution and transmission networks. Any large and ambitious integration of RES into the power system presents new challenges for grid owners and operators.

These challenges can be summarized:

* The remote location of RES changes the geographical distribution of power generation centers in relation to the load centers, such as large cities and industrial consumers.
* With a large share of the distributed power generation being connected to distribution networks, there is a risk that the transmission and distribution networks become congested and overloaded.

Wind and solar energy are intermittent energy sources and are affected by variable weather conditions. The integration of RES into the power systems will require large amounts of balancing power from conventional power plants being available. Long-distance transmission networks can integrate neighboring power markets with diversified power generation. Consequently, regional fluctuations can be balanced by the interconnectedness of power systems. The development of large dc supergrids is seen as another option to mitigate the challenges with the intermittency of RES. If mitigating actions are not taken, then the integration of large amounts of RES into the power systems may impair the stability and security of power supply.

Large integration of RES into the power generation is technically and economically feasible and has been documented in several international studies [6] A large contribution from renewable energy to the power system can be realized if the necessary steps are taken to maintain a high degree of security of supply and system security by developing and redesigning the entire value chain of the power system. The steps can be summarized:

* redesigning and reinforcing ac distribution and transmission networks;
* acquiring adequate amounts of primary and secondary reserves for balancing purposes;
* modifying forecasting methods used to operate transmission and power systems.

The transmission system's operation and balance management, power forecasting technologies, and cross-border flow management are among other issues to be considered and redesigned with the large integration of RES into the power system. In power systems, the power balance between generation and consumption must be continuously maintained and the balancing quality indicator is the power system frequency (Hz). If generation exceeds consumption, the system frequency rises; if consumption exceeds generation capacity, the frequency falls. It is the responsibility of the system operator to ensure that the power balance is maintained at all times. The intermittency of RES may impose unpredicted and immediate imbalances for the power system which can be balanced with spinning reserves provided by conventional generation resources or upcoming modern storage technologies, such as electrochemical batteries, flow batteries, or other storage technologies.

A VSC-HVDC system allows for the fully independent control of both the active and the reactive power flow within the operating range of the system. The use of VSC technology makes it possible to maintain the voltage and power quality on the connecting ac networks. The active power can be continuously controlled from full power export to full power import. Over the last decade, several VSC-HVDC links have been installed to increase the power transfer capacity between European countries. Many new European VSC-HVDC interconnector projects are under development, e.g. NordLink (a 1400 MW MMC-HVDC link between Germany and Norway) and the NSN Link (a 1200 MW MMC-HVDC link between the UK and Norway). The additional interconnection capacity will enable the better utilization of reserve and balancing capacities through power markets, potentially bringing down electricity prices while increasing the amount of renewables in the energy mix.

Enabling Technologies for the Large Scale Integration of RES into the Power System

The integration of RES into the power system will strongly rely on the availability of transmission and distribution networks. As indicated previously, one of the main barriers to the extensive integration of RES into the power system is the limited transmission capacity of existing ac networks. Transmission congestions can occur when the demand for a transmission path exceeds its reliable transfer capacity. It is obvious that any integration of large amounts of RES will require a large-scale expansion and upgrade of existing transmission and distribution networks.

Through grid expansion and reinforcement, the geographic diversity of RES can be exploited to smooth out intermittency caused by weather conditions. It is unlikely to have similar weather conditions over large geographic areas. Grid expansion and reinforcement will reduce grid congestion and increase the interaction and support between balancing areas. To meet the increasing load demands, new transmission lines have to be added to the system. However, owing to the possible negative impact on the environment, the expansion and installation of new transmission lines are often restricted and subjected to lengthy permitting and legal processes.

Resolving some of the existing constraints on ac networks' capacity requires new approaches. Grid owners and operators are utilizing innovative solutions to increase the power transfer capacity by integrating FACTS devices into the existing ac networks. Long-distance ac transmissions lines require the injection of sufficient amounts of reactive power to keep acceptable voltage profiles along the power lines. Static var compensators (SVCs) and static synchronous compensators (STATCOMs) provide the ac networks with the required reactive power compensation to make the ac grids more robust for the integration of large amounts of RES.

Integrating FACTS into the power system can improve and increase the transmission capacity and improve the stability of ac transmission systems. And integrating FACTS equipment into the existing ac networks improves the performance of renewable energy power plants in order to meet the grid code requirements. The MMC is considered as one of the enabling technology for FACTS and VSC-HVDC systems.

As many existing ac networks have reached their transmission capacity limits, a significant increase in transmission capacities is needed. One solution to this challenge could be to convert parts of the existing high-voltage alternating current (HVAC) power lines to HVDC power lines. VSC-HVDC transmission with the same voltage level as the peak voltage of the converted HVAC line increases the transmission capacity of existing corridors by a factor of three. As an example, an ac overhead transmission line with a line voltage of 230 kV and a transmission capacity of 480 MVA can be converted to a dc transmission system of +/− 320 kV with a transmission capacity of 1440 MW. Converting some of the existing ac overhead power lines to HVDC overhead power lines could potentially triple the transmission capacity.

In recent years, most of the European interconnector projects have been designed and realized with MMC-HVDC transmission technology. The MMC-HVDC terminals can be integrated into weak ac networks (e.g. offshore ac collector grids). In the event of ac network contingencies, MMC-HVDC schemes can support the network with fast power reversal and a broad range of ancillary services.

The MMC technology, recently adopted by major manufacturers of HVDC equipment, has demonstrated clear advantages in comparison with two-level VSC in terms of reduced losses and footprint. It is expected to achieve increased deployment in both offshore and onshore HVDC transmission systems as well as in future HVDC grids. This fast development has left a knowledge gap between the technology developers and the project developers as well as the academic world. This book aims to fill this gap by providing comprehensive and self-explanatory knowledge on the following topics:

- basic MMC properties and constraints;
- main-circuit design;
- converter dynamics and control strategies under steady-state and fault conditions;
- modulation and submodule energy-balancing methods;
- converter control under unbalanced grid conditions;
- application of MMC technology in HVDC transmission technology;
- control and protection of MMC HVDC under ac and dc network fault contingencies;
- multi-terminal dc grid (MTDC) grid control strategies and protection technologies;
- design and optimization of MMC-HVDC schemes for the export of wind energy from offshore wind power plants;
- recommended system studies for the development and integration of offshore HVDC schemes to offshore wind power plants;
- overview of available international MMC-HVDC standards and commissioning procedures.

References

[1] REN21, *Renewables 2015: Global Status Report*, http://www.ren21.net/wp-content/uploads/2015/07/REN12-GSR2015_Onlinebook_low1.pdf, accessed Mar. 30, 2016, 2015.

[2] Public Electricity Production from Fossil Fuels, IEA information paper in support of G8 plan of action.

[3] International Energy Agency, *Energy Technology Perspectives 2015*, IEA, https://www.iea.org/bookshop/710-Energy_Technology_Perspectives_2015, accessed Mar. 30, 2016, 2015.

[4] International Energy Agency, *World Energy Outlook 2015*, https://www.iea.org/bookshop/700-World_Energy_Outlook_2015, accessed Mar. 30 2016, 2015.

[5] N. Ahmed, S. Norrga, H.-P. Nee, et al., (2012) "HVDC SuperGrids with Modular Multilevel Converters: The Power Transmission Backbone of the Future," Proceedings of the 9th International Multi-Conference on Systems; Signals and Devices, SSD-PES 2012, Mar. 20-23 2012, Chemnitz, Germany.

[6] "Integration of renewable energy in Europe," https://ec.europa.eu/energy/sites/ener/files/documents/201406_report_renewables_integration_europe.pdf; "Design and operation of power systems with large amounts of wind power," https://www.ieawind.org/annex_XXV/Meetings/Oklahoma/IEA%20SysOp%20GWPC2006%20paper_final.pdf; "Integrating intermittent renewables sources into the EU electricity system by 2020: challenges and solutions," http://www.eurelectric.org/media/45254/res_integration_paper_final-2010-030-0492-01-e.pdf, "Large scale integration of wind energy in the European power supply: Analysis, issues and recommendations," http://www.uwig.org/eweastudy/051215_grid_report.pdf; "Integration of renewable energy sources in the German power supply system from 2015-2020 with an outlook to 2025: Dena study I &II," http://www.dena.de/fileadmin/user_upload/Projekte/Erneuerbare/Dokumente/dena_Grid_Study_II_-_final_report.pdf.

1

Introduction to Modular Multilevel Converters

1.1 Introduction

The purpose of this chapter is to explain the circuit topology and the operation of modular multilevel converters (MMCs), and put them into their context. To this end the chapter begins with a review of other voltage source converter (VSC) topologies, starting with the two-level converter. This review also includes other multilevel converters. This serves as a basis for a description and analysis of the MMC. Finally, a number of similar topologies are reviewed and compared to the MMC.

Until the 1970s, power electronics for conversion between alternating current (ac) and direct current (dc) was dominated by current source converters (CSCs) operating by natural commutation. The main application was rectification, to provide direct voltage for dc motors and industrial processes. Also, high-voltage direct current (HVDC) power transmission originally made use of CSCs. The available current valves, first thyratrons and mercury arc valves and later thyristors, lacked turn-off capability, and were therefore only suited for naturally commutated converters.

The advent of power semiconductors with turn-off capability, allowing for forced current commutation, paved the way for VSCs which can operate independently from the ac grid. These converters offer increased controllability and improved harmonic performance, which radically expanded the field of application.

At the same time several new challenges in terms of electric power conversion emerged. Concerns about the negative environmental impact of traditional energy sources, such as the burning of fossil fuels, have led to a desire to use renewable energy sources (RES), such as wind power and solar power. None of these uses the kind of large synchronous generators operating at constant speed traditionally employed for electricity generation. Therefore, there is a need for a power electronic interface, which in most cases is best implemented by VSC.

Similar concerns have also led to demands for energy savings. Within electrical drive systems very significant savings can be achieved by a transition to variable-speed operation.

Design, Control, and Application of Modular Multilevel Converters for HVDC Transmission Systems, First Edition.
Kamran Sharifabadi, Lennart Harnefors, Hans-Peter Nee, Staffan Norrga, and Remus Teodorescu.
© 2016 John Wiley & Sons, Ltd. Published 2016 by John Wiley & Sons, Ltd.
Companion Website URL: www.wiley.com/go/Sharifabadi/ModularConverters

In the past, pumps and fans, for instance, were usually operated at fixed speed with the fluid flow being controlled by throttling, which caused efficiency to be very poor at low flow rates. If the speed of the electric machine can be controlled overall power losses can be significantly reduced. Additionally, variable speed operation allows for improved control and precision in many industrial processes, contributing to increased automation and the improved quality of the product. The VSC fits these demands very well since it allows for convenient torque and speed control of synchronous machines and induction machines. The speed of dc motors could be controlled by CSCs, but such motors are less attractive because of the high maintenance cost of their brushes and commutators.

Also in transmission grid applications such as HVDC and FACTS (flexible alternating current transmission systems), the VSC has made inroads at the expense of thyristor-based naturally commutated converters. Several of the features of VSCs are attractive and sometimes also necessary in this field. The ability to impose a voltage on the ac side and thus control active and reactive power independently from each other is required in several applications. To these belong, for instance, HVDC connection of offshore wind farms as discussed earlier in this book.

From the beginning the prevailing VSC topology was the two-level converter and it is still the preferred solution for low-voltage applications. For systems operating at voltages beyond a couple of kilovolts, however, it is less useful since there are no available power semiconductors able to block voltages above this level. An option is to directly connect semiconductor elements in a series to create valves capable of withstanding higher voltages. Such concepts have been developed, but they tend to be complex since several problems need to be solved in order to ensure the operation and fault tolerance of such a valve. Also, this technology is not widely available, making it costly.

Instead, a transition to multilevel converter topologies offers better prospects for providing cost-effective power conversion at higher voltages. These topologies do not require direct series connection for increasing the operating voltage. Furthermore, the harmonic properties are much improved, so that the requirements on voltage and current distortion can be met without excessive switching losses.

The first multilevel topology to be used on a large scale was the three-level neutral point clamped (NPC) converter. It is still in frequent use for high-power motor drives. However, it cannot easily be extended to additional levels, which limits the field of application to the medium voltage range, at least if the series connection of semiconductors is not used to boost the voltage capability.

For operation at high voltage, using cascaded topologies that are based on series connection of converter *submodules* rather than semiconductors is a much more feasible alternative.

This chapter seeks to explain the development of cascaded converter topologies with a strong focus on MMCs for ac/dc conversion, since these are the main focus of this book. First, as a background, the two-level VSC is described. This is vital also to understand MMCs, since its basic building block, the two-level phase leg, is also used in the submodules of MMCs. The rest of the chapter is devoted to multilevel converters, starting with a general discussion of the consequences of the transition to multilevel operation. A brief treatment of diode-clamped converters is also given, focussing on the three-level variety, the NPC, since it is one of the most commercially viable multilevel converters. Thereafter, to introduce cascaded multilevel topologies, the concepts of converter submodules and submodule strings are presented and their operation is explained. This forms a basis for the treatment of the half-bridge MMC.

This topology and its operation and limitations are described in detail. Finally, a number of other cascaded converter topologies are reviewed and their properties are discussed in relation to those of the MMC.

1.2 The Two-Level Voltage Source Converter

The two-level converter is by far the most frequently used VSC for all applications with dc-side voltages up to approximately 1800 V. It is found in industrial drives, home appliances, automotive drive systems, as well as in low-voltage grid-connected converters (e.g. for integrating solar power into the power distribution grid). Furthermore, the two-level phase leg forms the basic building block also for the cascaded converter structures that are the main object of study in this book. For these reasons a thorough description of the two-level converter and its operation is given below.

1.2.1 Topology and Basic Function

Figure 1.1 shows the schematic diagrams of different two-level converters, only differing in terms of the number of phase legs. Like all VSCs, they are equipped with short-term energy storage, in the form of a dc capacitor, which will maintain the dc-side voltage approximately constant for a fraction of a fundamental cycle, regardless of the ac-side currents. The terminals of this capacitor form the dc terminals of the converter. To these terminals a number of phase

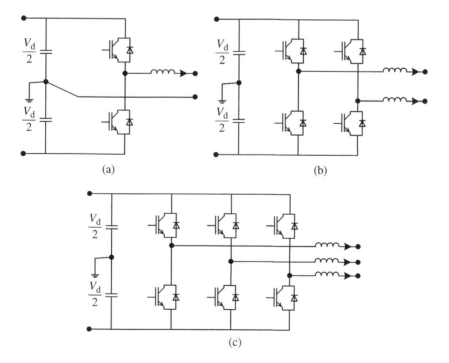

(a)   (b)

(c)

Figure 1.1 Two-level voltage source converters with different numbers of phase legs.

legs are connected in parallel. Each of these is equipped with two series-connected semiconductor valves capable of conducting current in both directions and blocking voltage in one direction by gate control. These are commonly implemented by a unidirectional controllable switch and an anti-parallel diode, although the switch and diode functions may be physically integrated, as is the case for metal-oxide-semiconductor field-effect transistors (MOSFETs) with a body diode. For a poly-phase converter, simply more phase legs are connected in parallel, as shown in the figure. The ac terminal is formed between the midpoints of the phase legs. Generally, an inductance is connected in series with the ac terminals, which will maintain the ac-side currents fairly constant in the short timeframe between commutations of a phase leg. Depending on the application, this inductance may be realized by a physical device, as in the case of a grid-connected converter with phase reactors. It may also be implemented indirectly, which is the case when an electric machine (generator or motor) or a transformer is connected to the converter ac terminal. Under such circumstances, the inductance results from the magnetic leakage fluxes of this device.

In Figure 1.2 the basic operation of the phase leg is explained in detail. The figure displays the different steps of a commutation cycle, which during operation is repeated at a frequency considerably higher than the ac-side fundamental frequency. Throughout the cycle, it is assumed that the ac-side current is directed out of the terminal as indicated by the arrow. Initially, in stage (a), the upper switch is conducting and the ac-terminal potential v_s equals that of the positive dc rail, $+V_d/2$. In stage (b), the upper switch is gated *off* and the current is forced to instead start flowing through the lower diode since this is the only available path for a current in the given direction. When the current has fully commutated to the lower diode, stage (c), the potential at the ac terminal instead coincides with that of the negative dc terminal $-V_d/2$. When this state has been achieved, normally the lower switch is also turned *on* so that a reversal of the direction of the phase current is possible. This is made after a certain delay, normally referred to as the *blanking time*. By inserting this delay, a situation where both switches are in the on-state and the dc link is shorted, albeit temporarily, can be avoided. Such a short-circuit can otherwise lead to a current surge which may damage the switches.

To make the ac outlet potential return to the positive dc rail the lower switch is first gated *off*. After the blanking time has passed the upper switch is again gated *on*, stage (d), whereby there will be a free current path from the dc-bus capacitor through this switch and the lower diode. This path only holds a low stray inductance, meaning that the current through it will increase rapidly, driven by the dc voltage. This rise implies that the current through the upper switch will increase, whereas that of the lower diode will decrease at the same rate. This becomes obvious by applying Kirchhoff's current law to the phase leg and keeping in mind that the ac-terminal current stays fairly constant during the commutation process. Finally, the lower diode turns off whereby the upper switch has replaced it as the path for the ac-side current. Thereby, the phase leg is back in the initial state and the sequence can be repeated.

The commutation sequence for the case where the ac-side current is instead flowing in the opposite direction (into the phase leg midpoint) is entirely analogous, but the current will instead be commutating between the upper diode and the lower switch.

It is often convenient to introduce a variable s to denote the switching state of a phase leg. It assumes the value $+1$ when the phase leg connects the ac outlet to the positive dc rail and -1 when instead the negative dc rail is connected, i.e.

$$v_s = s\frac{V_d}{2}. \tag{1.1}$$

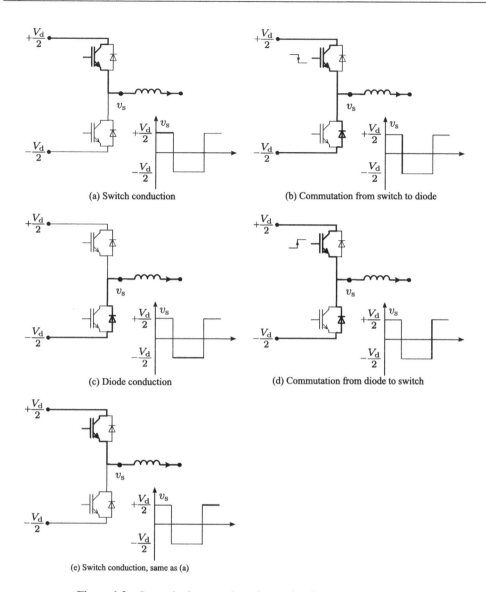

(a) Switch conduction

(b) Commutation from switch to diode

(c) Diode conduction

(d) Commutation from diode to switch

(e) Switch conduction, same as (a)

Figure 1.2   Stages in the operation of a two-level converter phase leg.

The switching states described above are used in the normal operation of the phase leg. It is obviously also possible to leave both switches in the *off*-state. Then the phase leg will behave like a diode bridge and the ac potential will be determined by the direction of the alternating current.

By employing the switching sequence described above the potential at the ac terminal, v_s, can be switched from one of the dc rails to the other at any desired instant and the converter can thus be made to behave as a controllable voltage source. Using repetitive switching and

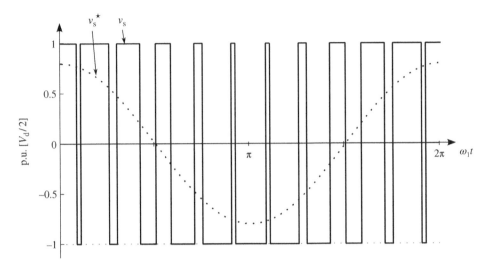

Figure 1.3   PWM of one phase leg.

appropriately timing the switching instances, a desired ac-side voltage can be achieved. This methodology is referred to as *pulse-width modulation* (PWM) since the outcome is that the width of the voltage pulses is varied. In Figure 1.3 it is explained how a desired reference voltage $v_s^\star$ can be realized by a converter employing PWM. In this figure the reference for the phase potential is assumed to be a sinusoid of frequency ω_1. Notably, the pulse width is varied so that the average of the switched phase voltage, taken over one cycle of the switching frequency, coincides with the reference voltage. Thus, when the reference curve assumes values close to $+V_d/2$, the ac terminal voltage will mostly be connected to the positive dc rail and, conversely, when the reference lies far below zero, it will instead be mostly connected to the negative rail. This way, the fundamental frequency component of the switched waveform will approach that of the reference. Because the output voltage is switched between discrete levels rather than varying continuously, the voltage will also contain a significant amount of higher-order harmonic components. However, owing to the high frequency, these will in most cases not cause significant currents or magnetic fluxes in equipment connected at the ac terminal. Therefore, in the first approximation they will not have any major impact on the system and can often be neglected. The issue of determining the switching instances is treated in more detail in Chapter 5, which covers modulation.

1.2.2   Steady-State Operation

The relationship between ac- and dc-side quantities will now be analyzed, with special emphasis on a converter with three phase legs that consequently can provide a three-phase alternating voltage; see Figure 1.4. As discussed above, the converter can be assumed to act as a controllable voltage source.

A key figure to be introduced at this stage is the *modulation index*, also sometimes referred to as the amplitude modulation ratio. It links the dc-side voltage to the fundamental component

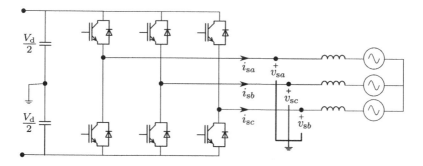

Figure 1.4 Three-phase two-level voltage source converter connected to symmetric load or source.

of the modulated voltage:

$$\hat{v}_s = m_a \frac{V_d}{2}.$$ (1.2)

Any static power converter needs to balance the active power flow on the dc and ac sides. This power balance can be formulated as

$$\frac{3}{2} \hat{v}_s \hat{i}_s \cos(\varphi) = V_d I_d.$$ (1.3)

Substituting Equation (1.2) into this expression gives an equation relating the currents on both sides of the converter:

$$3 m_a \hat{i}_s \cos(\varphi) = 2 I_d.$$ (1.4)

To allow for a simple analysis the phase voltages are assumed to be balanced and sinusoidal. Accordingly,

$$v_{sa} = m_a \frac{V_d}{2} \cos(\omega_1 t)$$

$$v_{sb} = m_a \frac{V_d}{2} \cos(\omega_1 t - 2\pi/3)$$ (1.5)

$$v_{sn} = m_a \frac{V_d}{2} \cos(\omega_1 t + 2\pi/3).$$

As already mentioned, the voltages of the converter will also contain high-order harmonics owing to their origin in PWM, but their impact can often be neglected.

Also the currents are assumed to be sinusoidal and balanced. This is, in most cases, a fair assumption given that the inductance, which is always connected at the ac terminal, will filter out higher-order harmonic components, leaving mainly the fundamental. Furthermore, they are lagging the voltage by an arbitrary power angle of φ.

$$i_{sa} = \hat{i}_s \cos(\omega_1 t - \varphi)$$

$$i_{sb} = \hat{i}_s \cos(\omega_1 t - \varphi - 2\pi/3)$$ (1.6)

$$i_{sc} = \hat{i}_s \cos(\omega_1 t - \varphi + 2\pi/3).$$

When φ is positive the converter will generate reactive power, whereas a negative value corresponds to the consumption of reactive power (assuming that $|\varphi| < \pi$ in both cases). The instantaneous power flow associated with one of the phases will amount to

$$p_{sa} = v_{sa}i_{sa} = m_a\frac{V_d}{2}\cos(\omega_1 t)\hat{i}_s\cos(\omega_1 t - \varphi) = m_a V_d\hat{i}_s\frac{1}{4}[\cos(\varphi) + \cos(2\omega_1 t - \varphi)]. \quad (1.7)$$

There is a constant term whose magnitude depends on the power angle between voltage and current. This term corresponds to the active power being exchanged between dc and ac sides. Also, there is a fluctuating term oscillating at twice the ac-side fundamental frequency. The presence of this term is not surprising, considering that both the current and the voltage show zero crossings with a frequency of twice the fundamental. At these zero crossings the instantaneous power will obviously be zero.

The presence of a phase shift that is common to both the phase current and the phase voltage will naturally result in the same phase shift in the instantaneous power as it amounts to the product of these quantities. Thus, if the phase voltages and currents are symmetrically phase shifted, as in Equation (1.5) and (1.6) this will also be the case with the three powers of the phases

$$p_{sa}(\omega_1 t) = p_{sb}(\omega_1 t - 2\pi/3) = p_{sc}(\omega_1 t + 2\pi/3). \quad (1.8)$$

Therefore, the power flow related to the other two phase legs can be written as

$$p_{sb} = m_a V_d\hat{i}_s\frac{1}{4}[\cos(\varphi) + \cos(2\omega_1 t - \varphi - 2\pi/3)]$$

$$p_{sc} = m_a V_d\hat{i}_s\frac{1}{4}[\cos(\varphi) + \cos(2\omega_1 t - \varphi - 4\pi/3)]. \quad (1.9)$$

The second-harmonic power fluctuation present in the phase quantities will thus cancel out in the total instantaneous power, which is obtained by summing p_{sa}, p_{sb}, and p_{sc}, and thus will be constant:

$$P_s = p_{sa} + p_{sb} + p_{sc} = \frac{3}{4}m_a V_d\hat{i}_s\cos(\varphi). \quad (1.10)$$

This result is also not really surprising considering that power remains constant in any balanced three-phase system and that there is no energy storage in the converter apart from the dc link capacitor. Hence, the only currents flowing into this capacitor will be higher-order harmonics resulting from the PWM. When either the phase currents or phase voltages are unbalanced, however, there will generally be second-harmonic pulsations in the capacitor.

Since only high-frequency harmonics will normally be present in the current fed into the capacitor, it can be small in size. The nominal energy storage capacity of the capacitor, which determines its size and cost, is computed as

$$E = \frac{CV_d^2}{2}. \quad (1.11)$$

Thus, it is proportional to the capacitance. Typical values for many grid-connected converters are 2–4 J of stored energy per kilowatt of rated converter power. This means that the energy stored in the dc link capacitor corresponds to the power converted during 2–4 ms of nominal operation. Thus, the total energy stored could only sustain the rated power during a fraction of a fundamental cycle even if the capacitor was entirely discharged. Such low values of energy storage are possible because only the higher-frequency current components are present in the current fed into the capacitor. These do not cause significant ripple in the voltage, bearing in mind that the reactance of the capacitor amounts to $1/\omega C$.

As will be evident in subsequent sections, in MMCs the dc capacitors need to absorb significant low-frequency harmonic current components. Therefore, these converters have greater demands in terms of stored energy in the capacitors.

1.3   Benefits of Multilevel Converters

Two-level VSCs are generally the most economical solution for low power ratings and low voltage, up to approximately 1 MVA. They have a simple structure with few components. For applications such as large industrial drives, FACTS, and HVDC transmission, where medium or high voltage needs to be handled, a number of shortcomings become evident, however.

- The ac-side voltages will contain significant harmonic components around multiples of the switching frequency. In converters using low-voltage power semiconductors high switching frequency can be used which mitigates this issue since only high-order harmonics will appear in the spectrum. However, for a converter in power transmission applications there is normally a desire to employ devices (typically insulated gate bipolar transistors (IGBTs) or integrated gate commutated thyristors (IGCTs)) with several kilovolts of blocking voltage. Each switching of such devices normally incurs high losses, implying that a switching frequency higher than approximately 1 kHz will cause excessive power losses. At low switching frequency more costly filters will be required to prevent the harmonics from causing negative effects in the equipment or the grid that is connected to the converter.
- The maximum blocking voltage of currently available power semiconductors applicable in VSCs (switches with turn-off capability and fast diodes) fall into the range of a couple of kilovolts. Thus, to realize a high-voltage converter for grid applications, a series connection of semiconductors has to be employed. This is related to a number of difficulties and no widely available technical solutions to achieve such series connection exist.
- In a two-level converter the phase voltage is always switched between the dc rails. In order to keep switching losses to feasible levels the switchings have to be fast, usually within a microsecond. At high direct voltages this implies that the voltage slope will be high, which imposes very significant stress on the insulation of any equipment connected to the ac terminal.

A transition to multilevel converter topologies can offer significant improvements of all these issues. Multilevel converters have more than one dc link capacitor, which means that several separate dc levels can be created inside the converter. The blocking capability requirements of the semiconductor valves is determined by the individual capacitor voltages rather than by the full dc link voltage. Therefore, the voltage rating and thereby the power rating of the converter can be increased without direct series connection of semiconductor elements.

The positive impact on the relation between the harmonic properties of the output voltage on the one hand and the switching frequency of the semiconductor valves on the other can be split into two largely independent parts. First, the output voltage waveform has more than two discrete levels. This is vital, since it implies that the amplitude of the harmonics can be reduced. Second, every semiconductor valve is not involved in every transition of the output voltage. This implies that the frequency at which the output voltage is changed can be increased without increasing the switching frequency. Thereby, the first harmonics in the spectrum will appear at higher frequency where they can be more easily filtered out.

In Figure 1.5 the impact of a transition from the use of two-level to multilevel converter topologies is illustrated in the time and frequency domains. In this figure the switching frequency of the semiconductor valves has been kept constant at 15 times the fundamental as the number of levels is increased from two to seven. In the first approximation the switching losses would thus remain constant in all cases. Both the phase voltage and the amplitude spectra of this signal are displayed for each number of levels. As discussed, the pulse frequency is also

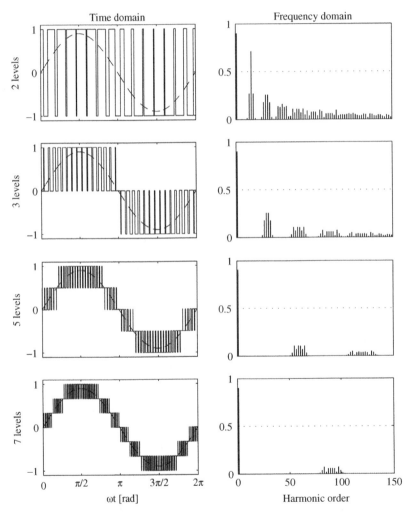

Figure 1.5 Impact of a transition to multilevel converters. The charts show the phase voltage of converters with two to seven levels in the time domain (left) and the frequency domain (right). In all cases the switching frequency is maintained constant at 15 times the fundamental frequency. The voltages are normalized by half of the pole-to-pole dc link voltage, i.e. the unit is p.u. $[V_d/2]$.

increased when a converter with a higher number of levels is used. The two-level waveform does not at all resemble a sinusoid, and its sole justification is that the use of PWM allows for a separation in the frequency domain between the desired fundamental and the undesired harmonics. However, with more levels the switched waveform will increasingly resemble the reference. In the frequency domain it is evident that more levels cause the harmonics to both appear at higher frequency and be of less magnitude. Both of these effects contribute to reducing the amount of filtering needed to achieve acceptable harmonic distortion. If the number of levels were to be increased further, an arbitrarily accurate approximation of the reference curve would result.

1.4 Early Multilevel Converters

This section will briefly describe a couple of multilevel VSCs that were introduced and developed before the MMC and other cascaded topologies, namely the *diode clamped topologies* and *flying capacitor topologies*. These have been extensively used for medium voltage applications such as high-power motor drives in industry, but also in some cases as high-voltage converters for power transmission applications. However, the number of levels generally cannot easily be extended beyond a few, which means that a direct series connection of power semiconductors is required for handling transmission-level voltages. As is the case for the two-level converter, they all have in common that a single dc-link capacitor, or a series connection of several capacitors, takes up the full pole-to-pole dc-side voltage.

1.4.1 Diode Clamped Converters

Diode clamped converters were first introduced by Nabae, Takahashi and Akagi in 1981 [1]; see Figure 1.6 showing three- and four-level variants. The dc link is split by series-connected capacitors creating additional dc levels. These topologies are also characterized by having diodes (called *clamping diodes*) providing additional paths for the current so that the ac terminal can be temporarily connected to these dc levels during operation. Thus, the creation of additional ac voltage levels is made possible.

The three-level variant is referred to as a *neutral point clamped* (NPC) converter since the clamping diodes in this case can link the ac terminal to the midpoint (which may be of neutral potential) of the dc link. This was also the name used in [1] which did not cover the extension to more than three levels. The NPC converter is the only diode clamped converter that has met significant commercial success. Although the concept is extendable to four or more levels, as shown in the figure, this has in most cases not been feasible in terms of cost. As the number of levels is increased the number of clamping diodes rises and, more importantly, the number of interconnects between the different valve elements goes up. This makes the mechanical design of the converter more complex. For a high-voltage converter this is particularly undesirable since in this case large mechanical clearances have to be observed, owing to the insulation requirements present. The three-level NPC converter, on the other hand, is being used extensively in medium-voltage applications, mainly high-power motor drives for industry. They are also used in railway traction converters and for the grid integration of wind turbines.

The different switching states of one phase leg of the three-level NPC converter are illustrated in Figure 1.7. States (a) through (c) are relevant to the case where the ac-side current is

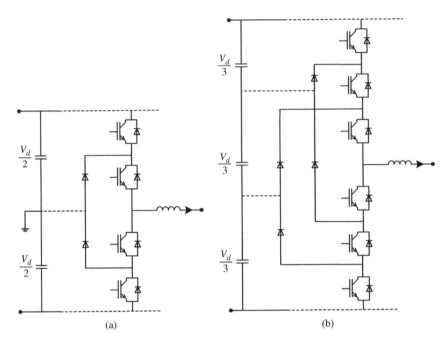

(a)   (b)

Figure 1.6 Diode clamped converter topologies. One phase leg shown of (a) a three-level (NPC) converter and (b) a four-level converter.

directed out of the phase terminal, whereas the states with opposite current direction are (d) through (f). The states where the phase outlet should be connected to either of the dc poles [(a), (c), (d) and (f)] are principally the same as in the two-level converter. Either both upper switches (for $v_s = +V_d/2$) or both lower switches (for $v_s = -V_d/2$) are gated on to achieve these states. To enter states (b) and (e) that connect the ac terminal of the phase leg to the dc link midpoint ($v_s = 0$) both inner valves S_2 and S_3 are gated on, whereas the outer switches are kept in the off-state.

During a fundamental cycle the phase potential should generally alternate between zero and $+V_d/2$ during the half-cycle when the reference is positive, and between zero and $-V_d/2$ during the other half-cycle (see also the graph corresponding to three-level modulation in Figure 1.5). From Figure 1.7 it is, for instance, evident that alternation between zero voltage and $+V_d/2$ is achieved by switching between states (a) and (b) when the direction of the alternating current is out of the phase leg, and between states (d) and (e) otherwise. As is the case for the two-level converter, there is no need to know the current direction to decide the switching pattern, as long as a blanking time delay is provided between the turn-off and turn-on of switches during a transition. For instance, states (b) and (e) are both achieved by having the inner switches conducting and the outer switches blocking, and the ac-side current direction determines which of these states is assumed.

Notably, as long as the dc link capacitors are well balanced, in none of the switching states does any valve need to block more than $V_d/2$ (i.e. half of the dc-link voltage). Thus, using semiconductor switches and diodes with a given blocking voltage and current rating, the three-level

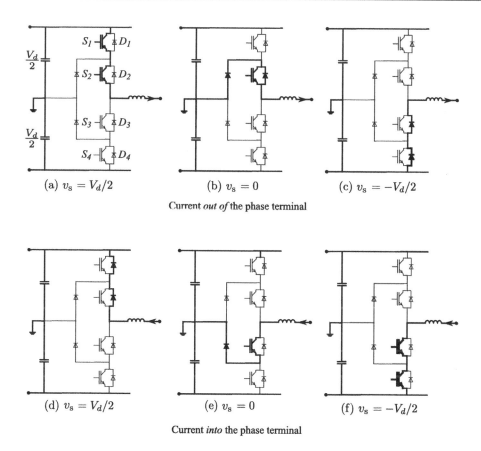

Figure 1.7   Switching states of the three-level neutral point clamped converter.

NPC converter can handle twice the direct voltage and hence twice the power of a two-level converter.

A weakness of the three-level NPC converter in HVDC applications is that in operation at high modulation index and with mainly active power flow ($\cos \varphi = 1$ or $\cos \varphi = -1$) the losses dissipated in the power semiconductors will be unevenly distributed. The outer switches S_1 and S_4 will experience both high switching losses and high conduction losses. The inner switches, on the other hand, will see mainly conduction losses since they stay in the conducting mode when transitions between states (a) and (b) and between states (e) and (f) take place. Hence, in an operating point that is typical for an HVDC converter the outer valves may see significantly higher losses than the inner valves. For this reason the valves need to be overrated, or otherwise semiconductors with a different rating would need to be used for the inner and outer valves respectively. Neither of these alternatives is attractive from a cost point of view.

To mitigate these issues another topology, labeled the *active neutral point clamped* (ANPC), converter has been introduced [2]. It differs from the normal NPC converter in that the clamping diodes have been replaced by valves of the same kind as the main valves (i.e. with a switch and an anti-parallel diode); see Figure 1.8. This opens up a few more possibilities in terms

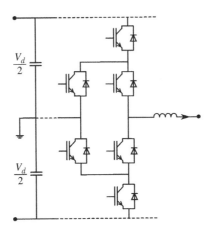

Figure 1.8 Three-level active neutral point clamped converter. One phase leg shown.

of control. More specifically, a three-phase ANPC phase leg offers more ways of implementing the zero-voltage state. This allows for a more even distribution of the losses among the semiconductors.

The ANPC converter has been used in two significant VSC HVDC projects, the *Cross-Sound Cable* and the *Murray link* [3, 4].

1.4.2 Flying Capacitor Converters

A further multilevel converter topology is the so-called *flying capacitor converter*, first presented in 1992 [5]. The topology is shown in Figure 1.9, which displays schematics of a generic phase leg with N levels (a) as well as a complete three-phase three-level converter (b). Like other multilevel topologies, it is based on using extra capacitors holding controlled direct voltages to provide the additional voltage levels. In this case the capacitors have no connection to a common dc link, and are therefore referred to as *flying capacitors*.

Figure 1.10 shows the different switching states of a three-level phase leg for the case where the ac-side current is directed out of the ac terminal at the midpoint of the phase leg. The voltage of the flying capacitor should be maintained at half of the pole-to-pole dc-side voltage (which is V_d). As evident from the figure, two of the states are again identical to the switching states of the ordinary two-level phase leg. These are labeled (a) and (d) and connect the ac terminal to either of the dc rails, whereby the terminal voltage becomes $+V_d/2$ or $-V_d/2$. By using other combinations of switches in the off- and on-states it is possible to instead force the current through the ac terminal to also flow through the flying capacitor. Thereby, the terminal voltage is zero and a third level is achieved. This becomes possible since in this case the flying capacitor voltage is added to, or subtracted from, the dc pole potentials. These states, labeled (b) and (c), are redundant in the sense that they produce the same null ac potential. However, the current direction through the flying capacitor is different in the two states. This makes it possible to balance the capacitor voltage by choosing the appropriate state whenever zero voltage should be imposed. With the direction of the ac-side current as indicated in the figure, state (b) will be chosen when the capacitor voltage is below the desired value, and state (c)

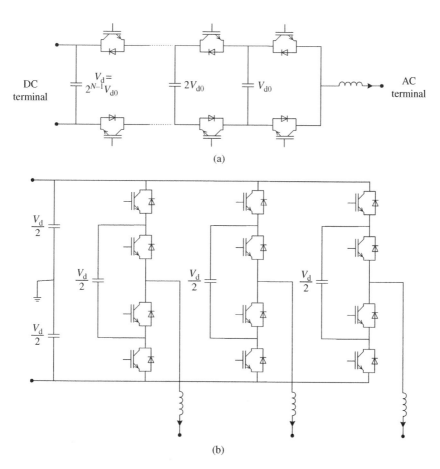

(a)

(b)

Figure 1.9 Flying capacitor converters, (a) one phase leg (generic schematic for an arbitrary number N levels), (b) three-phase three-level converter.

otherwise. At opposite direction of the ac-side current the states are interchanged. Thus, it is always possible to correct a deviation in the flying capacitor voltage in the next switching cycle. It should therefore generally only contain harmonics from the switching frequency and upwards. These are low in amplitude, owing to the low impedance of the capacitor at these frequencies. Thus, the existence of redundant zero-voltage states allows for maintaining the stored energy in the flying capacitor essentially constant over time.

As mentioned, Figure 1.10 shows the current paths during the possible switching states during the half-cycle when the ac-side current is directed out of the terminal. For the opposite case (into the terminal), the current paths in the phase leg are the same, but the current flows through diodes in the valves where it flew through switches in this figure, and vice versa.

In terms of voltage rating requirements, each valve in the three-level flying capacitor converter has to block half of the dc-bus voltage. Therefore, the required voltage blocking capability of each semiconductor device is the same as in the three-level NPC converter, and thus half of that of a two-level converter with the same rating.

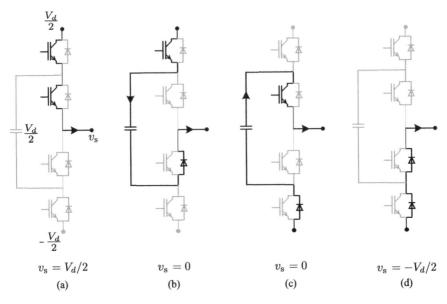

$$v_s = V_d/2 \qquad\qquad v_s = 0 \qquad\qquad v_s = 0 \qquad\qquad v_s = -V_d/2$$

$$\text{(a)} \qquad\qquad\qquad \text{(b)} \qquad\qquad\qquad \text{(c)} \qquad\qquad\qquad \text{(d)}$$

Figure 1.10 Switching states of a three-level flying capacitor converter phase leg. AC current directed out of the phase leg midpoint.

As seen in Figure 1.9(a), this topology can be extended to an arbitrary number of levels, and a correspondingly increased voltage handling capability, by adding more capacitors and semiconductor valves. Thus, it would be possible, at least in theory, to design a high-voltage converter (e.g. for hundreds of kilovolts) without the need for direct series-connection of power semiconductors. This would be the case if the number of levels were extended to the point where the voltage a single valve needed to block in the off-state were lower than the feasible blocking capability of available devices (i.e. a couple of kilovolts). However, such a circuit would be very complex, owing to the many flying capacitors that would connect valves at different points, creating many meshes. This would result in difficulties for the mechanical design of the converter, not least because the insulation requirements which stipulate that parts at different electrical potential need to be physically separated to avoid flashovers. Furthermore, it would require a large number of capacitors rated for different voltages, which would add significant cost. Notably, the total amount of stored energy in the capacitors would increase with the number of levels implemented. The capacitor cost, which is driven by the energy that can be stored, would therefore be higher for a converter with many levels. Therefore, a high-voltage converter, (e.g. for HVDC applications) would likely need to be implemented using a relatively limited number of levels and thereby would require valves capable of blocking high voltage. Thus, the valves would be implemented by series-connecting power semiconductor elements, although fewer devices would be in series compared to the case of a two-level converter of the same voltage rating.

Commercially, the flying capacitor converter appears to have met limited success compared to the NPC converter. It was marketed by the manufacturer Alstom for MV motor drive applications under the trade mark *Symphony*. Adoption by other vendors may have been impeded by certain patents, held by Alstom, that cover this topology.

1.5 Cascaded Multilevel Converters

The previous section treats two classes of multilevel topologies: diode clamped and flying capacitor converters. These are suitable for medium voltage applications (i.e. where the output voltage is a couple of kilovolts). The voltage can to a certain extent be increased without series connection of semiconductors, and the additional voltage levels improve the balance between harmonic performance and switching losses. For high-voltage applications they exhibit significant shortcomings, however. The number of levels cannot easily be extended, since this leads to very complex circuits where the commutation paths, where the current is switched to alter the output voltage, involve large parts of each phase leg.

This section introduces a further possibility, namely *cascaded multilevel converters*. These are based on the series-connection of converter elements, *submodules*, which each contain both of the core components of any VSC–dc capacitors and unidirectional semiconductor valves. They therefore represent a more truly modular approach where the commutation circuits are internal to the submodules. In general, no other effort than adding more submodules is required to increase the number of levels.

First, different kinds of submodules are discussed, and how these can be connected into strings to achieve more voltage levels and handle higher voltage. Later, the MMC with half-bridge submodules is discussed in detail. This converter permits ac/dc conversion with any number of levels in the voltages and it is the main topic of study in this book. Finally, a couple of other cascaded converters will be described and their operation explained.

1.5.1 Submodules and Submodule Strings

The fundamental building blocks of all cascaded multilevel converters are switching submodules (also referred to as *cells* or *chain links*), derived from the basic two-level phase leg described in Section 1.2. In Figure 1.11 schematics of the two basic alternatives are shown: half-bridge and full-bridge submodules. The half-bridge submodule consists of one two-level phase leg parallel to a dc capacitor that will maintain a direct voltage. The external terminals of the submodule are formed by the phase leg midpoint on the one hand and one of the dc capacitor terminals on the other hand. Two possible switching states are possible. In the first, called *bypass*, the switch in the valve parallel to the external terminal is conducting and the terminal voltage is zero. In the other state, labeled *insertion*, the valve in series with the submodule capacitor is conducting, implying that the voltage at the terminal equals the capacitor voltage. The output voltage of the submodule can be described in terms of a switching function s as

$$v = sV_c. \tag{1.12}$$

The switching function assumes the value 0 in the bypass state and 1 when the submodule is inserted. An additional state, labeled *blocking*, can be archived by leaving both switches in the non-conducting state. In this state the voltage at the terminals will depend on the direction of the current since only the diodes may conduct. In one direction the voltage will be zero and in the other the capacitor voltage will be present. This state is not used in normal operation, only during start-up and certain emergency conditions.

The full-bridge submodule has two phase legs connected in parallel to the same capacitor, and the external terminals are formed by the midpoints of the two legs. The normal switching

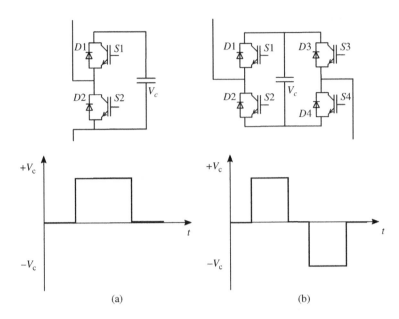

Figure 1.11 Converter submodules: half-bridge submodule (a) and full-bridge submodule (b).

states of the legs give rise to four different switching states for the entire submodule. Using the switching functions for the two legs the submodule voltage can be written as

$$v = (s_1 - s_2)V_c. \tag{1.13}$$

The switching functions again assume the values 0 or 1. With $s_1 = 1$ and $s_2 = 0$

$$v = +V_c$$

and with $s_1 = 0$ and $s_2 = 1$

$$v = -V_c.$$

Unlike the half-bridge, the full-bridge is thus able to provide a bipolar voltage. The two states $s_1 = s_2 = 0$ and $s_1 = s_2 = 1$ will both bypass the submodule and thus result in zero output voltage. However, the current through the submodule will follow different paths in these states. Either through the same node as the positive or the negative capacitor terminal. Therefore, in practical operation, both of these states will normally be used alternately in order to achieve even loading of the semiconductor valves and symmetric distribution of power losses.

Also the full-bridge submodule may be blocked (i.e. all of the switches are gated off). In this case it will behave as a diode bridge, and the terminal voltage will equal the capacitor voltage ($+V_c$) regardless of the current direction.

In conclusion, the converter submodules described can behave as controllable voltage sources, as long as the voltage of the submodule capacitor is maintained sufficiently constant. The half-bridge submodule can only provide a unipolar voltage and is therefore only suited to producing a voltage with a dc component. The full-bridge, on the other hand, is bipolar and can deliver a pure alternating voltage or a combination of ac and dc components.

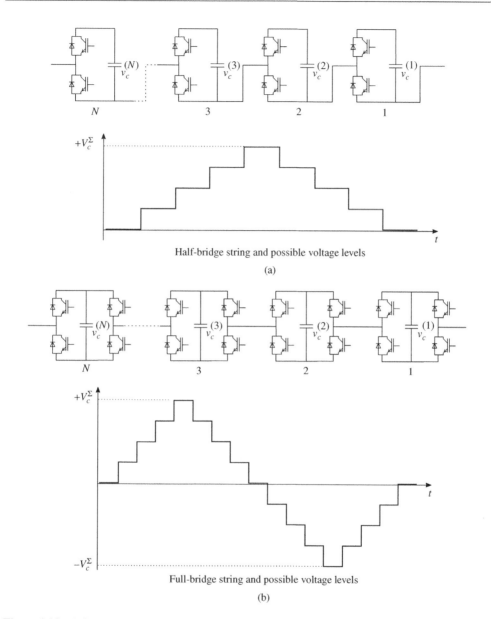

Figure 1.12 Submodule strings and their possible voltage levels: (a) half-bridge string and (b) full-bridge string.

In both of the discussed submodule types the capacitor has to be of significant size as it should maintain an approximately constant direct voltage while carrying significant currents at low harmonic order, usually fundamental and second.

Series connection of the submodules discussed above into *submodule strings* allows for extending the voltage that can be handled; see Figure 1.12. Also, the number of levels the terminal voltage can assume obviously increases with the number of series-connected submodules. A string of N half-bridge submodules can take values between zero and NV_c (i.e. a total of $N + 1$ levels). A full-bridge string with the same number of submodules can provide all voltage levels between $-NV_c$ and $+NV_c$, including zero, which implies a total of $2N + 1$ levels. In a converter intended for high voltage, strings of hundreds of submodules may be required to handle the voltage. Under such circumstances the voltage provided may be seen as approximately continuous since the size of one voltage level (i.e. the submodule capacitor voltage) is very small in comparison to the total voltage provided by the submodule string. As is further elaborated in Chapter 5, using more submodules also enables significantly reduced harmonic content.

For the operation of any submodule or submodule string it is mandatory that the power flow is balanced over time (i.e. that there are no constant terms in the power transferred at the terminals). Such a constant term would result in an uncontrolled increase or decrease in the capacitor energy, and thus also in the capacitor voltage. Accordingly,

$$\int_{t}^{t+T} vi d\tau = 0, \tag{1.14}$$

where T is the duration of a fundamental frequency period. In the frequency domain this expression translates into

$$V_d I_d + \frac{1}{2} \sum_{h=1}^{\infty} \hat{v}_h \hat{i}_h \cos \varphi_h \equiv 0, \tag{1.15}$$

where V_d and I_d are the dc components of the voltage and current, respectively, whereas $\hat{v}_h$ and $\hat{i}_h$ are the magnitudes of the voltage and current harmonic components of order h and φ_h represents the phase angle difference between them.

An exception to these rules occurs if the submodule is equipped with some sort of energy storage significantly larger than the submodule capacitor. There has, for instance, been proposals put forth in the literature for connecting batteries in parallel to the submodule capacitors. This would be a way of creating a variable energy storage with an ac terminal. In such a case there can be a nonzero term in the power flow to or from the string, amounting to the power exchange with the battery.

Fulfilling Equations (1.14) and (1.15) will ensure that the stored energy of the submodule string does not change over a fundamental cycle. Generally, different forms of closed-loop control are required to maintain this balance.

For a converter based on submodule strings it is usually convenient to split the power balancing problem into two parts, which are addressed separately. The first relates to the overall energy exchange with the string and thus with maintaining constant total energy in it. This is equivalent to keeping the sum of the capacitor voltages v_c^Σ at a desired set point. There will generally be a certain inevitable variation during a fundamental cycle, owing to low-frequency alternating currents injected into the capacitors. This control feature is, therefore, aimed at regulating the average voltage over a cycle.

The second part concerns the balancing of the capacitor voltages of the individual submodules *within* the string. Owing to the fact that the submodules are not all inserted or bypassed at the same time, the capacitor voltages will differ slightly between the submodules. This control is local to each string and its purpose is generally to maintain the individual capacitor voltages, v_c^k, at equal fractions of the total capacitor voltage, v_c^Σ, i.e.

$$v_c^k = \frac{v_c^\Sigma}{N}. \tag{1.16}$$

As long as the power balance is maintained, a submodule string can behave as a fairly ideal voltage source within boundaries defined by the total capacitor voltage. Appropriate modulation of the switching pattern can implement the desired low-frequency and direct voltage components at the terminals of a string. The total capacitor voltage is a key factor in determining the cost of the string. It will determine the rating and thus the cost of the power semiconductors as well as the submodule capacitors. In an MMC, generally a direct voltage component as well as an alternating voltage component need to be provided simultaneously by each string. It is, therefore, of interest to understand how limits in terms of the ability to generate these voltage components are related to each other, and to the capacitor voltage. Or stated otherwise: what total submodule capacitor voltage V_c^Σ is required to produce the voltages required for a certain conversion task? This is the case particularly since the capabilities of full-bridge and half-bridge strings differ considerably in this regard.

It can easily be understood that to produce a voltage waveform containing certain low-frequency and dc components the submodule string has to be able to provide a short-time average voltage (over a carrier cycle) coinciding with this desired waveform. This is evident from Figure 1.5. The submodule string voltage must be able to assume values spanning the waveform to be reproduced at all times. Thus, it follows that a half-bridge string can produce a short-time-average voltage in the range $[0, V_c^\Sigma]$. With a dc component of V_d the half-bridge string can produce a peak alternating voltage with maximum amplitude

$$\hat{v}_1 = \frac{1}{2}\{V_c^\Sigma - |2V_d - V_c^\Sigma|\}. \tag{1.17}$$

This relation is visualized in Figure 1.13(a). Evidently, the ac component can never exceed the dc level in the output. That is, without a direct voltage no alternating voltage is possible. Furthermore, from Equation (1.17) it can be understood that it is uneconomical to design the submodule string for operation with $V_d < V_c^\Sigma/2$ since a lower total capacitor voltage can allow for the same ac and dc components to be produced.

On the other hand, the instantaneous voltage of a full-bridge string has to fall in the range $[-V_c^\Sigma, V_c^\Sigma]$. Therefore, in this case, the maximum ac voltage will instead relate to the dc component as

$$\hat{v}_1 = V_c^\Sigma - |V_d| \tag{1.18}$$

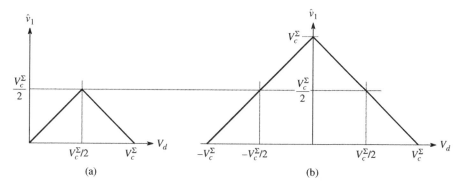

Figure 1.13 Capability of submodule strings to provide ac and dc voltage components given the same total submodule capacitor voltage V_c^Σ: (a) half-bridge string and (b) full-bridge string.

which is shown in Figure 1.13(b). Evidently, the capabilities either coincide with, or exceed, those of the half-bridge string. Only full-bridges can produce a pure ac component or a dc component with negative polarity. In case only an ac component is to be provided its magnitude can equal the full capacitor voltage, which is twice the maximum level possible with a half-bridge string. However, if the task is to produce a unipolar direct voltage with an ac component whose magnitude does not exceed the dc level, the two alternatives are equivalent. This can be understood by comparing the rightmost parts of the two charts in Figure 1.13.

The above reasoning assumes that the capacitor voltages are constant. Owing to the presence of harmonic currents fed into the submodule capacitors, there will generally be a certain fluctuation in the sum capacitor voltage (V_c^Σ) of a string during operation. This fluctuation may affect the capability to deliver a certain desired voltage. Depending on whether the peaks of the sum capacitor voltage coincide with the peaks of the reference voltage or not, the fluctuation may either reduce or increase the voltage capability, as explained in, for example, reference [6].

1.5.2 Modular Multilevel Converter with Half-Bridge Submodules

Until the year 2000 no pure cascaded converter topologies for ac/dc conversion were available. High-voltage VSCs (e.g. for HVDC applications), were implemented using direct series-connection of the semiconductor devices in order to withstand the voltage. Converters for reactive power compensation using full-bridge submodules were in use [7], but these lacked a dc terminal and were therefore not suited for ac/dc conversion, see also Section 1.5.3. There existed cascaded converters for ac/dc conversion, but these required individual isolated feeding of the submodules by transformers, and were not truly modular [8].

A major breakthrough was made by the introduction of the MMC with half-bridge submodules by Prof. Marquardt in 2002 [9]. It puts submodule strings to use in an ingenious way so that ac/dc conversion is made possible while maintaining the important benefits of cascaded converters in terms of voltage scalability and harmonic performance. These features have in a relatively short time made it one of the most used cascaded converter topologies.

Also, very intense research and development efforts, both in academia and in industry, have taken place. A search of the scientific database IEEE Xplore® made in the year 2015 using the exact search string "modular multilevel" yielded more than one thousand results. Notably, all of those results were published after the mentioned initial publications from 2002 so they have all been created within a time span of approximately 13 years. Also, the adoption within industry of this technology has been rapid. The manufacturer Siemens AG first employed the MMC for HVDC transmission, and the first major application was the Trans Bay Cable link installed in the US in 2010 [10]. Currently, all major vendors of HVDC systems offer solutions based on the modular multilevel topology.

1.5.2.1 Topology and Basic Function

In the previous section converter submodules and submodule strings were discussed. It was concluded that they can be made to behave as controllable voltage sources, as long as there is no net power exchange with a string over time, and as long as the energy balance between the submodules is maintained.

Figure 1.14 shows the schematic of an MMC for ac/dc conversion. The overall circuit structure resembles that of a three-phase two-level converter in that there are three phase legs with

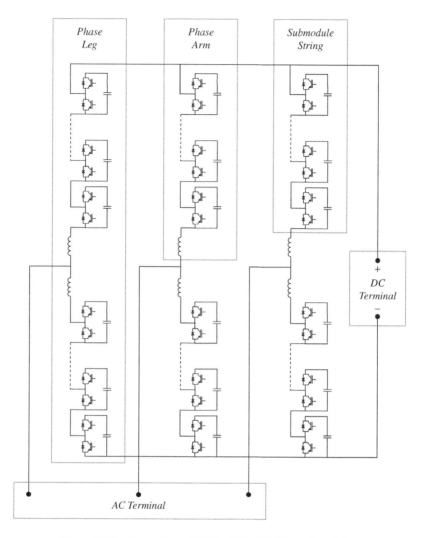

Figure 1.14   Three-phase MMC with half-bridge submodules.

a midpoint connection each. Together, these midpoints form the ac terminal of the converter. However, instead of the valves found in the phase arms of the two-level converter, the MMC has submodule strings. Therefore, the mode of operation is also fundamentally different. The circuit also has small inductors in each phase arm, which also implies an important difference from the two-level converter, where the inductance of the phase legs should be kept low to allow for rapid current changes during the commutations of the phase legs. In the MMC this is not needed since the commutations are internal to the submodules and there should be no such rapid changes of the phase arm currents. Instead, reactors are necessary in the phase arms since direct parallel connection of the voltage sources implemented by the submodule strings should be avoided. Without them, potentially high transient currents could occur between the phase legs. However, given that the voltages of the submodule strings can be controlled with great precision, particularly if the number of submodules is high, these inductors generally do not need to be large.

Notably, numbers of phase legs other than three are also possible, which allows for connection to ac networks with other numbers of phases. For a converter connected to a single-phase network (e.g. a catenary network for railway power supply), two phase legs can be used instead. In such a case, the single-phase ac terminal is obtained between the midpoints of these phase legs.

In short, the operation of the converter is based on controlling the six submodule strings in such a fashion that the combination of voltages they provide results in the desired voltages being imposed on the dc and ac sides of the converter. Figure 1.15 explains how this is possible. The figure shows how the upper and lower submodule string voltages of a phase leg change during the course of one fundamental cycle. The *sum* of these voltages is maintained approximately equal to the desired pole-to-pole dc-side voltage V_d. Simultaneously, the *difference* in voltage between the upper and lower strings is varied. This way the potential at the ac outlet can be set to any point between the dc rails, neglecting the voltage drop across the relatively small arm inductors. In particular, an alternating voltage can be imposed at the outlet. It is also obvious that in a converter with several phase legs a symmetric set of alternating voltages can be provided while all legs maintain the same direct voltage. Thus, the converter can behave as a controllable voltage source not only with respect to the ac side but also as seen from the dc side. This differs considerably from the two-level converter and the other multilevel converters described thus far, which function as current sources behind a capacitor as seen from the dc terminal. Given that any number of submodules can be used, both ac-side and dc-side voltages of the MMC can be controlled with great precision.

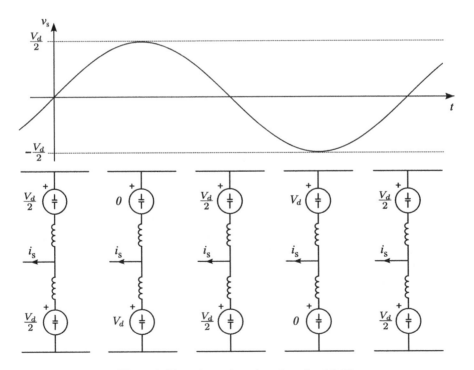

Figure 1.15   Voltages in a phase leg of an MMC.

Since the converter imposes voltages at both its terminals, the corresponding currents will depend on these voltages, and on the ac and dc networks connected. On the dc side it is a fair assumption that a relatively constant direct current can be maintained since the converter can provide a constant direct voltage. Likewise, if the converter is connected to a symmetric ac network and imposes symmetric voltages on it, symmetric ac-side currents will result. In Figure 1.16 the typical paths of these currents inside the converter are shown. It is seen how the dc-side currents split between the phase legs, and that the same current passes through both upper and lower arms. Also, the ac-side currents split between upper and lower phase arms. Since they are symmetric, they will sum to zero in the positive and negative dc poles and thus no ac components will appear on the dc side. It is obvious that the submodule strings of each phase arm need to provide direct as well as alternating voltage components while both direct and alternating currents flow through them.

The switching processes inside the MMC normally only involve one submodule at a time. Compared to the two-level converter, which switches the full pole-to-pole voltage at once, this implies a great improvement. In addition to the many benefits in terms of harmonic performance obtained by multilevel operation, the fact that only a small fraction of the direct voltage is switched simplifies the design of the converter. The steepness of the voltage flanks (dv/dt) during switchings is considerably reduced. Therefore, the impact of stray capacitances becomes much smaller, simplifying the mechanical design of the converter. Also the requirements on the insulation of equipment connected to the converter, such as transformers, can be significantly relaxed thanks to the lowered dv/dt.

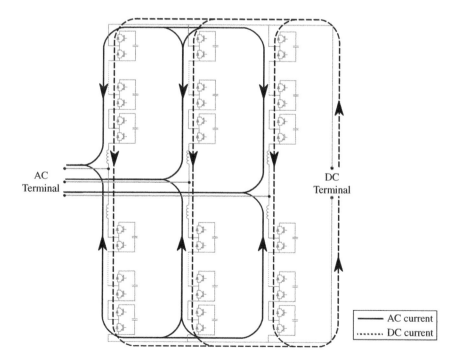

Figure 1.16 Paths for ac and dc current components within an MMC.

Furthermore, having the paths for the commutating currents internal to the submodules rather than involving large parts of the converter implies very significant benefits. The shortened commutation paths make it easier to keep the stray inductances low enough to avoid overvoltages during switchings. Also, reduced stray elements mean that faster switchings can be permitted, implying that the switching losses are reduced.

In terms of electromagnetic interference (EMI), the fact that the commutation loops are shortened and that the voltage steepness is reduced also has a very positive impact. Generally, less effort is required for limiting radiated EMI emissions in the design of an MMC than with a two-level converter of a similar rating.

1.5.2.2   Decoupling of AC and DC Sides

In this section a general framework for analysing the operation of the MMC is described. The circuit equations governing the converter are decoupled by using a linear variable transformation linking the currents and voltages of the phase arms to quantities that are more convenient for describing the operation. At this stage the submodule strings in each phase arm are treated as controllable voltage sources, as discussed in Section 1.5.1.

In Figure 1.17 an equivalent circuit of the converter is displayed. The voltages provided by the submodule strings of the phase arms are labeled $v_{l\phi}$ (lower arm, phase leg ϕ) and $v_{u\phi}$ (upper arm, phase leg ϕ), where index ϕ is substituted with any of the phase symbols a, b, or c. The corresponding currents flowing through the phase arms are labeled $i_{l\phi}$ and $i_{u\phi}$. The external circuits on the ac and dc sides are modeled in such a way so as to allow for common-mode as well as differential-mode voltages and currents on both sides.

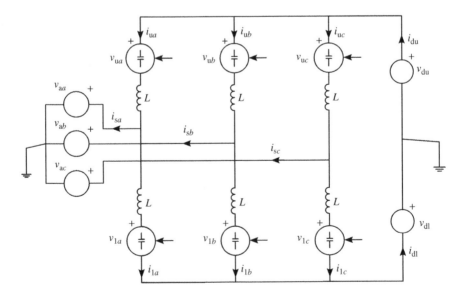

Figure 1.17   Simplified equivalent schematic of an MMC. Submodule strings depicted as controllable voltage sources.

The mentioned linear transformation is defined as

$$v_{s\phi} = \frac{v_{l\phi} - v_{u\phi}}{2}$$

$$v_{c\phi} = \frac{v_{l\phi} + v_{u\phi}}{2}$$

$$i_{s\phi} = i_{u\phi} - i_{l\phi} \tag{1.19}$$

$$i_{c\phi} = \frac{i_{u\phi} + i_{l\phi}}{2}.$$

A physical interpretation of these quantities is that $i_{s\phi}$ are the ac-side phase currents and $v_{s\phi}$ the inner emf (electromotive force) of each phase leg driving these currents. Correspondingly, $i_{c\phi}$ are the average currents of each phase leg, whereas $v_{c\phi}$ are the voltages driving these currents. The main part of these latter voltages and currents is the dc component, accounting for the active power flow through the converter. However, $i_{c\phi}$ also contains any currents circulating between the phase legs and $v_{c\phi}$ any voltages driving these circulating currents. The inverse of the transformation is obtained simply by solving for the phase-arm quantities:

$$v_{u\phi} = v_{c\phi} - v_{s\phi}$$

$$v_{l\phi} = v_{c\phi} + v_{s\phi}$$

$$i_{u\phi} = i_{c\phi} + \frac{i_{s\phi}}{2} \tag{1.20}$$

$$i_{l\phi} = i_{c\phi} - \frac{i_{s\phi}}{2}.$$

The voltage equation for a loop comprising an arbitrary phase leg and the dc link can be written as

$$v_{u\phi} + v_{l\phi} + L\frac{di_{u\phi}}{dt} + L\frac{di_{l\phi}}{dt} - v_{du} - v_{dl} = 0. \tag{1.21}$$

Substituting the inverse of the transformation (1.20) into this equation, and simplifying, yields

$$v_{c\phi} + L\frac{di_{c\phi}}{dt} - v_{dc} = 0, \tag{1.22}$$

with v_{dc} representing the average of the upper and lower dc-link voltages (i.e. $v_{dc} = (v_{du} - v_{dl})/2$). Correspondingly, the voltage equations for a loop going from the midpoint of the dc link to the midpoint of the ac side reads

$$v_{du} - v_{u\phi} - L\frac{di_{u\phi}}{dt} - v_{a\phi} = 0 \tag{1.23}$$

and

$$-v_{dl} + v_{l\phi} + L\frac{di_{l\phi}}{dt} - v_{a\phi} = 0. \tag{1.24}$$

By summing these two equations, and again substituting the inverse of the employed linear transformation (1.20) into the resulting equation the following is obtained:

$$v_{s\phi} - \frac{L}{2}\frac{di_{s\phi}}{dt} - v_{a\phi} - v_{ds} = 0. \tag{1.25}$$

The symbol v_{ds} represents any imbalance between the upper and lower dc-side voltages. It will be close to zero in most cases. Finally, applying Kirchhoff's current law to the positive and negative dc rails, respectively, of the circuit in Figure 1.17 we obtain

$$i_{du} = i_{ua} + i_{ub} + i_{uc}$$
$$i_{dl} = i_{la} + i_{lb} + i_{lc}. \tag{1.26}$$

Summing and subtracting these equations, respectively, and yet again employing the definitions given by (1.20) we get

$$i_{dc} = i_{ca} + i_{cb} + i_{cc} \tag{1.27}$$

and

$$i_{ds} = i_{sa} + i_{sb} + i_{sc}. \tag{1.28}$$

In most practical cases there is no connection from the dc link midpoint, implying that the latter current, i_{ds}, is zero.

Equations (1.25) and (1.28), on the one hand, correspond to an equivalent schematic diagram governing all ac-side currents and the common-mode dc-side current, shown in Figure 1.18(a). Equations (1.22) and (1.27), on the other hand, correspond to an equivalent schematic diagram governing the circulating currents and the differential-mode dc-side current; see Figure 1.18(b). The dominant part of the latter current is generally the dc component, which accounts for the active power flow between the dc and ac sides.

1.5.2.3 Steady-State Operation

The circuit behavior of the MMC as seen from the ac and dc terminals has been explained by the circuit decoupling described above. The insights gained will now be put to use to explain the basic operation of the converter. In particular, ac/dc conversion under stationary and symmetric conditions will be discussed. Furthermore, the implications for the internal energy fluctuations

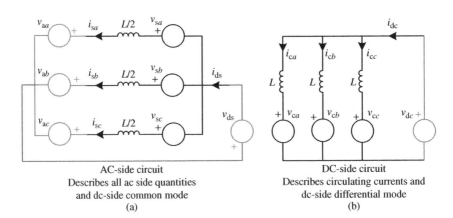

AC-side circuit
Describes all ac side quantities
and dc-side common mode
(a)

DC-side circuit
Describes circulating currents and
dc-side differential mode
(b)

Figure 1.18 Decoupled equivalent circuits of the converter for the ac and dc sides. Parts related to the converter highlighted.

inside the converter phase arms will be analyzed. Since symmetry is assumed, the operation of the different phase legs will be identical apart from a phase shift of 0 rad, $-2\pi/3$ rad or $+2\pi/3$ rad applying to all quantities of the respective leg. Therefore, the analysis can be made on a per-phase basis and then the indices a, b, and c, representing the three phases, can be consistently dropped.

The converter should impose a constant direct voltage, V_d, between the two dc poles. It is assumed that a constant differential-mode dc-side current, $i_{dc} = I_d$, is flowing as a consequence. In practice, a closed-loop controller is generally required to adjust the dc-side voltage so that this will be the case, something that is discussed in more detail in Chapter 3. However, since the aim at this stage is to understand the steady-state behavior, this is not taken into account. Although the decoupled circuit model allows for unbalance dc-side currents and voltages, these are assumed to be zero since they are generally not present in normal operation of the system (i.e. v_{ds} and i_{ds} are set to zero). In most cases there is a transformer connecting the converter to the ac grid that lacks a midpoint connection, meaning that the path for i_{ds} is broken up.

From Figure 1.18(b) it is evident that to fulfill the objectives the common-mode phase-leg voltages should equal half of the desired pole-to-pole dc voltage, i.e.

$$v_c = \frac{V_d}{2}. \tag{1.29}$$

This choice will also eliminate any currents circulating between the phase legs since there are no voltages driving such currents.

As discussed above, unlike the two-level converter, the MMC does not provide a physical, measurable, ac-side voltage. Instead, it behaves like a set of voltage sources, v_s, behind an inductance, amounting to $L/2$; see Figure 1.18(a).

$$v_s = \hat{v}_s \cos(\omega_1 t). \tag{1.30}$$

Obviously, the voltages for the three phases a, b, and c are phase-shifted by 0 rad, $-2\pi/3$ rad, and $+2\pi/3$ rad, respectively. Since the ac network is presumed to be symmetric, a symmetric set of currents will, as a consequence, flow in the three connections of the ac terminal. Thus,

$$i_s = \hat{i}_s \cos(\omega_1 t - \varphi). \tag{1.31}$$

Notably, the power angle φ can be negative, implying inductive behavior of the converter. As in the case of the two-level converter, power balance between ac and dc sides needs to be sustained since the converter does not possess a large energy storage. It will again read

$$\frac{3}{2}\hat{v}_s\hat{i}_s \cos(\varphi) = V_d I_d. \tag{1.32}$$

Also, in the same fashion as with the two-level converter, the modulation index is defined to link the magnitudes of the ac-side and dc-side voltages, although the ac-side voltage is replaced by the inner emf v_s since there is no explicit ac-side voltage in the MMC,

$$m_a = \frac{\hat{v}_s}{V_d/2}. \tag{1.33}$$

Again, this also permits a reformulation of the power balance as

$$3m_a\hat{i}_s \cos(\varphi) = 2I_d. \tag{1.34}$$

The next issue is to determine the submodule string voltages and currents. These can be obtained directly by inserting the desired terminal quantities v_c, v_s, i_c, and i_s into Equation (1.20). For any phase leg the upper and lower submodule string voltages are obtained as

$$v_u = \frac{V_d}{2} - \hat{v}_s \cos(\omega_1 t) \tag{1.35}$$

$$v_l = \frac{V_d}{2} + \hat{v}_s \cos(\omega_1 t). \tag{1.36}$$

Thus, each phase arm provides a dc voltage component as well as an ac voltage component, where the dc components form a common mode, whereas the ac components are in anti-phase. Likewise, the arm currents will be

$$i_u = \frac{I_d}{3} + \frac{1}{2} \hat{i}_s \cos(\omega_1 t - \varphi) \tag{1.37}$$

$$i_l = \frac{I_d}{3} - \frac{1}{2} \hat{i}_s \cos(\omega_1 t - \varphi). \tag{1.38}$$

A very interesting observation at this stage is that the conversion between ac and dc apparently occurs *within* the six submodule strings and that the ac and dc quantities are present simultaneously at their two terminals. Thanks to the topology of the converter, and the opposing signs of the currents and voltages, these quantities are separated so that pure dc quantities appear at the dc terminal and pure alternating voltages and currents appear at the ac terminal. The MMC can in fact be perceived as six ac/dc converters each handling one-sixth of the power.

An important issue is the choice of the sum capacitor voltage of the submodule strings. As noted in Section 1.5.1, for a half-bridge submodule string with total capacitor voltage V_c^Σ the alternating-voltage magnitude range is maximized when the direct voltage is half of this value. Therefore, the sum capacitor voltage of the submodule string in each arm is set equal to the dc-side voltage:

$$V_c^\Sigma = V_d. \tag{1.39}$$

Slight deviations from this choice are possible, and may be beneficial under certain circumstances, but in most cases this is the preferred capacitor voltage set point. It is evident that the semiconductor valves of an MMC phase leg will need to handle a total capacitor voltage of $2V_d$. This is twice the corresponding value for a two-level converter with the same dc-side voltage. Still, the total semiconductor rating is not necessarily doubled since the current rating of the semiconductors can be lower. Later in this section, a more detailed treatment of component ratings is made.

The power exchange with each of the submodule strings will now be considered. It is of interest since it governs the average capacitor voltage ripple. See also [11] for a more in-depth treatment of this topic. Instantaneous power expressions of the submodule strings of a phase leg can be found by multiplying Equations (1.35) and (1.37) and using the power balance (1.34). After some simplification they can be written as

$$p_u = \frac{1}{8} V_d \hat{i}_s [2\cos(\omega_1 t - \varphi) - m_a^2 \cos(\omega_1 t)\cos(\varphi) - m_a \cos(2\omega_1 t - \varphi)] \tag{1.40}$$

for the upper arm and

$$p_l = -\frac{1}{8} V_d \hat{i}_s [2\cos(\omega_1 t - \varphi) - m_a^2 \cos(\omega_1 t)\cos(\varphi) + m_a \cos(2\omega_1 t - \varphi)] \tag{1.41}$$

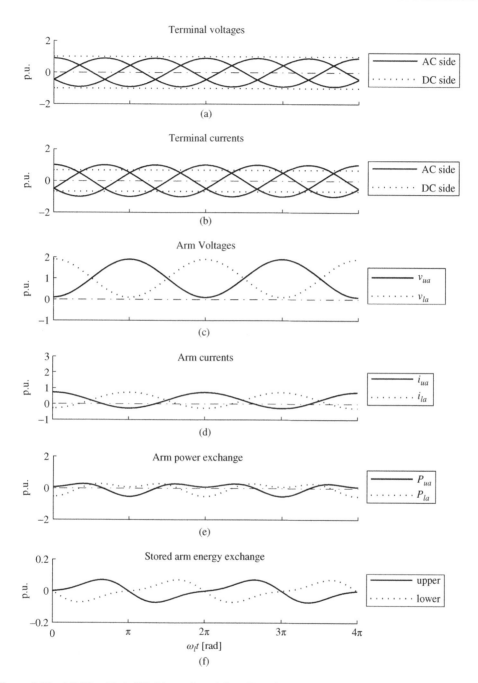

Figure 1.19 MMC with half-bridge submodules. Waveforms during a few cycles of steady-state operation with $m_a = 0.9$ and $\varphi = 0$.

for the lower arm. A first (rather trivial) observation from these expressions is that upholding the power balance at the converter level leads to power being balanced on the arm level since there is no constant term. That is, the condition for stable operation of the submodule strings is fulfilled according to Equation (1.14). Furthermore, it is clear that there will be significant low-frequency power pulsations in the submodule strings. First, there is an oscillation at the ac-side fundamental frequency. These terms have opposite signs in the expressions for the upper and lower phase arm, meaning that fundamental frequency power moves back and forth between the arms of the leg during operation. Moreover, there is a second harmonic term, proportional to the modulation index. This latter term is common to both submodule strings and thus corresponds to energy exchange with the entire phase leg. It relates to the second-harmonic power oscillation present in any single-phase voltage source, see Equation (1.7), which was derived for a two-level converter. For the two-level case these pulsations normally cancel out between the phase legs and therefore do not give rise to voltage ripple in the dc link. In an MMC, on the other hand, both the fundamental and second harmonic power ripple normally need to be absorbed by the submodule capacitors resulting in significant voltage ripple. For this reason, the total energy storage capacity of the capacitors, which also determines the size and cost of the capacitors, needs to be considerably larger in the MMC. For an application intended for operation at low frequency, such as a motor drive, this issue becomes particularly critical. As the impedance of the capacitors is increased ($X = 1/\omega C$) it will not be economically feasible to absorb the fluctuation by the submodule capacitors according to Equations (1.40) and (1.41). Instead, methods have to be found for redistributing energy within the converter using circulating currents; see [12]. Figure 1.19 shows steady-state waveforms of the variables discussed above from an MMC operating with only active power. The capacitor energy fluctuation is normalised with regard to the energy converted during one fundamental cycle.

1.5.2.4 Limits of Operation

The treatment of the MMC with half-bridge submodules has so far assumed that the submodule strings behave like ideal voltage sources and that there is voltage present at the dc terminal of the converter at all times. As discussed in Section 1.5.1, there are limitations to the capability of the strings to provide voltage, which influence the operating range of the converter; see Figure 1.13. In particular, the ac-component amplitude in the voltage can never exceed the dc component. For the MMC, this implies that the peak ac-side phase voltage may never exceed half of the dc-side pole-to-pole voltage. In other words, the modulation index is limited to the interval $m_a \in [0, 1]$. For a three-phase system, where there is no direct connection between the dc side and the ac side, a zero-sequence component can be added to the phase voltage references to extend the modulation index to $2/\sqrt{3} \approx 1.155$. Still, when the network connected at the dc terminal cannot sustain a voltage, the alternating voltage also has to be zero.

An extreme consequence of this circumstance occurs when the dc-side terminal of the converter is shorted. In a converter used for HVDC transmission this situation arises whenever a short-circuit fault on the dc line occurs. This is particularly common with dc overhead lines, and is usually caused by a lightning strike to the line. Since all ac-side inner emfs will then be zero, the converter loses control of all currents. This can also be understood by studying Figure 1.20, which shows an equivalent schematic diagram of an MMC with the dc-side shorted. Notably, the unipolar voltages that can be provided by the submodule strings would only add to the dc-side short-circuit currents, and the submodules thus have to be blocked or

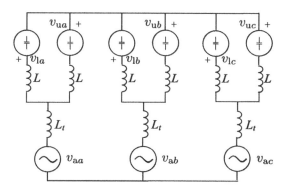

Figure 1.20 Schematic of an MMC at dc-side short-circuit.

bypassed. As a result, the converter will appear as a three-phase short-circuit from the ac side and essentially behave as a rectifier of the short-circuit currents. This state will generally persist until circuit breakers on the ac side open, which generally takes a few cycles (i.e. several tens of milliseconds). Therefore, the short-circuit currents will be determined by the short-circuit power of the ac grid and the overall ac-side reactance of the converter. The point-of-wave of the fault instance will determine the maximum instantaneous current, with the worst case being

$$I_{sc} = 2\sqrt{2}\frac{V_a}{Z_a + Z_s} \tag{1.42}$$

where Z_a is the short-circuit impedance of the grid and Z_s represents the overall ac-side impedance of the converter when blocked. This latter quantity is composed of the arm inductance and any reactance in series with the converter. The second part in most cases consists of the short-circuit reactance of a transformer connecting the converter to the grid. Notably, the short-circuit current will not be split between the upper and lower arms, as the ac-side current does during normal operation.

The dc short-circuit behavior presents several important issues for the design and fault handling strategy of the system. First, the peak short-circuit currents, according to Equation (1.42) will generally be much higher than the normal operating currents; see Figure 1.21, which shows a typical ac-side phase current during a symmetric fault. These currents will flow through diodes in the converter, and may well damage these if appropriate protective measures are not taken. Such measures can be implemented in several ways. First, the series reactance Z_s can be increased to limit the fault currents (e.g. by increasing the arm inductances). Also, a fast bypass switch (typically implemented by a thyristor) can be connected in parallel to each submodule. At a dc-side short-circuit these thyristors are fired and thus divert the currents away from the diodes. This method makes use of the high surge current capability of thyristors. Finally, the surge-current capability of the diodes may be dimensioned to handle the fault currents. This generally implies increasing the active area of the diodes significantly. All of the methods imply added cost, and often a combination is the most cost-effective solution. In Chapter 9 a more in-depth treatment of converter protection concepts including handling of dc-side short-circuits is provided.

A further issue is the fact that the converter, as discussed, will short also its ac terminal, which is undesirable. Obviously, no active power exchange is possible when a dc short-circuit

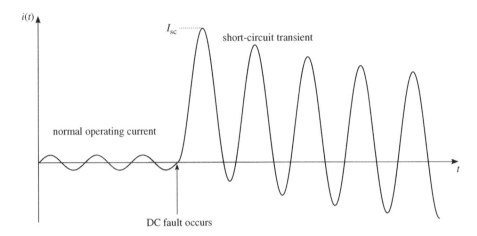

Figure 1.21   One of the ac-side currents during a stiff dc-side short-circuit fault in a half-bridge MMC.

fault occurs, but the behavior is similar to a three-phase symmetric grid fault until the ac break-ers have cleared it. It may have detrimental effects on the operation of the ac grid such as a temporary voltage sag or worse.

A final matter that must be taken into account with regard to the handling of dc-side short-circuits is related to the wider operation of the dc system. The circumstance that current control is lost will generally also delay the recovery of the dc network. This is particularly serious for converters connected to an HVDC grid. If one or several converters feed short-circuit current into the fault, it will be more difficult to disconnect the faulty part.

Connecting fast dc breakers that can interrupt short-circuit currents on the dc side can solve most of the mentioned problems. If such switchgear can separate the converter from the dc lines in a very short timeframe (a few milliseconds), there will be no currents fed into the dc network. Also, the converter can stay in operation feeding reactive power into the ac grid. Other strategies for handling dc-side faults are based on using alternative converter topologies that can limit or control dc fault currents. Several such topologies have been proposed and some are described in the section 1.5.3.

1.5.2.5   Component Rating Issues

This section briefly treats the fundamental rating requirements of the half-bridge MMC based on the discussion of the converter topology and its operation in the previous sections. A more in-depth coverage of component rating follows in Chapter 2.

It is generally a fair assumption that the cost of semiconductors in an MMC is strongly related to the sum of the submodule capacitor voltages and the peak current in the arms. The semiconductor valves need to withstand the capacitor voltage when they are in their off-state, which will define their blocking capability requirements. As discussed previously, the capaci-tor voltage will generally contain significant ripple components. These will, to a certain extent, drive up the peak voltage during operation, which must also be taken into account when dimen-sioning both the submodule capacitors and the semiconductors. Therefore, the total blocking voltage of all semiconductor valves in an MMC is generally more than twice that of a two-level converter of similar rating.

As regards the current rating, the low switching frequency at which the MMC can operate means that the switching losses are low in comparison to those of a two-level converter. Therefore, the devices will in most cases not be thermally constrained (i.e. the maximum junction temperature will not determine the required rating). Instead, the needed semiconductor current rating will be defined by the safe operating area limitations of the devices, meaning that the peak instantaneous current during operation is the critical parameter. For steady-state operation it can be obtained by identifying the maximum value during a cycle of the arm currents according to Equations (1.37) or (1.38):

$$\hat{i}_{u,l} = \frac{I_d}{3} + \frac{\hat{i}_s}{2}. \tag{1.43}$$

By using the energy balance equation (1.34) to link the ac- and dc-side currents the expression can be rewritten as

$$\hat{i}_{u,l} = \hat{i}_s \left(\frac{1}{4} m_a \cos \varphi + \frac{1}{2} \right). \tag{1.44}$$

Sinusoidal and stationary conditions are assumed. The part within brackets is always below unity, which implies that the peak arm current is lower than the peak ac-side phase current. Hence, the current-wise rating of each semiconductor valve can generally be lower in an MMC than in a two-level converter of similar rating, where the valves always have to switch the full ac-side current. This will, to a certain extent compensate for the fact that the overall sum of the device blocking voltages is at least doubled. The maximum instantaneous voltage the submodule string has to deliver will determine the required total capacitor voltage. This quantity will in its turn decide the number of submodules. It can be estimated as

$$\hat{v}_{u,l} = \hat{v}_s + \hat{V}_d = \hat{v}_s \left(1 + \frac{1}{m_a} \right). \tag{1.45}$$

for $|\cos \varphi| = 1$, which is the worst case.

According to the reasoning above, the cost of semiconductors and capacitors in a submodule string is, in the first approximation, proportional to the maximum instantaneous values of the current and the voltage at its terminals. Therefore, the product of the peak string voltage and current can be used as an indicator of the cost, and thus help in understanding how different design choices affect it. To this end, the *semiconductor power* P_{sc} is defined as

$$P_{sc} = k_c \hat{v}_{u,l} \hat{i}_{u,l}. \tag{1.46}$$

The coefficient k_c represents the number of semiconductors in a submodule. It assumes the value 2 for a half-bridge MMC and 4 for a full-bridge MMC. In case comparisons between converters with different number of phase arms should be made, k_c should also involve the number of phase arms. For an MMC intended for active power transfer, P_{sc} can be obtained using Equations (1.44) and (1.45) as

$$P_{sc} = \hat{v}_s \hat{i}_s \left(1 + \frac{1}{m_a} \right) \left(\frac{1}{2} + \frac{1}{4} m_a \right). \tag{1.47}$$

After some simplification, the semiconductor power can be rewritten in terms of the phase-leg apparent power $(S = \hat{v}_s \hat{i}_s/2)$, as

$$P_{sc} = k_c S \frac{1}{2} \left(m_a + 3 + \frac{2}{m_a} \right). \tag{1.48}$$

The extrema of this function are now sought. Differentiating the expression within brackets with respect to m_a and setting the resulting expression to zero yields

$$1 - \frac{2}{m_a^2} = 0. \tag{1.49}$$

The only feasible solution is $m_a = \sqrt{2}$. The second derivative is positive for all valid values of m_a, so there is a minimum at this point. However, the dependency on m_a is rather weak for applicable values of m_a. This is evident from Figure 1.22, which shows the normalized semiconductor power as a function of m_a.

Importantly, the reasoning above about component ratings will only give an indication of the actual requirements. In practice, faults and other abnormal operating modes that may be encountered will often determine the necessary ratings of the semiconductors and passive elements. Therefore, practical dimensioning of MMCs, and other power converters, will generally require detailed calculations and simulations that take into account different fault cases as well as the properties of the control system. However, the stresses incurred during contingencies can often be sufficiently considered by using the steady-state values and include a margin to account for the additional rise in currents and voltages.

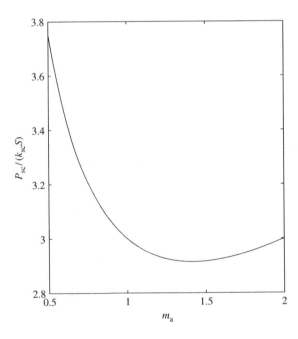

Figure 1.22   Semiconductor installed power versus modulation index for an MMC.

One important such abnormal case, which may impact the rating of components, is a dc-side short-circuit. As discussed above, it leads to immediate loss of the capability to control the currents. The converter will appear as a short-circuit from the ac side and as a rectifier of the short-circuit currents. These currents may exceed the normal operating case several times, depending on the size of the arm reactors and the reactance in series with the converter (such as the leakage reactance of a transformer connecting the converter to the grid), according to Equation (1.42). They will flow in the anti-parallel diodes of the valves parallel to the terminal of each submodule [diode $D2$ in Figure 1.11(a)]. Although the surge current capability of diodes tends to be superior to that of semiconductor switches, such as IGBTs, the current spike encountered may make it necessary to increase the active area of these diodes. Otherwise, other measures can be implemented to protect the diodes, as discussed in Chapter 2, but this will also come at additional cost.

1.5.3 Other Cascaded Converter Topologies

As discussed in the previous sections, the conventional MMC with half-bridges offers several benefits over the two-level converter for high-power applications. Thanks to its scalability, it has removed the upper limit in terms of voltage for VSCs in power grids. Also, the topology offers an excellent combination of harmonic properties and low power losses. Still, the topology suffers form some significant drawbacks. Like the two-level converter, it loses control of both ac-side and dc-side currents in case of a dc-side short-circuit; see Section 1.5.2, which makes it less useful in HVDC grids, at least when not equipped with fast dc-breakers. Also, the amount of stored energy in the submodule capacitors is large, leading to a high capacitor cost.

A number of other modular converters have been proposed prior to, and after, the development of the half-bridge MMC. For certain applications these show significant advantages. This section, therefore, briefly describes some of these topologies and their merits and drawbacks in comparison to the half-bridge MMC.

1.5.3.1 STATCOM with Cascaded Full-Bridges

There are many applications where reactive power compensation is desired, for instance to compensate for non-resistive or fluctuating loads. It allows for better utilization of the grid since the amount of reactive power that may displace the useful active power is reduced; see Figure 1.23. Also, in case it is used to compensate for a rapidly fluctuating load such as an *electric arc furnace* (EAF), used in the steel industry, it can mitigate the fast grid voltage variations (flicker) that would otherwise result. In case fast response times are desired, a VSC connected to the grid is often used, whose sole purpose it is to generate or absorb reactive power. Such a device is usually referred to as a *static synchronous compensator* (STATCOM). They may be installed separately, which is often the case for industrial installations, or as part of a larger installation for reactive power compensation. Early STATCOMS were often implemented by two-level converters or three-level NPC converters (see Section 1.4.1) connected to the grid by step-up transformers [13, 14]. Generally, several converters had to be connected in parallel through the transformers to handle the great power required. Also high-voltage three-level NPC converters were used for this purpose [15]. This way parallel connection could be avoided, and in some cases the converter could be connected directly to the grid without a

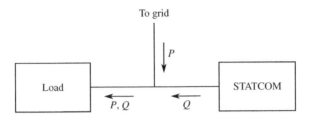

Figure 1.23 Basic principle of reactive power compensation. In case a power source is compensated the active power, P, will be negative.

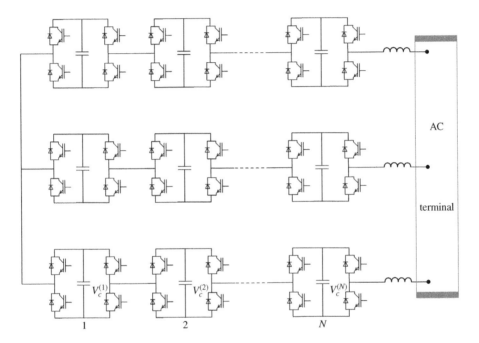

Figure 1.24 Cascaded full-bridges converter for providing reactive power.

transformer, thus reducing the equipment cost. As in the case of two- and three-level convert-ers for HVDC, direct series-connection of power semiconductor elements is required in these cases to handle the high voltage at the converter terminals. This implies challenges in terms of voltage sharing between the semiconductors, static as well as dynamic. Also, semiconductor switches with guaranteed short-circuit failure mode (see Section 2.1.6) are required to ensure fault-tolerance.

A more straightforward way to provide a VSC capable of delivering reactive power at the scale required in power grids and for compensating large industrial loads is shown in Figure 1.24. H-bridge submodule strings are connected in either delta- or wye-connection forming a three-phase ac terminal; see Figure 1.25. In series with each string a small reactor (in per-unit terms) is connected, to limit harmonic currents and to allow for control of the arm

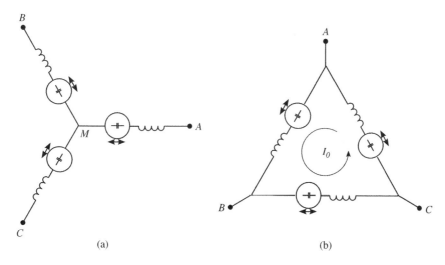

Figure 1.25   Cascaded full-bridges STATCOM in (a) wye and (b) delta configuration.

currents. This is likely to be the simplest cascaded topology. As evident from the figure, the converter only has an ac terminal, since the only task of the converter is to provide or absorb reactive power. Cascaded full-bridges converters of this kind date back to the 1970s with an early reference being a patent filing from MIT [16]. This filing appears to cover mainly single phase systems and the application as a STATCOM was not foreseen. In 1996 a three-phase version of this converter intended as a STATCOM was presented [17], likely for the first time. For practical use the converter topology was first adopted by the manufacturer Alstom [7] for a STATCOM that should form part of a larger FACTS installation. This early utility application of the topology made use of gate-turn-off (GTO) thyristors to implement the four valves of each submodule. The low switching frequency made possible by the cascaded approach was particularly valuable given the high switching losses of the GTO thyristors. The concept has since been taken up by several manufacturers and is currently in widespread industrial use [18].

The benefits over two-level converters are fundamentally the same as with other cascaded converters. A great number of levels can be achieved in the output voltage, enabling acceptable harmonic performance even at very low switching frequencies. The modular approach gives scalability, and any desired rating in terms of reactive power can be achieved by adding more submodules and thus increasing the operating voltage. The voltage can then be adjusted to suit the grid by a transformer. This is different from the case of HVDC transmission, where the dc-side voltage and current have to be adapted to suit the current handling capability of the HVDC transmission cable or overhead line. Furthermore, vital fault tolerance can be implemented by including redundant submodules in the strings. However, as in the case of other cascaded converters, there are a number of drawbacks. Most significantly, the fact that the dc capacitors are distributed in the phase arms impedes power exchange between the submodule strings of the phases. This is especially critical for a converter that should be able to compensate for loads that may be severely unbalanced, such as EAF. In a two-level converter, power exchange between the phases occurs easily since the phase legs lack energy storage and are all connected to a common dc link; see Section 1.2.2.

The energy pulsation, which has to be managed by the submodule capacitors, can be derived as follows for balanced steady-state operation. First, the voltage across the string and the current through it are assumed to be sinusoidal, with the current lagging the voltage by a power angle, φ:

$$v(t) = \hat{v}\cos(\omega_1 t)$$
$$i(t) = \hat{i}\cos(\omega_1 t - \varphi). \tag{1.50}$$

The overall power exchange with the string is as usual obtained by multiplying the instantaneous string voltage and current

$$p(t) = \frac{1}{2}\hat{v}\hat{i}[\cos\varphi + \cos(2\omega_1 t - \varphi)]. \tag{1.51}$$

The first term is constant, so the requirement of power balance over time in this case mandates that the only feasible power angles are $\varphi = -\pi/2$ rad and $\varphi = \pi/2$ rad, which will eliminate this term. It corresponds to either the pure consumption or the pure generation of reactive power, respectively. This is expected, since the converter has no dc terminal at which active power could be exchanged. In either of these cases there will be a second harmonic fluctuation that will result in a submodule capacitor ripple common to all submodules. This is analogous to the conditions of the half-bridge MMC, but the MMC also has a fundamental frequency component in the ripple. The energy fluctuation under these circumstances can be found by integrating Equation (1.51) over time:

$$e(t) = \frac{\hat{v}\hat{i}}{4\omega_1}\sin(2\omega_1 t - \varphi). \tag{1.52}$$

As mentioned, the phase arms of a full-bridge STATCOM may either be configured in wye connection or delta connection, as seen in Figure 1.25. In terms of power rating these two cases are obviously equivalent. That is, the amount of reactive power that can be provided given submodule strings with certain voltage and current ratings is equal. Only the base impedance of the system (i.e. ratio of rated voltage to rated current of the converter) will differ, by a factor of three. This can easily be realized by observing that at symmetric conditions the line-to-line voltage V_Δ and the phase currents I_Δ in the delta-connected case will relate to the corresponding submodule string quantities V, I as

$$V_\Delta = V$$
$$I_\Delta = \sqrt{3}I, \tag{1.53}$$

whereas for a wye-connected converter the corresponding relations will be

$$V_Y = \sqrt{3}V$$
$$I_Y = I. \tag{1.54}$$

Thus, with given submodule string ratings,

$$\frac{V_Y}{I_Y} = 3\frac{V_\Delta}{I_\Delta}. \tag{1.55}$$

The wye-connected configuration is therefore more beneficial when there is a desire to match a certain voltage at the terminals by the least number of submodules. As discussed, however, in many cases the voltage can be adapted to the grid by a transformer. Then, the terminal voltage of the converter matters less. Apart from this obvious difference, there are a few other aspects to be considered when choosing between the wye and delta configurations. As mentioned, a STATCOM may be required to produce a negative-sequence current in order to compensate for a load that can be severely unbalanced, such as an EAF. This will generally make it difficult to prevent imbalances between the overall capacitor energies of the phase arms. That is, the sum capacitor voltages, V_c^Σ, of the three phases will tend to diverge rapidly. However, by using the delta configuration, it is possible to inject a zero-sequence current component, I_0, that can be used to counteract such a divergence, as indicated in Figure 1.25(b). By properly adapting the magnitude and phase displacement of the zero-sequence current, power redistribution between the submodule strings may be performed to that effect, as described in, for example, [19].

On the other hand, use of the wye-connected STATCOM offers the possibility of adding zero-sequence (common-mode) components to the string voltages. If the midpoint M is not connected, or if there is a transformer without a connected midpoint, such components do not affect the external behavior of the converter. However, a zero-sequence voltage at the fundamental frequency can be used to redistribute energy between the phase arms in a similar way as a zero-sequence current can be used with the delta-connected circuit. Thus, there is a possibility for balancing the capacitor energies during operation under unbalanced conditions also using the wye-connected STATCOM. Furthermore, third-harmonic zero-sequence injection in order to allow for an increase of the fundamental output voltage is possible. In a similar fashion as in the case of, for example, a two-level converter such a component at the third harmonic order can allow for an increase of the modulation index from $m_a = 1.0$ to $m_a = 2/\sqrt{3}$. This leads to a corresponding increase of the power rating of the converter, thus reducing the cost per unit of reactive power.

1.5.3.2 Full-Bridge MMC

The shortcomings of the ordinary MMC with half-bridges in terms of dc-side fault handling capability can be overcome by replacing the half-bridges with full-bridges. The bipolar voltage that can be provided by full-bridge strings permits the converter to control both ac-side and dc-side currents even in the case of a stiff dc-side short-circuit. This can be understood by considering Figure 1.20 and replacing the half-bridge strings by full-bridge strings. Under such circumstances the converter will topologically resemble a cascaded-submodule wye-connected STATCOM, as described in the previous section, where the submodule strings of each phase leg appear connected in parallel. Given that no direct voltage has to be imposed by the arms during a dc-side fault, the peak ac-side emf that can be provided by the converter then rises to the sum of the submodule capacitor voltages V_c^Σ. It normally equals the dc-side voltage, V_d, and at a nominal modulation index of $m_a = 1.0$ it exceeds the peak value of the driving voltage of the ac grid twofold. Therefore, it is more than sufficient for controlling the currents.

Furthermore, in normal operation the use of full-bridges can allow for an extension of the possible operating range as discussed in Section 1.5.1. Equations (1.17) and (1.18), visualized in Figure 1.13, define the reachable operating regions in terms of alternating and direct voltage components, at a given sum capacitor voltage, V_c^Σ, for a single submodule string with either half-bridges or full-bridges. Given that for an MMC the dc-side voltage equals the sum of the

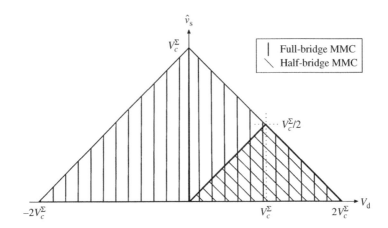

Figure 1.26 Operating region in terms of alternating and direct voltage of half-bridge and full-bridge MMCs given a sum capacitor voltage per arm of V_c^Σ. Impact of capacitor voltage fluctuation is not taken into account.

output voltages of the strings in the upper and lower arms, whereas the ac-side inner emf v_s is half of the difference between these–see Equation (1.19)–the corresponding regions for an entire MMC built from either type of submodules can be constructed; see Figure 1.26. The difference between the submodule types lies in the region where $V_d < V_c^\Sigma$. In this region a converter equipped with full-bridges can provide an ac-side emf higher than half of the direct voltage ($\hat{v}_s > V_d/2$). Thus, infinite modulation index is in theory possible.

It should also be noted that the operating region of the full-bridge converter is symmetric with regard to the $\hat{v}_s$-axis so that the direct voltage can also be negative. This is an attractive feature, for instance in HVDC applications, where dc polarity reversals may occur, such as when interoperability with CSC technology is required.

Importantly, the reasoning above is only valid when the overall capacitor voltages remain approximately constant. In practice the capability to provide voltage components will be affected by the ripple in the capacitor voltages, although this is generally a second-order effect. As long as the average capacitor energy is in the range of several tens of Joules per kilowatt of rated power, this extension or reduction of the operating area will not be dominant (at 50 or 60 Hz operating frequency).

On the same topic, since the operating range can be extended, it can also be of interest to investigate the impact on the fluctuation of the stored arm energies, and thus on the capacitor voltage ripple, in points outside the range possible with the conventional MMC. In particular, by increasing the modulation index beyond 1.0 the capacitor energy variation can be significantly reduced [20]. By observing Equations (1.40) and (1.41) it is evident that with $m_a = \sqrt{2}$ the terms corresponding to power exchange at the fundamental frequency will cancel out when only active power is exchanged on the ac side ($\varphi = 0$ rad or $\varphi = \pi$ rad). The expressions for the power exchange with the upper and lower submodule strings in a phase leg will both take the following form:

$$p_u = p_l = -\frac{1}{8} V_d \hat{i}_s m_a \cos(2\omega_1 t - \varphi). \tag{1.56}$$

Thus, under these circumstances, the power exchanged between the upper and lower phase arms disappears. Only the inevitable fluctuation at the second harmonic frequency remains, since it is related to the power exchange with the entire phase leg. Figure 1.27 shows steady-state waveforms of an MMC operating in this point. The dc-side voltage and current have been set to 1.0 p.u., as in Figure 1.19, to allow for comparisons. The magnitude of the ac-side emf increases to 1.41 p.u. and the currents on the ac side are reduced by the same factor to preserve power balance. As predicted by Equation (1.56), the curves corresponding to the power exchanged with the submodule strings will coincide and only contain a second harmonic component. This also applies to the energy fluctuations in the submodule capacitors [chart(f)] since these quantities are time integrals of the power exchanged. Comparing Figures 1.19 and 1.27, it is evident that the capacitor energy fluctuation is reduced by approximately half under the given circumstances. Therefore, a full-bridge MMC intended for operation at elevated modulation indices could potentially be designed with smaller submodule capacitors. However, this only holds at pure active power exchange at the ac side. At significant production or absorption of reactive power the energy ripple will increase as the fundamental-frequency variation will be present, as explained in [20].

At this point it should be noted that the power angle φ refers to the active and reactive power exchanged with the inner emf of the converter. This will deviate slightly from the reactive power exchanged with the entire converter since half of the arm inductance appears in series with the inner emf as seen from the ac side–see Figure 1.18(a)–and this inductance consumes a certain amount of reactive power.

In terms of semiconductor expenditure the full-bridge MMC appears less attractive. For a converter intended for operation in the same fashion as a half-bridge MMC, that is with modulation index below 1.0 (or below $2/\sqrt{3}$ with third harmonic injection), the required over-all semiconductor active area will in the first approximation be doubled. For a given dc-side voltage the same number of submodules is required, but each submodule has two phase legs instead of one, whereas the maximum instantaneous current will be the same. However, since the converter can control the currents during dc-side short-circuits, certain savings are possible in terms of the rating of the diodes. As mentioned in Section 1.5.2, for a half-bridge MMC the diodes in the valves parallel to the submodule terminals will need to conduct the ac-side short-circuit currents in case the dc-side voltage disappears. In case no other means for protecting these diodes against surge currents are present in the system, this translates to a significant over-rating of the diodes, which can possibly be avoided in a full-bridge converter.

Under the same operating conditions, the semiconductor conduction losses will be doubled since the current will always pass though two valves in each submodule. However, the average switching frequency of the valves can in theory be halved, maintaining the same output voltage distortion and submodule capacitor-voltage ripple. This can, for instance, be achieved by only switching one of the phase legs and permanently keeping one valve of the other phase leg in the conducting state. Therefore, the overall semiconductor switching losses will remain the same as in the half-bridge case.

Since the full-bridge MMC can be operated at peak alternating voltages higher than the direct voltage, the semiconductor rating requirements of a converter designed for such operation are also of interest. However, at a given dc-side voltage (which is most often the case in HVDC applications) any increase in the ac-side emf magnitude will entail an increase in the required total capacitor voltage (i.e. that more submodules are required), according to Equation (1.18). Maintaining the same power, the ac-side currents will be reduced while the dc-side

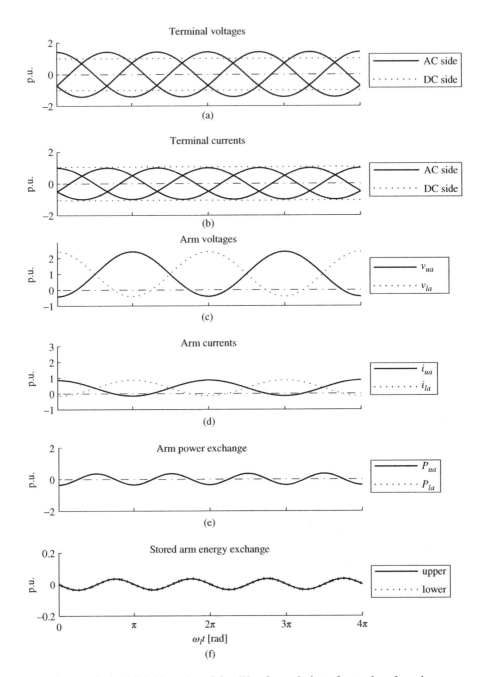

Figure 1.27 MMC with full-bridge submodules. Waveforms during a few cycles of steady-state operation with $m_a = \sqrt{2}$ and $\varphi = 0$. Note how the fundamental-frequency term of the submodule string energy ripples disappears at this modulation index.

current is unchanged. To understand the overall influence on the semiconductor expenditure, the previously derived Equation 1.48 in Section 1.5.2 is useful. It links the semiconductor cost, measured as the product of the peak submodule string current and voltage, to the modulation index at $\cos \varphi = 1$. It was found that the cost has a minimum for $m_a = \sqrt{2}$ (i.e. in the region only reachable by a full-bridge converter). A plot illustrating the relationship can be found in Figure 1.22. The difference in P_{sc} when m_a is increased from 0.9 to $\sqrt{2}$ is only 5%. The possibility of operation at elevated modulation indices, enabled by full-bridges, can therefore not compensate for the fact that the number of semiconductors is doubled with full-bridges ($k_c = 4$ instead of 2). The total semiconductor expenditure is, therefore, approximately doubled for a converter using full-bridges, regardless of how it is operated.

Behind this reasoning also lies the assumption that a transformer is present at the ac terminal, whose turns ratio can be adapted to suit the chosen output voltage of the converter. Also in terms of power losses, designing an MMC for operation at elevated modulation indices does not appear to offer any benefits. For applications where the dc voltage may be chosen freely, such as back-to-back HVDC interconnections, operating the full-bridge MMC with lower dc voltage may, however, be beneficial.

1.5.3.3 Alternate-Arm Converter

The quest for converter circuits with the capability of blocking the dc-side current at faults has also resulted in other solutions. The so-called *alternate-arm converter* (AAC) is shown in Figure 1.28. Its arms combine strings of full-bridge submodules and valves of the kind found in a normal two-level converter, as shown in the figure. These latter valves are labeled *director switches*. It was proposed by the manufacturer Alstom Ltd [21], which has also filed several patent applications related to the topology and its control. The full-bridge submodule strings

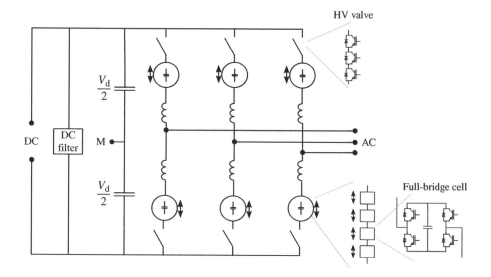

Figure 1.28   Schematic diagram of the AAC.

give this converter the capability of controlling the currents during dc-side faults. Indeed, during a stiff dc-side short-circuit, when $V_d = 0$, the equivalent schematic will resemble that of the full-bridge MMC, under the same conditions.

The basic mode of operation of a phase leg is based on having the director switches each conduct during half of the fundamental cycle. The upper switch conducts during the positive half-cycle, and during this interval the upper submodule string provides the ac-side emf. Thus, the string voltage should be $V_d/2 - v_s$ during this time. As earlier, V_d and v_s represent the dc-side voltage and the ac-side emf, respectively. Conversely, the director switch in the lower phase arm is put in the on-state during the half-cycle with negative ac-side emf, and the corresponding lower submodule string provides a voltage to the ac side during this period, (i.e. $V_d/2 + v_s$). Figure 1.29 shows current and voltage waveforms during a couple of fundamental cycles.

The operation thus resembles the standard two-level converter in the sense that the two arms of the same phase alternately conduct, hence the name of the converter. This is in contrast to the MMC, where the two submodule strings of a phase leg implement voltage sources that conduct current continuously. A further important difference compared to the MMC is the behavior towards the dc side. The MMC is a multilevel VSC seen from both the ac and dc networks to which it is connected. Each phase leg of the AAC instead acts as a current source as seen from its dc terminal, consistently injecting one half-cycle of the alternating current waveform into one of the dc poles and the other half-cycle into the other pole. This follows from the operating mode of the converter where the director switches only permit current to flow into one of the dc poles at a time. Like the two-level converter, the AAC therefore requires a capacitor across the dc terminal to smoothen the direct voltage. However, there is a significant difference in terms of the harmonic content of the currents injected into the dc link. A three-phase two-level converter mainly produces high-frequency components, resulting from the switching process, since the second-harmonic fluctuations cancel out between the phase legs, as explained in Section 1.2. The AAC, on the other hand, will inject significant current harmonics at multiples of six times the fundamental (6th, 12th, 18th, etc.). This occurs since the AAC phase legs inject half-cycle current waveforms into the dc poles, whose sum over the three phases is periodic at six times the fundamental frequency, as evident from Figure 1.29(b).

As discussed previously the power flow of any submodule string always needs to be balanced over time so that no net energy exchange occurs during an entire fundamental cycle. One of the consequences is that the power flow at the ac and dc terminals of the converter also needs to be balanced over a fundamental cycle. Therefore, the direct coupling between ac and dc side currents that exists in the AAC implies that there will also be a fixed ratio between the ac-side and dc-side voltages to maintain power balance. The average power flow at the ac terminal of one of the phase legs, as usual, equals

$$P_a = \frac{1}{2}\hat{v}_s\hat{i}_s \cos(\varphi) \tag{1.57}$$

since the ac-side current and voltage can be assumed to be sinusoidal. Keeping in mind that the current injected into the positive dc terminal has the waveform of a half-cycle of a sinusoid, phase shifted by φ rad, the corresponding dc-side expression is

$$P_d = V_d\frac{1}{\pi}\int_0^\pi \hat{i}_s \sin(x - \varphi)dx, \tag{1.58}$$

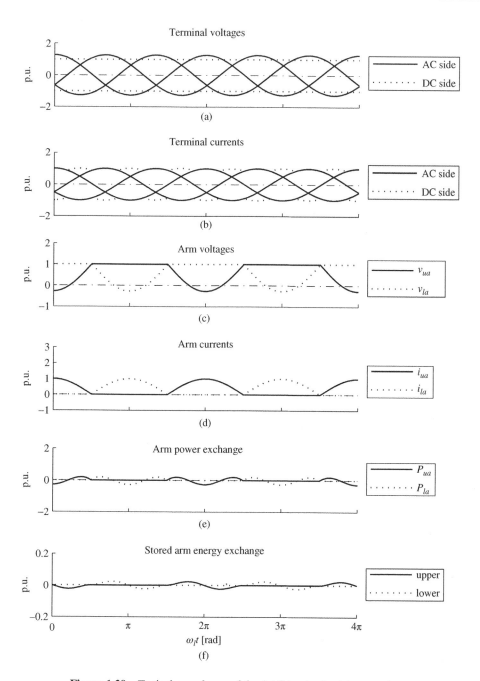

Figure 1.29 Typical waveforms of the AAC in steady-state operation.

which by evaluation of the integral can be rewritten

$$P_d = \frac{2}{\pi} V_d \hat{i}_s \cos \varphi. \tag{1.59}$$

During the other half-cycle the ac-side current is injected into the other dc pole resulting in the same average power flow. Equating the right-hand sides of Equations (1.57) and (1.59) and dividing by $\hat{i}_s$ and the power factor, which appear on both sides, yields

$$\hat{v}_s = \frac{2}{\pi} V_d. \tag{1.60}$$

The operating range is thus limited, and the converter only works in this specific operating point, commonly labeled the *sweet-spot*. Outside the sweet-spot there will be no natural balancing of dc- and ac-side power flows, leading to uncontrolled drift of the submodule capacitor voltages and collapse of the converter operation. Notably, the sweet-spot corresponds to a modulation index of

$$m_a = \frac{4}{\pi} \tag{1.61}$$

which is higher than what can normally be achieved by a two-level converter or a half-bridge-based MMC. As noted in the previous section, an MMC with full-bridges can in theory operate with any modulation index.

The modulation index normally stays fairly constant during the operation of a grid-connected converter since the alternating and direct voltages in most cases do not depart significantly from their nominal magnitudes. This is in contrast to the case of a motor drive, where alternating voltage levels from zero and upwards may be required, depending on the speed of the motor, to maintain constant flux linkage. Still, even for a grid-connected converter, a certain flexibility in terms of the voltage ratio is required to compensate for ac and dc voltage variations and also to allow for reactive power interaction, which requires that the converter emf differs in magnitude from the grid voltage. Thus, the fixed voltage ratio would render the AAC useless for most practical applications. For this reason a few methods have been developed for balancing the power flows also at other modulation indices.

First, it is possible to introduce a short overlap period during which the director switches of both the upper and lower arms are conducting simultaneously. During such a period, a current will flow through the entire phase leg thus exchanging energy between the upper and lower arm submodule strings and the dc terminal. The overlap periods lead to additional losses since more current will flow through the phase arms. Also, the amount of dc-side harmonic distortion will increase, owing to the overlap periods since they result in short current bursts being injected into the dc terminal. Notably, the overlap period will need to be longer the farther from the sweet-spot the converter is to be operated.

Another balancing strategy is based on having a controllable zero-sequence current at the third harmonic frequency flowing on the ac side [22]. In order for this to be possible the converter can be connected to a delta-wye transformer with the wye-connected windings toward the converter. To provide a path for the zero-sequence currents the midpoint formed by these windings is connected to the dc-link midpoint (denoted M in Figure 1.28). Since a transformer is generally connected between the converter and the ac grid, this arrangement does not necessarily imply any significant additional cost. The third-harmonic currents do not give rise to any average power exchange with the dc link. Also, since they have zero phase sequence they will circulate in the grid-side delta winding of the transformer and thus not transfer any power to or from the ac side. Instead, they will cause power interaction between the phase arms. This is due to the fact that the director switches connect each phase arm during half a fundamental cycle, corresponding to 1.5 cycles at the third harmonic, which will give a net contribution to altering the charge of the submodule capacitors of the concerned arm. A disadvantage of this method is that the third harmonic component will cause a voltage ripple across the dc link capacitors. This ripple is particularly unattractive from a harmonics mitigation perspective since it is of common-mode nature and therefore especially prone to cause EMI. It is therefore likely to require additional filtering to allow for compliance with distortion limits.

It can be concluded that the discussed balancing methods can indeed compensate for the unbalanced power flow caused during operation away from the sweet-spot. However, they both rely on using additional currents. These currents will increase power losses and potentially drive up the rating requirements of the semiconductors. Furthermore, as mentioned earlier, they cause additional harmonic distortion at the dc side of the converter which may necessitate additional filtering to comply with norms and regulations.

As discussed in Section 1.5.2 the half-bridge MMC suffers from large energy pulsations in the phase arms during operation, which implies that the overall stored energy in the submodule capacitors needs to be large to limit the voltage ripple. This can to a certain extent be mitigated by using full-bridge submodules thanks to the possibility of operating with modulation indices far beyond one, whereby the energy pulsations occurring between upper and lower phase arms can be reduced. For the AAC the arm power pulsation can be derived by assuming sinusoidal ac-side quantities, whereby the upper arm voltage and current is

$$v_u = V_d/2 - \hat{v}_s \sin(\omega_1 t)$$
$$i_u = \hat{i}_s \sin(\omega_1 t - \varphi). \tag{1.62}$$

during the interval when the upper director switch is conducting ($0 < \omega_1 t < \pi$). Multiplying these expressions, and some simplification, yields the following equation for the power exchanged with the upper arm during the concerned interval:

$$p_u = \frac{V_d}{2}\hat{i}_s \sin(\omega_1 t - \varphi) - \frac{1}{2}\hat{v}_s\hat{i}_s[\cos(\varphi) - \cos(2\omega_1 t - \varphi)]. \tag{1.63}$$

During the other half-cycle ($\pi < \omega_1 t < 2\pi$) the current, and thus the power, is zero. The corresponding equations for the lower phase arm will be the same but phase shifted by half of a fundamental cycle. In the sweet-spot, according to Equation (1.60), Equation (1.63) simplifies to

$$p_u = \frac{1}{2}\hat{v}_s\hat{i}_s \left\{ \frac{\pi}{2} \sin(\omega_1 t - \varphi) - \cos(\varphi) - \cos(2\omega_1 t - \varphi) \right\}. \tag{1.64}$$

As evident from this equation the power fluctuation to be handled by the submodule capacitors contains a fundamental frequency term and a second harmonic term, like in the MMC. In addition there is a constant term. However, since the current is only present in the arm during the interval $0 < \omega_1 t < \pi$ the power flow can average to zero. Furthermore, the constant and sine terms will to a significant extent cancel out during the period. Therefore, the energy fluctuation and thus the capacitor voltage ripple is reduced. As evident from Figure 1.29(f) the capacitor energy fluctuation is low in comparison to the MMC. The choice of operating point and normalization of the signals used in Figures 1.19, 1.27, and 1.29 were made so as to allow for comparisons. A comprehensive analysis of the capacitor energy pulsations of the

AAC confirms these findings [23]. Thus, potentially the AAC can be designed with less capacitive energy storage than the MMC. However, operation far from the sweet-spot, employing the discussed means of balancing, will generally increase the energy fluctuations, and a proper analysis of each case is required.

In terms of overall power semiconductor expenditure and power losses the AAC is generally perceived as falling between the half-bridge MMC and the full-bridge MMC [24]. In the first approximation the required maximum output voltage of all full-bridge submodules per arm is half the pole-to-pole dc-side voltage. Thus, the total capacitor voltage per arm has to be rated for this value as well, which is half of what is required for a half-bridge MMC. However, since each full-bridge has two phase legs, the number of semiconductor elements for the submodule strings will be similar. The director switches, on the other hand, need to withstand the ac-side peak phase voltage, which causes extra semiconductor expenditure. Furthermore, to implement the director switches direct series connection of semiconductor elements is generally required since there are no available elements that can withstand the voltages encountered in a converter for transmission applications. This represents a step away from a truly modular approach to designing converters, and will generally imply additional costs. For such, series-connected, devices there is always a need to provide even voltage sharing, both static, during blocking, and dynamic, during the switching processes. Possibly, the dynamic voltage sharing can be simplified by the circumstance that the switchings mainly occur at zero voltage but the valve design also has to take into account abnormal operating cases when zero-voltage switching cannot be fulfilled.

As regards current rating, the peak arm current will also in this case likely determine the rating of semiconductors of both the full-bridge strings and the director switches. It will amount to the peak ac-side current, which in most cases is more than in the half-bridge MMC; see Equation 1.44.

A final issue that needs consideration is the harmonic properties of the converter. The inner ac-side inner emf is a multilevel voltage much like that of the MMC. Therefore, the need for filtering on the ac side should be minimal. In many applications the arm reactors together with the leakage inductance of a transformer, connecting the converter to the grid, will be sufficient to keep the ac-side distortion within permissible limits. However, unlike the MMC, which also acts as a voltage source toward the dc side, the AAC, as mentioned, resembles a diode bridge seen from the dc terminal. Therefore, the dc-side current contains harmonic components at multiples of six times the fundamental. Therefore, to comply with the stringent requirements concerning harmonic distortion in HVDC applications, the AAC likely also requires a filter, in addition to the dc capacitor, to prevent these harmonics from propagating to the network connected on the dc side, as indicated in Figure 1.28.

In summary, the AAC represents a possible alternative to the MMC for HVDC applications. It appears to be seriously considered for this purpose by the manufacturer GE, formerly Alstom. The component expenditure and the power losses appear favorable, especially when the capability of blocking dc faults is taken into account. In particular, the size of the submodule capacitors could be reduced compared to the MMC given that the overall energy fluctuations of the submodule strings are smaller.

However, the fixed ratio between direct and alternating voltages (the sweet-spot) implies an important restriction to the operation of the converter. The proposed methods for allowing operation away from the sweet-spot also cause certain drawbacks in terms of increased losses and harmonic distortion. Furthermore, the AAC is not modular to the same extent as the

MMC. In addition to the submodules it requires series-connected valves capable of blocking high voltage, as well as a relatively large dc capacitor dimensioned for the full pole-to-pole dc-side voltage. Also, the harmonic distortion at the dc side is considerably higher than that of the MMC.

1.6 Summary

The two-level converter enjoys extremely widespread use for low-voltage motor drives as well as for grid connected applications, such as the integration of RES and uninterruptible power supplies. This topology benefits from its simplicity, and the easy power transfer between phase legs which leads to minimal requirements on the size of the capacitor energy storage.

In the voltage range beyond a few kilovolts two-level converters become harder to implement because of the lack of power semiconductors with sufficient blocking capability. Direct series connection of devices to handle this issue is possible, but it is an expensive technology that has never gained widespread use. Instead, multilevel converters are in most cases more suited for these higher voltages. These split the switched direct voltage among several dc capacitors so that semiconductor series connection can be avoided. Also, the relationship between switching frequency and harmonic distortion of the output voltage is much more favorable. This is due to both the increase in the number of voltage levels and the decoupling of the pulse frequency from the switching frequency.

For medium-voltage applications up to several kilovolts diode-clamped topologies currently dominate the market. These share the benefit of not requiring large dc capacitors with the two-level converter. However, extension to more than three levels is cumbersome, owing to the many clamping diodes required, and the fact that these all need to be connected to the common dc link. Therefore, diode-clamped converters are not suited for applications in power transmission and distribution systems, where tens or hundreds of kilovolts are to be handled.

At such levels of voltage instead cascaded converters, employing series-connection of converter submodules, are rapidly gaining market share. This is happening at the expense of CSCs that previously reigned supreme in this power range. A number of important factors reinforce this development.

First, the scalability in terms of voltage, and thereby power, is excellent. The voltage rating can be increased by simply adding more submodules. Since the commutation paths are internal to the submodules, the impact of stray elements is not more critical than in a low-voltage converter. Furthermore, since the number of levels can easily be increased, the switching frequency can be reduced to only a few times the fundamental while still fulfilling requirements concerning harmonic distortion. This implies that semiconductor devices of high blocking voltage can be used without excessive switching losses. Moreover, the modular design implies benefits in terms of design and manufacturing, since the converter largely consists of a number of identical units.

On the negative side, cascaded converters generally suffer from considerable internal energy fluctuations at low frequency, which have to be absorbed by the submodule capacitors. This implies that the total energy stored in the capacitors needs to be much larger than in, for instance, a two-level converter. The cost of the capacitors is increased, and the large stored energy also implies challenges for the handling of internal short-circuit faults. The energy pulsations occur at multiples of the fundamental frequency, which is particularly critical for a converter that should be able to operate at low frequency (e.g. in a motor drive application).

Generally, additional measures have to be taken to balance the energy pulsations between the phase arms in these cases.

The introduction of the MMC with half-bridge submodules was a major step forward, since it was the first truly submodule-based converter to allow ac/dc conversion. It has sparked intense development efforts, leading to the commercialization of this technology within less than ten years. Currently, several HVDC links are already in operation that rely on MMCs for the conversion between ac and dc. This means that VSCs can compete successfully with CSCs also in the high-power range. One weakness, however, is the loss of current control that occurs whenever the direct voltage is absent.

The possibilities of cascaded converters have also been exploited in other ways. Particularly the full-bridge STATCOM has found many practical applications. MMCs with full-bridges are also seriously considered for HVDC applications thanks to the capability of maintaining current control even during dc-side short-circuit faults. Full-bridges also offer a radically increased operating range, but the semiconductor cost and the power losses will be considerably higher.

In the coming chapters various aspects of the design, operation, and applications of the MMC are described and explained in more detail.

References

[1] A. Nabae, I. Takahashi, and H. Akagi, "A new neutral-point-clamped PWM inverter," *IEEE Transactions on Industry Applications*, vol. IA-17, no. 5, pp. 518–523, 1981.

[2] B. Bijlenga, "HVDC device for converting between alternating voltages and direct current voltages," US Patent 6 480 403, 2002.

[3] B. D. Railing, J. J. Miller, P. Steckley, G. Moreau, P. Bard, L. Ronström, and J. Lindberg, "Cross-Sound Cable project: Second generation VSC technology for HVDC," in *Proceedings of the Cigré Session*, Paris, France, 2004.

[4] I. Mattsson, A. Ericsson, B. D. Railing, J. J. Miller, B. Williams, G. Moreau, and C. D. Clarke, "Murraylink: The longest underground HVDC cable in the world," in *Proceedings of the Cigré Session*, Paris, France, 2004.

[5] T. Meynard and H. Foch, "Multi-level conversion: High voltage choppers and voltage-source inverters," *Proceedings of the 23rd Annual IEEE Power Electronics Specialists Conference*, 1992.

[6] K. Ilves, A. Antonopoulos, L. Harnefors, S. Norrga, L. Ängquist, and H.-P. Nee, "Capacitor Voltage Ripple Shaping in Modular Multilevel Converters Allowing for Operating Region Extension," *Proceedings of the 37th Annual Conference of the IEEE Industrial Electronics Society (IECON 2011)*, 2011, pp. 4403–4408.

[7] J. D. Ainsworth, M. Davies, P. J. Fitz, K. E. Owen, and D. R. Trainer, "Static VAr compensator (statcom) based on single-phase chain circuit converters," *IEE Proceedings: Generation, Transmission and Distribution*, vol. 145, no. 4, pp. 381–386, 1998.

[8] P. W. Hammond, "A new approach to enhance power quality for medium voltage AC drives," *IEEE Transactions on Industry Applications*, vol. 33, no. 1, pp. 202–208, 1997.

[9] R. Marquardt, A. Lesnicar, and J. Hildinger, "Modulares Stromrichterkonzept für Netzkupplungsanwendung bei hohen Spannungen," *Proceedings of the ETG-Fachtagung, Bad Nauheim, Germany*, 2002.

[10] T. Westerweller, K. Friedrich, U. Armonies, A. Orini, and D. Parquet, "Trans Bay Cable: World's first HVDC system using multilevel voltage-sourced converter," *Proceedings of the Cigré Session*, Paris, France, 2010.

[11] K. Ilves, S. Norrga, L. Harnefors, and H.-P. Nee, "On energy storage requirements in modular multilevel converters," *IEEE Transactions on Power Electronics*, vol. 29, no. 1, pp. 77–88, 2014.

[12] A. Antonopoulos, L. Ängquist, S. Norrga, K. Ilves, L. Harnefors, and H.-P. Nee, "Modular multilevel converter ac motor drives with constant torque from zero to nominal speed," *IEEE Transactions on Industry Applications*, vol. 50, no. 3, pp. 1982–1993, 2014.

[13] S. Mori, K. Matsuno, T. Hasegawa, S. Ohnishi, M. Takeda, M. Seto, S. Murakami, and F. Ishiguro, "Development of a large static var generator using self-commutated inverters for improving power system stability," *IEEE Transactions on Power Systems*, vol. 8, no. 1, pp. 371–377, 1993.

[14] C. Schauder, M. Gernhardt, E. Stacey, T. Lemak, L. Gyugyi, T. Cease, and A. Edris, "Development of a ±100 MVAr static condenser for voltage control of transmission systems," *IEEE Transactions on Power Delivery*, vol. 10, no. 3, pp. 1486–1496, 1995.

[15] T. Larsson, R. Grünbaum, and B. Ratering-Schnitzler, "SVC Light: A utility's aid to restructuring its grid," *Proceedings of the IEEE Power Engineering Society Winter Meeting 2000*, vol. 4, pp. 2577–2581, Singapore, 2000.

[16] R. H. Baker and L. H. Bannister, "Electric power converter," US Patent 3 867 643, Feb. 18, 1975.

[17] F. Z. Peng, J.-S. Lai, J. McKeever, and J. VanCoevering, "A multilevel voltage-source inverter with separate DC sources for static var generation," *IEEE Transactions on Industry Applications*, vol. 32, no. 5, pp. 1130–1138, 1996.

[18] M. Pereira, D. Retzmann, J. Lottes, M. Wiesinger, and G. Wong, "SVC PLUS: An MMC STATCOM for network and grid access applications," *Proceedings of the 2011 IEEE PowerTech*, Trondheim, Norway, 2011.

[19] F. Peng and J. Wang, "A universal STATCOM with delta-connected cascade multilevel inverter," *Proceedings of the IEEE 35th Annual Power Electronics Specialists Conference*, vol. 5, pp. 3529–3533, 2004.

[20] K. Ilves, S. Norrga, and H.-P. Nee, "On energy variations in modular multilevel converters with full-bridge submodules for AC-DC and AC-AC applications," *Proceedings of the 15th European Conference on Power Electronics and Applications (EPE)*, Lille, France, 2013.

[21] D. Trainer, C. Davidson, C. Oates, N. MacLeod, D. Critchley, and R. Crookes, "A new hybrid voltage-sourced converter for HVDC power transmission," *Proceedings of the Cigré Session*, Paris, France, 2010.

[22] F. Moreno, M. Merlin, D. Trainer, K. Dyke, and T. Green, "Control of an alternate arm converter connected to a star transformer," *16th European Conference on Power Electronics and Applications*, Lappeenranta, Finland, 2014.

[23] M. Merlin, T. Green, P. Mitcheson, F. Moreno, K. Dyke, and D. Trainer, "Cell capacitor sizing in modular multilevel converters and hybrid topologies," *Proceedings of the 16th European Conference on Power Electronics and Applications*. Lappeenranta, Finland: IEEE, 2014.

[24] S. Norrga, X. Li, and L. Ängquist, "Converter topologies for HVDC grids," *Proc. 2014 IEEE International Energy Conference (ENERGYCON)*, pp. 1554–1561. Dubrovnik, Croatia: IEEE, 2014.

2

Main-Circuit Design

2.1 Introduction

The purpose of this chapter is to provide an understanding of various design choices for different parts of the main circuit of modular multilevel converters (MMCs). This includes power semiconductors, submodule capacitors, arm inductors, submodule design, redundant submodules, auxiliary power supplies, and start-up procedures.

The design of the main circuit of an MMC involves several steps, ranging from dimensioning of main-circuit components to issues with auxiliary power supplies and start-up procedures. The chapter starts with a presentation of the power semiconductors used in MMCs today, and possible future alternatives. In Section 2.3, the submodule capacitors are described. These capacitors are voluminous and costly, and as such they deserve considerable attention. The following section, i.e. Section 2.4, describes arm inductors. Section 2.5 describes different possible submodule configurations. This is important because the choice of submodule configuration influences various properties, such as fault-handling capability, losses, and the size of the submodule capacitors. In the same section the designs of three different commercial realizations of MMCs are also described briefly. In Section 2.6, the choice of main-circuit parameters is discussed. This involves choice of power semiconductor devices, choice of the average submodule capacitor voltage, number of submodules, submodule capacitance, and arm inductance. The handling of redundant and faulty submodules is treated in Section 2.7. Depending on the choice of power semiconductor devices in the submodules, and whether the submodules contain a distributed control unit, different amounts of auxiliary power supply are necessary. This becomes a major task when higher auxiliary power supplies are required. In Section 2.8, therefore, this problem is described, along with suggestions for possible realizations found in the literature. In Section 2.9, a brief description of start-up procedures is given. Finally, a short summary of the chapter is given in Section 2.10.

Design, Control, and Application of Modular Multilevel Converters for HVDC Transmission Systems, First Edition.
Kamran Sharifabadi, Lennart Harnefors, Hans-Peter Nee, Staffan Norrga, and Remus Teodorescu.
© 2016 John Wiley & Sons, Ltd. Published 2016 by John Wiley & Sons, Ltd.
Companion Website URL: www.wiley.com/go/Sharifabadi/ModularConverters

2.2   Properties and Design Choices of Power Semiconductor Devices for High-Power Applications

The intention of this section is to provide an insight into the main physical properties of power semiconductor devices without going into all details of the underlying semiconductor physics. The only part penetrating slightly deeper into semiconductor physics is Section 2.2.8 on silicon carbide power semiconductors. This is the last subsection of this section, and the reader who is only interested in the power semiconductor technology of today can skip this subsection.

This section starts with a brief historical overview of the development toward modern power semiconductors, starting with the mercury arc valve. In Section 2.2.2, the basic conduction properties of silicon are described. In the next subsection, i.e. Section 2.2.3, the blocking capability of a p–n junction is described along with the choices of doping densities and thicknesses of the semiconductor layers. Thereafter, in Section 2.2.4, the conduction properties of semiconductors are discussed. It is explained that pure or moderately doped silicon layers have far too weak conduction properties for high-voltage devices if only the resistivity of the material itself is considered. The concept of carrier injection is introduced, and its effect on conductivity is explained. Also the effects of the carrier lifetime are discussed, with respect to both the voltage drop and the switching properties. It is emphasized that the choice of carrier lifetime is a delicate trade-off between conduction and switching characteristics of a power device. In the case of the MMC, this optimization results in a very different set of device parameters than for other voltage-source converters. The reason is that the switching frequency per device can be chosen much lower for the MMC. Examples are given showing how this favors high-voltage thyristor-based devices, which have lower ON-state voltage drops than other devices. Next, in Section 2.2.6, different alternatives for packaging are presented along with the consequences for the application. In Section 2.2.7, the reliability of power semiconductor devices is treated, and finally in Section 2.2.8, silicon carbide power devices are described.

2.2.1   Historical Overview of the Development Toward Modern Power Semiconductors

During the first half of the 20th century, high-voltage ac transmission had reached such a maturity that it was the obvious choice for building meshed high-voltage transmission systems. The historical battle between ac and dc transmission was almost forgotten and the common opinion was that essentially all onshore transmission should be based on high-voltage ac overhead lines. In some specific cases, however, high-voltage ac transmission could not be used. Two such cases are long transmissions across the sea and interconnections of unsynchronized ac grids.

The mercury arc valve had already been invented, in 1902, by Peter Cooper Hewitt. It was used as a rectifier for speed control of dc motors, battery charging, and electroplating for metal surface treatment. The use was, however, limited to voltages below a few kilovolts because of problems with *arc-back* (or backfire), a phenomenon where the valve conducts in the reverse direction when the applied voltage is negative. During the 1930s and 1940s, Uno Lamm at ASEA in Sweden did pioneering work which finally resulted in the graded-electrode mercury arc valve [1]. In Figure 2.1 a photograph of a mercury arc valve converter is shown.

Using this technology, the first mercury arc valve high-voltage direct current (HVDC) transmission system was put into service in 1954. This 96 km long 20 MW/100 kV single-cable link

Figure 2.1   Photograph of a mercury arc valve (Courtesy of ABB).

with sea return connected the island of Gotland with the mainland of Sweden. A photograph
of a mercury arc valve for this HVDC link is shown in Figure 2.2. In 1941 a similar project
was also initiated in Germany, but because of the war the project was never put into operation
in Germany. Instead, the equipment was moved to the Soviet Union and put into operation
almost at the same time as the Gotland HVDC link.

This was the start of the mercury arc valve HVDC transmission era, which lasted until the
mid-1970s. The mercury arc valves could successfully fulfill the conversion tasks, but the high
costs of the valves in combination with drawbacks such as limited voltage handling capability,
reliability problems, and concerns regarding the handling of the poisonous mercury prevented
a massive growth of the technology.

Starting with the invention of the thyristor by the research group led by William Shockley
in 1950, the power semiconductor technology developed in parallel with the high-voltage
mercury arc valves. The first commercially available thyristors were introduced in 1956
by General Electric. In the late 1960s the development of high-voltage thyristor valves for
HVDC transmission began, and in 1972 the first thyristor-based HVDC link was put into
service by General Electric.

This was the Eel River link in Canada, and from 1977 no new mercury arc valve systems were installed. The thyristor era was initiated, and because of comparatively low cost,
efficiency, and reliability, the thyristor-based HVDC transmission technology has enjoyed continuous growth ever since its introduction forty years ago.

The semiconductor technology was, however, also used for other power conversion tasks
than HVDC transmission. Already in the early years of the mercury arc valve, rectifiers were
used for speed control of dc motors. During the 1950s and 1960s, both thyristor and transistor

Figure 2.2   Photograph of the Gotland HVDC link mercury arc valve (Courtesy of ABB).

technology developed, ranging from gate turn-off (GTO) thyristors, to bipolar junction transistors (BJTs) and metal-oxide-semiconductor field-effect transistors (MOSFETs) [2], which were all used for variable-speed motor drives.

After intense research activities by several research groups, H. W. Becke and C. F. Wheatley Jr. filed a patent on a structure they called "power MOSFET with an anode region" in March 1980 [3]. This structure is basically what today is called the insulated-gate bipolar transistor (IGBT). In parallel with Becke and Wheatley Jr, J. B. Baliga had been working on the same concept, which was published in 1982 in *Proceedings of the IEEE*. In this publication, however, the device was called 'insulated gate rectifier', a name which would change several times over the next decade.

The term *insulated gate* refers to the fact that the gate of the device is formed in the same way as in a MOSFET, i.e. by means of an oxide layer. This implies that the device is voltage controlled, in contrast to a BJT, which requires a base current as long as the device conducts a main current (collector current). The benefit of having a voltage-controlled device is that the power consumption of the gate driver is substantially lower than for a current-controlled device.

This may seem insignificant, but in high-voltage converters the gate drivers operate at very high voltages, and providing the necessary supply voltage at several hundred kilovolts is not a trivial task [4]. In Section 2.8 special attention is given to this issue. Usually, such power supplies are costly, and the higher the power, the higher the cost. The low gate-power requirement of the IGBT is one of the main reasons why the IGBT replaced the GTO and BJT in motor-drive applications during the 1990. The IGBT could also switch significantly faster than the GTO and BJT, which meant that the switching losses could be reduced or that a higher switching frequency could be used.

During the late 1990s, the IGBT was introduced in voltage-source converter-based HVDC applications. The preceding development phase had involved several technology steps. First, a packaging technology that permitted series-connected stacks of IGBTs was developed. Second, a gate driver in combination with passive elements for simultaneous switching of all IGBTs in the stack was developed. Finally, a new converter type had to be designed and dimensioned along with new control systems and protection schemes.

The first voltage-source converter-based HVDC transmission was put into service by ABB in 1997 [5] in Hällsjön, Sweden. This was a test system for evaluation purposes. In 1999 the first commercial IGBT-based HVDC transmission link was delivered. This was the new Gotland link interconnecting the island of Gotland with the mainland of Sweden. The converters were two-level voltage-source converters (VSCs) with pulse-width modulated ac-side voltages. In many ways these new HVDC converters set a new standard for what an HVDC transmission system could accomplish.

The main difference compared to thyristor-based HVDC transmission is the high controllability which comes from the fact that an IGBT can not only turn on a current at a certain instant, like the thyristor. It can also turn off a current at a desired instant. As a consequence, an IGBT-based HVDC converter can control both active and reactive power independently. Especially for weak grids, this is very beneficial because the ac-side voltage can be controlled accurately by injecting (or extracting) a precisely controlled amount of reactive power at any instant.

Another benefit is that IGBT-based converters can operate without the presence of generators on the ac side (black-start capability). With IGBT technology, the application area of HVDC in general has been widened to cover many cases where thyristor-based HVDC cannot be used. A problem, however, with the two-level VSC technology was that the converter losses were considerably higher than for thyristor-based converters.

The main reason for this is that an approximately 15-times-higher switching frequency was used for the two-level IGBT VSCs than for the line-commutated thyristor converters. For more than a decade various ways to reduce the switching losses were evaluated, but no major breakthroughs were in sight until the MMC entered the stage.

Even though Rainer Marquardt had already presented the idea at international conferences 2003 [6], [7], not many people realized that this technology would totally change the whole HVDC transmission field until at least five years later. The MMC permits much lower switching frequencies per device, which means that the IGBTs can be optimized for minimum conduction losses. It is this optimization, and a few other issues, that are investigated in more detail in this chapter. Power devices for HVDC transmission must be designed for minimum losses but, as will be evident from the discussion below, this involves delicate trade-offs, especially between low ON-state voltage drops and fast switching.

2.2.2 Basic Conduction Properties of Power Semiconductor Devices

A semiconducting material has electrical conduction properties that are intermediate to conductors and insulators. This means that the conduction properties of a pure semiconductor material are inferior to electrical conductors by several orders of magnitude. If, for instance, a current density of 100 A/cm^2 flows through a pure silicon wafer with a resistivity of 2000 Ωm and a thickness of 1 mm, a voltage drop of approximately 2 MV would be observed. Such a voltage drop would make the silicon wafer useless in any electric power application.

The same holds for the insulation properties of a pure semiconductor material compared to an electrical insulator. Despite this, semiconductors are very successful when applied to solid-state power conversion. The reason is that the electrical properties of a semiconductor can be varied by orders of magnitude, for instance by means of doping and injection. An attempt to illustrate this is performed below.

2.2.3 P–N Junctions for Blocking

When used in power semiconductor devices, semiconducting crystals are always doped. In the case of a semiconducting crystal based on silicon, this means that atoms of dopants from groups V and III of the periodic table have replaced some of the silicon atoms in the semiconductor lattice.

If dopants from group V are used, a so-called n-type semiconductor has been formed, and if dopants from group III are used, the resulting semiconductor is said to be a p-type semiconductor. In the former case the excess charge carriers are electrons which have a negative (n) charge. In the latter-case, the excess carriers are so called holes which are positive (p).

A p–n junction is said to exist when a p-type region and an n-type region are adjoined to each other inside the semiconductor crystal. This can be achieved either by growing one type on top of the other using epitaxy or by changing the doping of the outermost layer of a base material through diffusion or implantation of dopants of the opposite kind compared to the base material.

An important feature of a p–n junction is that it permits current flows only in the direction from the p side to the n side. In the other direction, the p–n junction blocks the current flow. The consequence is that the p–n junction must block the voltage of the circuit which would otherwise drive the current through the junction. This blocking property is possible up to a maximum voltage, called *breakdown voltage*, which depends on the levels of doping, i.e. the relative content of dopants in the p region and the n region.

At the breakdown voltage a p–n junction is usually exposed to a so-called avalanche breakdown, where ever more charge carriers are generated. This occurs at a specific maximum field strength, E_{bd} (approx. $3 \cdot 10^5$ V/cm for silicon), in the junction and is unfortunately usually destructive. Many power MOSFETs are, however, dimensioned to withstand avalanche breakdown (avalanche rating), and they are tested in so-called unclamped inductive switching experiments [8].

Also in other modern power semiconductor devices (especially high-voltage IGBTs and diodes) attempts have been made to enable a certain degree of resilience against operation above the breakdown voltage. This is called *self-clamping*, and the main feature of the design is to avoid local current crowding at the event of an avalanche breakdown. The reader who desires a full description of this design choice is referred to [9] and [10].

Before explaining how the p–n junction should be dimensioned in order to avoid a breakdown, the concept of the space charge layer has to be introduced. The space charge layer is created on both sides of the junction as a consequence of carriers diffusing from one side to the other. Electrons diffuse from the n side to the p side, leaving positive dopant ions. Similarly, holes from the p side diffuse over to the n side leaving negative dopant ions on the p side. As shown in Figure 2.3, a positive space charge is created on the n side of the junction, and a negative space charge is created on the p side. The diffusion currents generating the space charge layer are, however, balanced by drift currents driven by the electric field between the

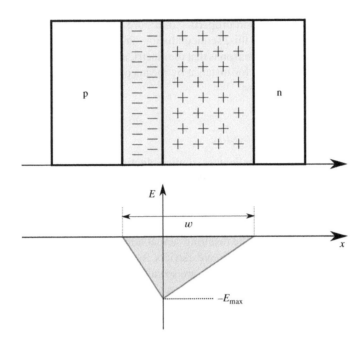

Figure 2.3   Space charge layer of a p–n junction.

positive charges on the n side and the negative charges on the opposite side. As a consequence, an equilibrium state between the two processes is established.

A closer analysis of the electric field, E, created by the space charges on the n side reveals that

$$\frac{dE}{dx} = \frac{qN_d}{\epsilon},$$ (2.1)

where N_d is the density of donors, q is the electron charge, and ϵ is the permittivity of silicon. On the p side

$$\frac{dE}{dx} = -\frac{qN_a}{\epsilon},$$ (2.2)

where N_a is the density of acceptors. The total voltage drop across the space charge layer can be obtained if E is integrated over the width, w, of the space charge layer. Accordingly,

$$V_{sc} = \int_w E(x)\, dx.$$ (2.3)

Typically, the n side of the junction has a significantly lower doping level than the p side. This means that w is practically located entirely on the n side. When a voltage is applied across the junction, the width of the space charge layer is changed. If the applied voltage has a polarity such that the positive terminal of the voltage source is connected to the n side and the negative terminal to the p side, the junction is said to be *reverse biased*.

In this case w is increased compared to the case when no voltage was applied to the junction. Regardless of the magnitude of this voltage, the gradients of the electric field on both sides of the junction will be constant, as given by Equations (2.1) and (2.2). This means that a high reverse voltage will cause both E_{max} and w to assume high values compared to the case when no voltage is applied to the junction. If the applied reverse voltage is sufficiently large, E_{max} may exceed the breakdown electric field, E_{bd}. This would cause a breakdown of the junction. In order to prevent this from occurring, two measures are taken. First, the doping level on the n side of the junction is chosen such that

$$N_d \propto \frac{1}{V_{bd}}, \tag{2.4}$$

where V_{bd} is the voltage at which the junction suffers a breakdown and N_d is the chosen doping level on the n side. The voltage V_{bd} is set to be higher than the rated voltage of the junction, so that various tolerances of the design would not force the junction to breakdown at voltages below the rated voltage. The second measure to prevent breakdown is to choose an appropriate thickness, w, of the low-doped n layer such that the space charge layer can be accommodated well within the layer. Accordingly,

$$w \propto V_{bd}. \tag{2.5}$$

In silicon w would be chosen to be approximately 100 μm for a junction with a voltage rating of 1000 V. In many modern high-power bipolar power semiconductor devices the low-doped layer which should accommodate the space charge layer in the blocking state does not have a thickness that would be sufficient to cover the whole space charge layer. Instead, a so called n-type buffer layer with higher doping is introduced between the low-doped layer and the p layer (anode or collector). The benefit of such a design is that a lower forward voltage drop can be achieved because of the reduced thickness of the low-doped n− layer. The reader who seeks a detailed description of the dimensioning of w including the buffer layer is referred to [10].

2.2.4 Conduction Properties and the Need for Carrier Injection

A semiconductor device for power electronics should not only be able to block high voltages. It should also have good conduction properties. As already stated above, a semiconductor has relatively poor conduction properties compared to a conductor. In the case of pure silicon, a voltage of 2 MV was required to achieve a current density of 100 A/cm^2 through a silicon wafer with the thickness 1 mm. If the same wafer would have an n doping of 10^{20} m^{-3}, the voltage drop would be reduced to 500 V.

This is a dramatic reduction, but the voltage drop is still approximately two orders of magnitude too high for practical applications. The dependence on the doping level, however, indicates that high doping levels in combination with thin layers may result in acceptable voltage drops.

Recalling Equations (2.4) and (2.5), it is found that high doping levels in combination with thin layers correspond to a junction that is dimensioned for a low breakdown voltage. In practice, therefore, doped silicon has sufficient conductivity if the rated voltage of the device does not exceed a few hundred volts. This property is exploited in power MOSFETs. For higher voltages, however, the layer that should accommodate the space charge layer has to be thicker and more lightly doped, resulting in unreasonably high voltage drops, like the 500 V in the example above.

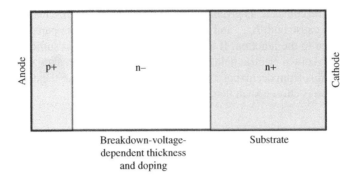

Figure 2.4   The semiconductor layers of a power diode.

The key to achieving sufficiently high conductivities for thick structures with low doping levels is *injection* of charge carriers. This mechanism is also referred to as *conductivity modulation* as it can increase the conductivity of a low-doped semiconductor layer by orders of magnitude. An example of a structure using carrier injection is a normal power diode. For reasons that are explained below, a power diode is not merely a p–n junction. Typically, three layers are necessary, as shown in Figure 2.4. The anode is formed by a highly doped p+ layer, which is capable of injection holes to an adjacent n layer. On the other side of the junction there is a low-doped n– region. This region has to accommodate the space charge layer when the diode is reverse biased, and consequently the thickness of this layer is dependent on the voltage rating of the device. The next layer is the highly doped n+ layer on the cathode side, which should inject electrons into the n– layer. The manufacturing process, however, starts from the substrate on the cathode side (n+) and then the n– layer is grown epitaxially. Finally, the p+ layer is diffused into the anode side.

In Figure 2.5 the charge densities in the three layers of a typical power diode are shown. As already mentioned, the highly doped p+ and n+ regions at the anode and cathode sides of the power diode serve as sources of charges that diffuse into the n– region. This diffusion of charges from both sides of the n– region creates a *plasma* of charges ($p = n$ for charge neutrality) with substantially higher carrier densities than those obtained from the doping (p_{n0} and n_{n0}).

As indicated above, approximately two orders of magnitude is typically what is necessary in order to achieve sufficiently low voltage drops for high-voltage devices. From a conduction perspective, the highly doped and thin p+ and n+ layers are unproblematic. In the p+ region on the anode side the hole density p_{p0} is of the order of 10^{18} m^{-3}, which yields very good inherent conduction properties. The electron density n_{p0} obtained from the doping is too low to contribute significantly to conduction. However, some of the electrons diffuse over to the p+ region from the n– region. These electrons are denoted by the electron density n_p.

The corresponding charge densities of the highly doped n+ region on the cathode side are denoted n_{n0} ($= 10^{19}$ m^{-3}), p_{n0}, and p_n. Owing to the high value of n_{n0}, this layer also has good inherent conduction properties.

The intermediate n– region, however, clearly benefits from the injection of holes from the p+ region and electrons from the n+ region. The property to maintain a charge plasma is also

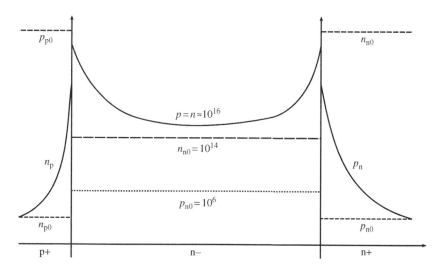

Figure 2.5 Charge densities of a power diode in cm^{-3}.

the most significant property of a typical bipolar device, in contrast to unipolar devices, where the conduction mechanism relies entirely on drift with one type of charge carrier (electrons or holes), usually electrons.

The expressions for the exact values of the charge densities in the three layers will not be presented here, as this would require a more detailed description, which would reach far beyond the scope of this introduction to power semiconductors. The excess charges in the n− region, i.e. the total amount of charges in excess of n_{n0} in the volume of the n− region, are denoted Q_F. This quantity is related to the forward current, I_F, of the power diode through the approximate relation

$$I_F \approx \frac{Q_F}{\tau}, \tag{2.6}$$

where τ is the carrier lifetime. This relation between stored charge and forward current is important for the conduction mechanism because, as a consequence of it, the voltage drop across the n− layer is approximately inversely proportional to the carrier lifetime. Experiments on high-voltage thyristors show that the total ON-state voltage drop minus a constant voltage term is almost inversely proportional to the carrier lifetime [11].

The tuning of the carrier lifetime is usually one of the last process steps in the fabrication of a power device. Typically, it is performed by diffusion of gold or irradiation of highly energetic electrons. Typical values of carrier lifetimes are given in Table 2.1.

The strong influence on the voltage drop would indicate that the carrier lifetime should be chosen as high as possible. In Figure 2.6 the forward current density J_F as a function of the forward voltage drop V_F of a power diode for two different values of carrier lifetime are shown. Note that both the magnitude and the slope of the forward voltage drop are influenced by the carrier lifetime.

Unfortunately, the carrier lifetime also has implications on the switching properties of a high-voltage device. The value must, therefore, be chosen with care, weighing conduction properties against switching properties.

Table 2.1 Carrier lifetimes.

	τ
Pure silicon	several seconds
Silicon substrate	>100 µs
High-voltage thyristors	15–100 µs
Fast thyristors	2–5 µs
IGBTs and diodes	0.1–10 µs

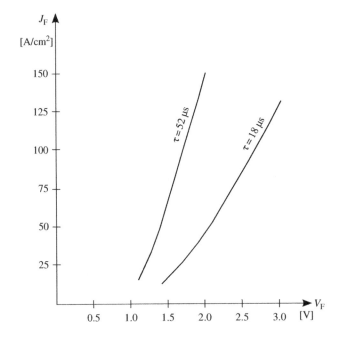

Figure 2.6 The forward current density J_F as a function of the ON-state voltage drop V_F of a power diode for two different values of carrier lifetime.

In this context, the MMC is different from most other VSC topologies because the switching frequency per device can be chosen very low without sacrificing the quality of the output voltage waveform. This clearly indicates that the ideal power device for an MMC is a slow-switching high-voltage device, where the main design efforts are put into making a device having a minimum ON-state voltage drop. Even though such a device would exhibit comparably high switching losses, the total losses would still be lower than for a fast-switching counterpart.

An investigation regarding the choice of power semiconductors for MMCs is presented in [12]. To allow comparison of devices with different voltage ratings, the ON-state voltage drop was normalized with respect to the direct voltage of the dc capacitor of the submodule.

In Figure 2.7 the relative voltage drop as a function of the forward current I_F is shown for seven different power devices (labeled A–G). The device types and voltage ratings are given

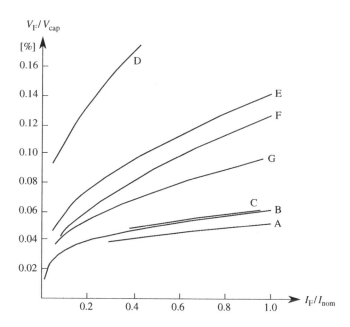

Figure 2.7   Relative ON-state voltage drops of the seven different power devices presented in Table 2.2.

Table 2.2   Main parameters for the devices A–G
shown in Figure 2.7.

Device	Type	V_{max} [V]	I_{nom} [A]
A	IGCT	4500	2100
B	IGCT	4500	1700
C	IGCT	6500	1290
D	IGBT	1700	2400
E	IGBT	3300	1500
F	IGBT	4500	1200
G	IGBT	6500	500

in Table 2.2. The influence of the choice of power semiconductors is very significant, and the dominating position of the IGBT in MMCs has been questioned in favor of thyristor-based devices like the integrated gate-commutated thyristor (IGCT) [13], which is a further development of the GTO. Note especially that the high voltage drop of the 1700 V IGBT (Device D) is beyond comparison with the IGCTs (Devices A–C). Also the other IGBT devices (Devices E–G) have substantially higher voltage drops than the IGCTs.

It is, however, an interesting observation that IGBTs with higher voltage ratings have lower relative voltage drops than IGBTs with lower voltage ratings. However, even the IGBT with the highest voltage rating, i.e. Device G, has a significantly higher relative voltage drop than any of the IGCTs. The main reason for this difference in forward voltage drop is that the IGBT has comparably poor injection of electrons from the emitter side, which results in a lower charge density of the plasma on the emitter side [14]. In Figure 2.8 this difference is shown.

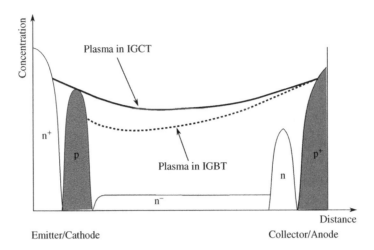

Figure 2.8 Comparison of the charge plasma during conduction for IGBTs and IGCTs. Approximate doping concentrations for the various layers of the devices are given as reference.

This property is an inevitable consequence of the IGBT device structure, which lacks a highly doped n region on the emitter side of the lowly doped drift region. In the location where the electron current from the channel bends down vertically toward the collector no electron injection source except the MOS-channel is available.

A development of the IGBT with lower voltage drops than conventional IGBTs is the injection enhanced insulated gate bipolar transistor (IEGT) [15]. The main idea of this structure is to increase the injection of electrons by means of using a deep trench gate which generates an accumulation layer [16] of electrons that can serve as an electron emitter. It has also been shown that it is important to let the width of this accumulation layer be large in comparison to the width of the source (emitter) [17]. From measurements it is clear that the forward voltage drops of IEGTs are close to those of IGCTs.

2.2.5 *Switching Properties*

In order to achieve good conduction properties a high-voltage power device relies on carrier injection in combination with a suitable carrier lifetime. The excess charge in the lowly doped n− region of the device, which is beneficial for the conduction mechanism, is detrimental for the switching processes. In power diodes this excess charge gives rise to a phenomenon called *reverse recovery*, which is characterized by a reverse current through the diode before turn-off. In IGBTs or IGCTs the excess charge of the n− layer instead gives rise to a *tail current* [18]. In both cases the phenomenon causes significant switching losses. From this point of view it would make sense to minimize the carrier lifetime, because this would reduce the excess charge. This would, however, increase the voltage drop when the device conducted a forward current. In Figure 2.9 an illustration of the concepts of reverse recovery and tail current is given. The reverse recovery charge, Q_{rr}, is shown as the hatched area between the current plot and the time axis.

Dynamic effects related to the build-up of charge plasma at turn-ON can also be observed for most high-voltage power semiconductor devices. This appears as an increased voltage drop in

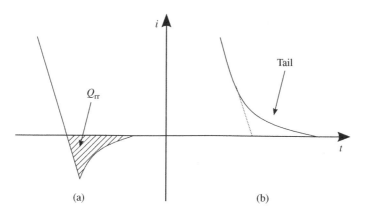

Figure 2.9   (a) Reverse recovery (b) tail current.

the initial phase of a conduction interval, and is referred to as *forward recovery* or *conductivity modulation lag* [18].

As already mentioned above, a trade-off between conduction and switching properties must be made. Sometimes, especially for low-voltage devices ($\leq$1200 V), the device manufacturers offer different versions of their devices with names like *low-saturation* or *high-speed*. In the former case this corresponds to a device with a long carrier lifetime, whereas the second case corresponds to a short carrier lifetime. The exact values are, however, rarely given by the manufacturers.

2.2.6   Packaging

Power semiconductor devices are available in a multitude of packages. In high-power applications, such as MMC, so called power modules are used. They are available in two different types. The most common type is the industrial power module with and isolated base plate and a non-hermetic plastic cover. These modules are used in industrial motor drives and traction converters. Because of the comparably wide application range, the production volumes are high, and consequently comparably low costs can be reached. The second type is the hermetic press-pack which was first developed for thyristors and GTOs. Later, this package type was also developed for IGCTs and IGBTs.

Thyristors, GTOs, and IGCTs make use of an entire wafer for each device, and therefore these devices have very similar packages. The press-pack IGBT module, however, is a more complicated design because a multitude of chips have to be connected electrically and thermally. Below, isolated industrial modules and press-pack modules are described separately.

2.2.6.1   Industrial Isolated Power Modules

Isolated industrial IGBT power modules are available in various versions, and with various voltage ratings, and current ratings. For MMCs voltage ratings of 3.3 kV or 4.5 kV are most common, and the current rating is typically 1.2 kA or higher. A photograph of such a power module with the ratings 4.5 kV and 1.2 kA is shown in Figure 2.10. The particular module

Figure 2.10   Industrial isolated IGBT module with the ratings 4.5 kV and 1.2 kA.

shown in the figure contains one IGBT and one anti-parallel diode. Each of the devices is realized by a multitude of chips which are interconnected by means of copper strips, ceramic substrates with direct-bonded copper (DBC) surfaces, and bond wires.

The parts visible from the outside are the plastic cover, the base plate, and the screw terminals for the IGBT and the anti-parallel diode. The two smaller terminals that are shorted by a metal strip are the gate and auxiliary emitter (also called *Kelvin emitter*) terminals of the IGBT. These terminals are used for connecting the gate-drive unit. The reason for having a separate auxiliary emitter terminal for the gate-drive unit is to avoid inductive voltages caused by the emitter current. Such voltages would affect the gate circuit, and could potentially have an adverse effect on the switching performance.

The third small terminal is the auxiliary collector terminal, which is used mainly for short-circuit detection. Most advanced gate-drive units for IGBTs have a built-in short-circuit protection, and typically the *de-saturation method* is used [19], [20], [21]. The basic idea of this method is that if the IGBT is gated ON and has a high voltage drop, this is probably a sign of short-circuit, because the high current has forced the device out of saturation. If such an event is detected, the IGBT can be turned OFF safely, and thus an explosion [22] can be avoided.

The plastic cover is made of one of five different materials. These are polyphenylene sulfide (PPS), polybutylene terephthalate (PBT), polyphthalamide (PPA), polyamide (PA) and polyethylene terephthalate (PET). The latter three are often used for high-voltage modules. The materials are chosen based on the specific application requirements (e.g. traction, industrial, or transportation), and depending on various application-dependent standards.

The base plate of the power module in the figure is made of aluminum silicon carbide (AlSiC), but other manufacturers may also use nickel-coated copper base plates. The reason for using AlSiC base plates is to achieve better power-cycling capability. Comparing copper and AlSiC in Table 2.3 reveals that copper has a more than six times higher coefficient of thermal expansion (CTE) than silicon, whereas the same quantity for AlSiC is only 2.6 times higher than for silicon. The thermal conductivity (TC) of copper is, however, more than twice that of AlSiC. Power cycling is described briefly at the end of this subsection.

Table 2.3 Coefficient of thermal expansion and thermal
conductivity for various materials related to packaging of
power semiconductor devices.

Material	CTE [ppm/K]	TC [W/(m K)]
Si	2.6	150
SiC	4.0	490
Cu	16	385
Al	23	210
Mo	5.2	140
AlSiC	6.8	170
Alumina	6.7	25
AlN	4.6	175
Si_3N_4	2.6	90

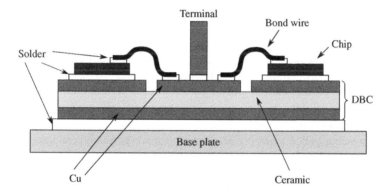

Figure 2.11 The different layers of an industrial isolated IGBT power module. Here, a DBC substrate
with a patterned top copper surface is shown. In many cases, instead, separate DBC substrates are used.
In this case, three substrates would have been used.

The different layers of the internal structure of an IGBT power module are shown in
Figure 2.11. On top of the base plate a DBC substrate is soldered. A ceramic tile is bonded to
the copper layers on the top and bottom by means of a high-temperature oxidation process. The
purpose of the ceramic layer is to electrically isolate the chips from the base plate. However,
the DBC substrate must also provide a path for the heat from the chip to the base plate.

The most common ceramics for this purpose are aluminum oxide (alumina) and aluminum
nitride (AlN), of which the latter has approximately seven times better heat conduction properties; see Table 2.3. The CTE for AlN is only 1.8 times higher than for silicon, whereas the
CTE for alumina is 2.6 times higher than for silicon. The main drawback of AlN is its comparably high cost. Another material used for ceramic substrates is Si_3N_4, which has a CTE that
matches that of silicon, and a TC between those of alumina and AlN. An advantage of Si_3N_4 is
that its bending strength and fracture toughness are significantly higher than for both alumina
and AlN.

The top copper layer of the DBC substrate can be etched in the same way as a printed circuit board in order to form electric circuits. These circuits can be used to interconnect the chips with each other and to make connections to the metal strips forming the external terminals of the power module.

The IGBT and diode chips are soldered on top of the patterned DBC surface. The IGBT chips are soldered with the collector side facing downwards toward the DBC substrate. The upper side of the IGBT chips have emitter and gate metal pads for bond wires. These pads are used when soldering the bond wires. The other ends of the bond wires are soldered onto the connection points on the patterned DBC substrate.

Metal strips forming the external terminals of the IGBT power module are soldered (or contacted by means of ultrasonic welding [23]) to their connection points on the DCB substrate. After all electrical interconnections have been made, the voids of the interior of the power module are filled with silicone gel in order to prevent partial discharges at metal edges inside the module and, at least to some extent, seal the system against moisture and atmospheric contaminants [24]. At the edges of metallizations of the chips and DBC substrates high electric fields may occur, and even if a good dielectric material is used around these edges, there is a risk of partial discharges, especially for high-voltage (e.g. 6.5 kV) power modules [25].

A significant reliability problem of industrial power modules is that both the bond wires and all soldering layers are subjected to stress caused by thermal expansion, especially *power cycling*, which is the result of varying self-heating and may cause device failure. The mismatch in thermal expansion between the silicon chip and the copper surface of the DBC substrate creates a shear force on the soldering layer between these surfaces. Owing to variations in the operating conditions of the module, the chip temperatures will vary, and each temperature cycle consumes a certain amount of lifetime.

The failure, which is called *soldering delamination*, starts as a small crack somewhere along the edge of the chip, often at a corner. For each temperature cycle, this crack extends, which results in a gradually impaired cooling capability. This causes an acceleration of the aging process. Finally, the chip lifts off from the DBC substrate, which causes an open circuit between the collector metallization of the chip and the collector terminal of the device. The same thing can occur in any of the soldering joints in Figure 2.11.

Regarding the bond wires, two different failure mechanisms are possible. The first is a soldering delamination of the solder joint between the chip and the bond wire. This is usually referred to as *bond-wire lift-off*. The second failure mechanism is caused by bending forces acting on the point of the bond wire where it bends and follows the surface of the chip. These bending forces will vary depending on the current through the bond wire, and finally the bond wire cracks because of material fatigue. This is commonly referred to as *bond-wire heel crack*.

The bond wires are also a weak spot regarding short-term surge currents. In the event of an internal short-circuit or a dc-side short-circuit of an MMC, excessive currents may flow through the bond wires. The excessive heat generated in the chip and bond wires may then melt the solder and cause a rapid bond-wire lift-off [26].

The remaining bond wires will then be subjected to even higher temperatures, which may cause a cascading effect of bond wires lifting off. Finally, there is no other path for the current than through an arc between the chip and the lifted bond wire. This arc will create a rapid pressure build-up and excessive heat, which results in an explosion which will destroy not only the module itself but also any other equipment in the vicinity of the module [22]. In cases with very high surge currents or extended durations of the surge current the bond wires may even evaporate along with the emitter metallizations, forcing the current into an arc which causes an explosion.

The first type of damage that can occur to adjacent equipment is that gases escape from the exploded housing. The second type is aerosols that are emitted out of the housing. These aerosols are composed of finely dispersed gel and rubber particles. The result of this emission is a sticky surface on adjacent equipment. Before this equipment can be used again costly cleaning operations may be necessary.

The third type of damage is caused by comparably heavy objects that are ejected out of the exploded module. These objects may by metal parts, parts of chips or whole chips, or plastic debris from the housing. The kinetic energy of these ejected objects may cause severe mechanical damage to adjacent equipment.

The above scenario is the worst possible failure in HVDC installations, and any possible measures to prevent it are, therefore, taken. One way of dealing with this problem is to make the submodule housing explosion proof [27]. Another way to solve the problem is to use power modules without bond wires. Such modules are called *press-pack power modules*, and are described in the following subsection.

2.2.6.2 Press-Pack Power Devices and Modules

The first press-pack device was the thyristor, which is typically a whole-wafer device. The main idea is that the silicon wafer should be pressed between two parallel rigid plates which serve as both heat sinks and electrical terminals. The wafer is consequently cooled from both sides, which is a considerable advantage because the cooling area is doubled.

When compared to an industrial IGBT power module, however, the main advantages are that a press-pack thyristor has no soldering layers that can delaminate and no bond-wires that can crack or lift off from the chip. Additionally, the casing is hermetic, which means that moisture or pollutants cannot enter the inside of the package and cause subsequent damage. For these reasons, a press-pack thyristor is the most reliable and rugged power device available today.

As such, it is also the ideal device for HVDC transmission applications, where reliability is the most important property at the system level. The same level of reliability can also be achieved with less reliable power devices, for instance by means of various kinds of redundancy, but such measures always come with an additional cost.

Figure 2.12 shows a cross-sectional view of a press-pack thyristor. In the middle, the silicon wafer is pressed between two molybdenum (Mo) discs, which are covered by a thin layer of rhodium (or sometimes with ruthenium) to prevent oxidization. The reason for using molybdenum discs between the silicon wafer and the copper pole pieces on the outside is that its CTE is only two times higher than that of silicon, whereas the CTE of copper is more than six times higher than that of silicon; see Table 2.3.

The pole pieces outside of the molybdenum discs are usually made of nickel-plated copper. If these pole pieces would be applied directly to the silicon wafer without the molybdenum discs in between, the silicon wafer would crack as soon as the wafer heats up, because of the elongating forces acting on the silicon wafer from the much more thermally expanding copper.

The edges of the silicon wafer are cut at specific angles in order to minimize the electrical field along the edge of the wafer. In this way, electric breakdown at the edge of the wafer can be avoided. This cutting process is called *beveling*. Outside the beveled edge of the wafer a silicone rubber ring is fitted as a passivation to prevent breakdown at the edge of the wafer.

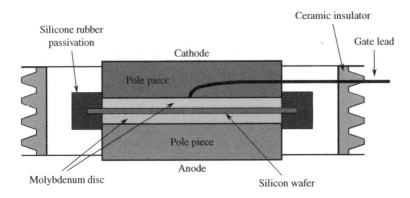

Figure 2.12   Cross-sectional view of a press-pack thyristor.

Figure 2.13   A 4.5 kV press-pack IGCT along with its integrated gate-drive unit.

On the outside a grooved ceramic housing protects the silicon wafer mechanically, and pro-vides sufficient creepage distance between the anode and cathode. Together with the metal lids on the anode and cathode sides, this housing creates a hermetic package which protects the wafer from moisture and pollutants.

From a packaging point of view, an IGCT has roughly the same characteristics as a standard thyristor. The only differences are the low-inductance bus bar system which interconnects the gate and that the gate is contacted by a ring-shaped electrode instead of at a point in the center. A photograph of a 4.5 kV IGCT along with its integrated gate-drive unit is shown in Figure 2.13. On the left-hand side the press-pack power device is seen, and on the right-hand side the gate-drive unit can be seen.

The array of circles seen in the middle of the figure are the capacitors of the output stage. These capacitors store the energy that is necessary to provide the high turn-off currents of the IGCT. Directly to the left of these capacitors a number of small black rectangles can be seen. These are the transistors of the output stage. The structure joining the output transistors with the power device is a low-inductance parallel-surface busbar system interconnecting the gate-drive unit with the power device.

Figure 2.14 StakPak IGBTs (Courtesy of ABB).

Press-pack IGBT modules are different from press-pack thyristor devices because they employ a multitude of chips instead of a whole wafer. This makes the design far more complicated. The primary goal of the design is to prevent a faulty device appearing as an open circuit. As already explained above, such a scenario is catastrophic because the current will generate an arc and a subsequent explosion (or fire). In Figure 2.14 a photograph of two press-pack IGBT modules is shown. The gray rectangular areas on the top of the modules are the collector surfaces of the subassemblies of the power module. Each subassembly contains a certain number of chips which are contacted electrically and thermally by means of individual spring systems for each chip.

In Figure 2.15 a simplified drawing of a spring system for an individual chip can be seen. The collector side of the IGBT chip is soldered onto a molybdenum base plate. As mentioned, from a CTE perspective a molybdenum base plate is more than three times better than copper; see Table 2.3. Molybdenum is also 30% better than AlSiC from this point of view, but the thermal conduction properties are actually almost 20% worse than for AlSiC.

On the emitter side of the chip, an individual spring unit for each chip provides the necessary force via an aluminum/silver (Al/Ag) platelet [28]. The latter achieves both electrical and thermal contact by means of pressure, while the chip is free-floating with respect to the Al/Ag platelet. This mechanical arrangement ensures that each chip has sufficient contact force even if the force is not perfectly uniform on the outside of the power module. This is an important feature if stacks with many series-connected press-pack modules are to be assembled.

In order to achieve, that a faulty device causes a short-circuit, the Al/Ag platelet plays an important role. In case of a fault, the faulty device will dissipate significantly more heat than in the normal case. This heat spreads to the Al/Ag platelet by means of thermal conduction, and

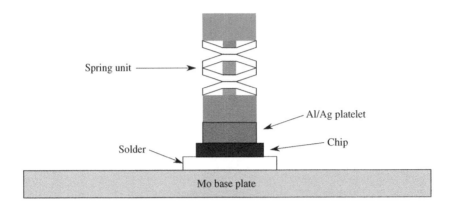

Figure 2.15   Mechanical arrangement of an individual chip of a press-pack IGBT module.

when the platelet has reached a sufficiently high temperature it starts to melt. When melting, the aluminum and silver form a stable alloy with the silicon of the faulty chip [29]. Silver and aluminum are good choices of metals for this purpose as they form low-melting-point eutectic alloys with silicon. The alloy formed by the platelet and the chip has to conduct all of the module's current. This current could be more than ten times greater than during normal operation, because only one single spring-chip unit would be conducting the same current as all parallel-connected chips would do during normal operation. This state has to be maintained until the next scheduled maintenance, which may not be for several months.

The gate contact of each chip is contacted via a spring contact. In order to achieve a long lifetime of this contact, the metallization of the gate pad makes use of three different metal layers, with silver as the top layer.

2.2.7   Reliability of Power Semiconductor Devices

Power semiconductor devices for HVDC applications are designed to have very high reliability. A common way to express the reliability of any component or system is with the so-called failures in time (FIT) rate. This number suggest how many failures are expected in 10^9 operating hours (corresponding to approximately 114,000 years). Typical values for power semiconductor devices for voltage-source-converter high-voltage direct current (VSC-HVDC) applications are of the order of 100, which is a remarkably low value. This number, however, depends on several factors, and in order to achieve such low FIT rates, a long list of requirements specified by the device manufacturer has to be fulfilled. The most important requirements are related to:

- maximum values (average and surge) of voltage and current, or safe operating area (SOA);
- gate driving conditions;
- required auxiliary circuits, for instance snubbers;
- mechanical fitting (mainly to the heat sink).

If these requirements are violated, it is likely that a certain amount of the remaining expected lifetime of the device is lost or, in the worst case, that the device fails immediately.

The violation of any of the requirements may trigger a multitude of failure mechanisms, ranging from obvious reasons of failure such as voltage breakdown and over-heating to more sophisticated phenomena like filamentation [30], dynamic avalanche, latch-up, thermo-mechanical problems (as described in Section 2.2.6), and cosmic-ray-induced failures [31]. Below, five items associated with reliability are discussed. These are: the SOA, cosmic-ray failures, failures associated with gate-driving conditions, influence of auxiliary circuitry, and issues related to humidity and environmental impact. Finally, an example from the literature of an estimation of the reliability of a complete 8 MVA inverter is presented.

2.2.7.1 Safe Operating Area

The safe operating area (SOA) of a power semiconductor device is determined by several mechanisms related to device physics. The most important are avalanche breakdown and over-heating, but filamentation, latch-up, and dynamic avalanche may also be important failure mechanisms influencing the SOA. In Figure 2.16, an SOA of a fictitious IGBT is shown. The rated values of voltage and current (V_{rated} and I_{rated}) are given in the data sheet of the device. The actual maximum values used in the converter are usually significantly lower in order to fulfill various requirements, and in order to have a certain safety margin.

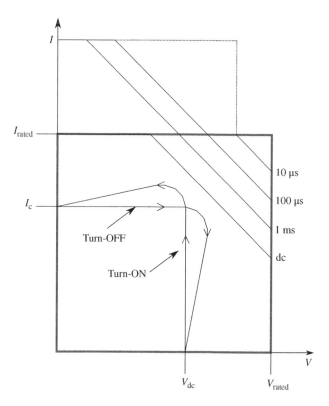

Figure 2.16   The SOA of an IGBT. V_{rated} and I_{rated} are rated values of voltage and current, respectively. I_C is the collector current during normal operation and V_{dc} is the dc-link voltage.

The collector current, I_C, may vary from zero up to the peak value at maximum load, and the dc-link voltage, V_{dc}, is chosen with respect to cosmic ray issues and the estimated maximum ripple of the submodule capacitor voltage. As in any hard-switching, VSC overshoots of the voltage and current are inevitable during the switching transitions. At turn-ON of an IGBT the current has an overshoot owing to the reverse recovery of the anti-parallel diode which is turning OFF, and during turn-OFF an over-voltage is observed across the IGBT that is turning OFF. Typical switching loci for turn-ON and turn-OFF are shown in Figure 2.16. Despite the overshoots in both voltage and current during switching, a typical design would still allow for considerable safety margins on top of the overshoots.

For short transients the SOA is rectangular, but for longer durations the upper right corner of the SOA gets cut off at ever larger portions. The main reason for this is high power dissipation, but also other mechanisms may influence this indentation. The line indicated with dc has to be interpreted with consciousness, because a very high power dissipation is obtained. Typically, the cooling system will not allow such a high power dissipation continuously without exceeding the maximum junction temperature. The limits for 10 μs are drawn with dashed lines, as these correspond to a separate short-circuit SOA. For short-circuit cases usually a lower voltage than that rated is specified, but the current is allowed to assume very high values (e.g. ten times the rated current).

2.2.7.2 Cosmic Ray

Problems with cosmic ray were first observed in locomotive converters operating at high altitudes. At first, the failures of the power semiconductors could not be explained, but after intense research the origin of the problem was found. In 1994, by comparing experiments performed in salt mines 140 m below the surface with measurements above the ground, it was explained that silicon power diodes can run into thermal destruction because of cosmic-radiation-induced effects [33], and in [31] eight different devices (diodes, thyristors, and GTOs) from three different manufacturers were investigated and it was found that the failure rate was independent of device type but strongly dependent on the voltage. An approximate function which could fit the observed results was also formulated. A conclusion from this work is that high-voltage devices must be operated at a voltage level which is significantly lower than the breakdown voltage in order to have a low FIT rate with respect to cosmic-ray-induced failures.

The basic mechanism of cosmic-ray-induced effects is described in [34]. An event is initialized by an incident particle (e.g. a neutron from outer space) that hits a silicon nucleus located in the space charge layer of a reverse-biased power semiconductor device. The exposed silicon nucleus decays into lighter ions having kinetic energies of the order of 10 to 100 MeV. This energy is deposited in the silicon lattice and creates a strongly localized and highly concentrated electron-hole plasma along the path of the ions. Consequently, high electric field strengths arise at the ends of these paths, and this results in impact ionization. Depending on the magnitude of the applied voltage, this may cause a massive carrier multiplication. The highly conductive plasma quickly extends through the entire device, leaving an almost short-circuited device.

In [32] a 4 kA/4.5 kV IGCT was investigated with respect to the same phenomenon, and in Figure 2.17 results retrieved from that publication are shown. The figure shows the cosmic-ray withstand capability as a function of dc-link voltage. From investigations like this, it is often concluded that a dc-link voltage of approximately 2.8 kV is suitable for 4.5 kV devices.

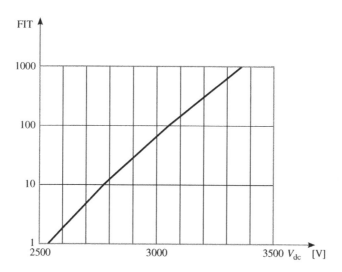

Figure 2.17 Cosmic-ray withstand capability of a 4 kA/4.5 kV IGCT as a function of the dc-link voltage. The data for this graph is retrieved from Figure 8 in [32].

2.2.7.3 Gate-Driving Conditions

The reliability of a power semiconductor device is also strongly dependent on how the gate is driven. One example is the choice of the gate-source voltage of an IGBT when it is in the ON-state. A high gate-source voltage would give a considerably lower ON-state voltage drop than a low gate-source voltage. An example is the IGBT module 5SNA1200G450350 (not unlike the module shown in Figure 2.10) by ABB which has an ON-state voltage drop of 2.5 V at 25 °C with a gate-source voltage of 17 V. If instead a gate-source voltage of 11 V is used, the ON-state voltage drop becomes 3.3 V. This yields an increase in conduction losses of 32%. The drawback of having a high ON-state gate-source voltage is that the short-circuit current gets much higher. The ON-state gate-source voltage must, therefore, be chosen with great care, considering both short-circuit handling and conduction losses.

Even more important is how the gate is driven through the switching transitions at turn-ON and turn-OFF. Fast transitions, on the one hand, with high gate currents yield low switching losses. Unfortunately, fast transitions may also cause undesired effects like overshoots in voltage and current, and severe oscillations. Additionally, very fast transitions may also increase the risk for filamentation, dynamic avalanche, latch-up, and electromagnetic interference. Slow transitions, on the other hand, not only cause high switching losses, but there is also a risk that noise from adjacent current paths may affect the switching transient. The switching transitions must, therefore, be chosen with great care considering a long list of requirements, and it is important that the switching transitions are always the same regardless of the operating conditions (unless the gate-drive unit is adaptive with respect to, for instance, temperature). From this discussion it becomes clear that the gate-drive unit is one of the most important parts of a high-power inverter. Sometimes it is even referred to as the *brain* of a converter. As such, the gate-drive unit is also one of the most important parts influencing the reliability of a high-power converter. The FIT rate of the gate-drive unit should, therefore, be of the same order of magnitude as the power semiconductor device it drives.

2.2.7.4   Auxiliary Circuitry

In order to guarantee a certain degree of reliability most high-power semiconductor devices need some auxiliary circuitry. These circuits could be snubbers and clamps in the case of an IGCT, grading resistors and resistive-capacitive networks for series-connected IGBTs, or special circuits for handling over-currents and start-up. Without these circuits, the specified low FIT rates of the power semiconductor devices cannot be guaranteed. Consequently, these circuits must also be reliable, and sometimes specially designed components have to be used in order to achieve a sufficient reliability. This also holds for auxiliary power supplies (as described in Section 2.8) for the gate-drive units and signal transmission for the gate-drive units.

2.2.7.5   Humidity and Other Environmental Impact

Humidity, gaseous contamination, and salty environments may also have an impact on reliability. Usually this is dealt with when choosing the packaging of the power semiconductor device. If the environment is too demanding, only a hermetic package may be a viable solution. However, in most locations where HVDC converters are situated, the environment is not very extreme, and the available countermeasures against humidity and salty environments can be handled satisfactorily.

The basic underlying problem with humidity and pollutants is aluminum corrosion and failing junction terminations. These processes have been monitored closely in [35], and it is clear that it is important to make sure that humidity and pollutants do not come in direct contact with the metallizations on the chip, especially those of the junction terminations. In this work it was also shown that the degradation accelerated by a factor of two by increasing the voltage from 65% of the rated voltage to 90% of the rated voltage.

2.2.7.6   Reliability of a Complete Converter

In [32] the reliability of a whole 8 MVA three-level inverter was studied. The study included all components of the inverter and included all known failure mechanisms. The authors emphasize that the numbers used in the calculation are based on many years of field experience. Three different realizations of the inverter were investigated: one GTO version, one IGCT version, and one IGBT version. The FIT rates for the complete inverter were 3960 for the IGCT version and 9120 for the IGBT version. The FIT rates of the devices themselves were 50 for the IGCT and 120 for the IGBT. The gate-drive unit of the IGCT had a FIT rate of 200, whereas that of the IGBT was 150.

2.2.8   Silicon Carbide Power Devices

From the presentation above it is clear that the properties of the power semiconductor devices are of utmost importance for any HVDC transmission project. The main properties to be evaluated are: reliability, cost, and losses. Even though there is still a continuous development of power semiconductor devices in silicon, it is not likely that disruptive technology improvements, which can dramatically change device properties, are possible.

The idea of achieving such disruptive improvements by replacing the existing silicon-based power semiconductor devices with new devices based on silicon carbide (SiC) is not new. During the 1990s, SiC was identified as the successor of silicon in HVDC applications because of the possibilities of much higher blocking voltages than what was possible in silicon, and because dramatic loss reductions were foreseen. Intensive research was initiated in Europe, Asia, and America, partly explicitly aiming at high-voltage low-loss devices for HVDC transmission applications. At this time, the main obstacle was that the base material for development of power devices—the SiC substrates—had far too many defects to permit series production with a reasonable production yield or to a reasonable standard. Additionally, the cost for these imperfect substrates was immensely high compared to the existing silicon substrates. This did not prevent, for instance, ABB from continuing the development of high-voltage diodes in SiC, because tremendous benefits were foreseen in a successful device.

Sometime at the beginning of the new millennium, however, an unpredicted drift phenomenon was observed. The forward voltage drop of the new SiC diodes was increasing over time when they were subjected to forward current conduction [36]. Later this phenomenon was identified as stacking faults related to dislocations in the crystal structure [37]. Owing to the nature of the problem, i.e. a deterioration of device parameters due to a recombination of charge carriers, the phenomenon was referred to as *bipolar degradation*. Later investigations revealed that the problem was very severe, and that any bipolar device would exhibit bipolar degradation unless a perfect crystal structure could be guaranteed. During the early years of the millennium, this was impossible except for devices with very small chip areas. Even today, this problem should not be underestimated if large-area devices are targeted.

Even though the problem with bipolar degradation effectively reduced the expectations on the SiC technology for high-voltage bipolar devices, unipolar devices could still be developed successfully. In fact, the first SiC Schottky diode was already reported in 1992 [38]. The first commercially avaliable SiC Schottky diodes were, however, introduced by Infineon as late as 2001. These diodes were highly appreciated for their excellent switching properties, having virtually no reverse recovery at all. They were used in power factor correction equipment together with silicon MOSFETs and IGBTs. The SiC age had started and switch-mode devices in SiC were already on their way.

2.2.8.1 Unipolar Devices

As explained in Section 2.2.3, any p–n junction has to be dimensioned for the maximum voltage it should be able to sustain without experiencing a breakdown. When the junction is reverse-biased a space charge layer is formed, in which the absolute value of the electric field increases linearly when going from the edge of the space charge layer towards the very boundary between the p and n regions of the junction, where the electric field has its maximum value, E_{max}. This is described mathematically by Equations (2.1) and (2.2) and visualized in Figure 2.3. An increase in voltage would cause a corresponding increase of both E_{max} and the width, w, of the space charge layer. At the maximum permissible voltage, E_{max} must not exceed the electric field, E_{bd}, at which a breakdown occurs. In order to fulfill these requirements, the doping density must be reduced if the junction should be dimensioned for a higher voltage. This reduction in doping density is associated with a corresponding increase in resistivity of the material. Additionally, by reducing the doping density, the width of the space charge layer increases. The total resistance, therefore, increases quadratically with the rated voltage. In practical devices the resistivity may even increase faster than quadratically if the voltage rating of the device is increased. In [10] it is shown that the ON-state resistance of a unipolar device should be expected to increase with the blocking voltage to the power of 2.5. Below, however, a simplified explanation is given.

When comparing the conduction processes of unipolar devices in silicon (Si) and SiC, the basics of *drift* in semiconductors has to to be reviewed. From [10], or any other textbook on power semiconductor devices, it is found that the drift current density in an n-type unipolar semiconductor can be written as

$$J_n = E_d q N_d \mu_n, \tag{2.7}$$

where E_d is the electric field driving the drift current, q is the electron charge, N_d is the doping density, and μ_n is the mobility of electrons. If this current density flows through a drift region with the width w, it would cause a forward voltage drop of

$$V_{on} = E_d w. \tag{2.8}$$

If this voltage drop is divided by the current flowing through the drift region, an expression for the resistance of the drift region is obtained. Thus, dividing Equation (2.8) with the product of Equation (2.7) and the area of the device, A, the resistance of the drift region is obtained as

$$R_{on} = \frac{w}{q N_d \mu_n A}. \tag{2.9}$$

Now, it becomes clear that the quantities affecting the conduction properties are interrelated to those affecting the blocking properties. The doping density, N_d, and the width of the drift region, w, must both be chosen with respect to the rated voltage of the p–n junction. For simplicity, it is assumed that $E_{max} = E_{bd}$ when the rated voltage, V_{bd}, is applied across the junction. This means that there is no margin between the rated voltage and the breakdown voltage of the device. If the electric field in the blocking state has the triangular shape given by Equation (2.1), and if the voltage applied across the junction in the blocking state is calculated using Equation (2.3), the width of the drift region can be written as

$$w = \frac{2V_{bd}}{E_{bd}}. \tag{2.10}$$

Note that w is inversely proportional to E_{bd}, which means that a material with a high value of E_{bd} will have a thin drift region.

The doping density, N_d, in the denominator of Equation (2.9) must also be chosen with respect to V_{bd}. As already stated by Equation (2.4), N_d must be chosen inversely proportional to V_{bd}. The relation to E_{bd} has, however, not been discussed. In order to find an expression for N_d, an expression for E_{max} is first established. This is achieved by means of integration of Equation (2.1) over the drift region. Thus,

$$E_{max} = \int_w \frac{q N_d}{\epsilon} \, dx = \frac{q N_d w}{\epsilon}. \tag{2.11}$$

Once again it is assumed that $E_{max} = E_{bd}$. Substituting E_{max} with E_{bd} in Equation (2.11) and solving for w yields

$$w = \frac{\epsilon E_{bd}}{q N_d}. \tag{2.12}$$

Now, N_d is found by combining Equations (2.10) and (2.12) and solving for N_d. Accordingly,

$$N_d = \frac{\epsilon E_{bd}^2}{2qV_{bd}}. \qquad (2.13)$$

Note that N_d is proportional to the square of E_{bd}. For a material with a high value of E_{bd}, therefore, very high doping densities can be used.

Finally, having the expressions for w and N_d, the resistance of the drift region can be obtained. Thus, substituting Equations (2.10) and (2.13) in (2.9) yields

$$R_{on} = \frac{4V_{bd}^2}{\epsilon \mu_n A E_{bd}^3}. \qquad (2.14)$$

This equation, which highlights the factors influencing the conduction properties of any unipolar power device, can also be found in [39]. It is clear that the ON-state resistance increases quadratically with the rated voltage of the device, which indicates that at some voltage the ON-state resistance would become unfavorably high such that bipolar devices would become more advantageous. However, what is even more interesting is the very strong dependence on E_{bd}. A quantity derived from Equation (2.14) is Baliga's figure of merit (BFOM) [40], which is a quantity used for comparison of relative ON-state resistances of unipolar devices based on different semiconductor materials. It is defined as

$$\text{BFOM} = \epsilon \mu_n E_{bd}^3. \qquad (2.15)$$

Mainly, as a consequence of the fact that E_{bd} is approximately ten times higher for SiC than for Si, ON-state resistances that are several hundred times lower for SiC than for Si are expected. A compilation of the most important material properties for Si and SiC along with BFOM for the two materials is presented in Table 2.4. From the table it is evident that the mobility of SiC is significantly lower than that of Si, and depending on the value of mobility and breakdown electric field, different values of BFOM are obtained. In some cases, only a fraction of the mobility value in Table 2.4 are obtained. Nevertheless, BFOM indicates that unipolar devices in SiC could totally change the performance of power electronic converters.

A remarkable feature of unipolar power semiconductor devices in general, and SiC versions in particular, is that the ON-state resistance can be reduced arbitrarily by increasing the total chip area. *The conduction losses of such a switch can, therefore, be chosen arbitrarily low provided that a sufficient amount of chip area is available.* This is an interesting design feature

Table 2.4 Material parameters for Si and SiC along with the normalized value of Baliga's figure of merit (BFOM).

Quantity	Si	SiC
E_{bd} [MV/cm]	0.30	3.5
μ_n [cm^2/Vs]	1400	1000
ϵ/ϵ_0	11.9	9.7
BFOM	1	925

for HVDC transmission, because additional cost spent on increased device area can be regained in life-cycle costs for the losses of the converter. These low ON-state losses can also be used in the reverse direction, because the channel of any unipolar device conducts in both directions [41], [42]. This has two distinct advantages. First, the voltage drop may be chosen to be much lower than that of a diode by increasing the chip area. Second, the anti-parallel diode may be entirely omitted, which may result in significant cost savings.

The first unipolar switch-mode device in SiC that became available as engineering samples was the lateral-channel JFET (junction field-effect transistor) [43], [44], which despite the name is a vertical device. Mainly, because of the lateral gate structure, this device did not possess the extremely low ON-state resistance predicted by BFOM. In several publications, however, it was shown that the device could switch very rapidly. Switching times of the order of 30–40 ns were reported [45], [46]. *As a matter of fact, any unipolar power semiconductor device could switch arbitrarily fast provided that sufficient gate power can be provided, and that the device is connected to a circuit that permits the desired switching speed.* A consequence of this is that arbitrarily low switching losses can be obtained if the parasitic elements of the circuit are sufficiently small.

Another distinctive feature of the SiC JFET is that it, typically, is a normally ON device. This means that the gate-drive circuit must provide a negative voltage at the gate in order to keep the device in the OFF-state. This property is seen as a great disadvantage by many design engineers, because a loss of power supply to the gate-drive circuit results in a short-circuit of the dc link if the device is connected in a bridge leg of a VSC (or in a submodule of an MMC). Especially at start-up, this may be very problematic because no power is usually available before the dc-link has been charged. Therefore, when the device was introduced, it was connected in a cascode configuration with a low-voltage Si MOSFET in order to achieve a normally OFF device. The drawback of this cascode connection was that the series-connected Si MOSFET increased the ON-state resistance and the SiC JFET lost some of its controllability from the gate. Another problem with using a silicon device in conjunction with a SiC device is that the high-temperature properties of SiC cannot be used. From an HVDC point of view, this may not be a serious limitation, but from a general applicability point of view, it is not desirable.

Another possible solution to the *normally ON problem* is to use a gate driver that can provide the necessary gate power without external power supply both at start-up and during continuous operation. Such a self-powered gate driver is presented in [47] and [48]. This driver concept indicates that the normally ON characteristic of the device can be handled by a combination of innovative system solutions. As a matter of fact, it may even be advantageous from a systems perspective, because a faulty submodule must appear as a short-circuit in order to provide a path for the arm current whatever has occurred on the submodule level.

The next SiC-based power transistor to appear as engineering samples was the BJT [49], [50], which despite its name has the main characteristics of a truly unipolar device. In the ON-state, the device is resistive like a unipolar device, and during switching no bipolar effects like the tail current can be observed. The unipolar character is evident from the comparison with a silicon carbide JFET presented in [46]. An important difference compared to the JFET is that the BJT is current controlled. For a given collector current, I_C, the base driver must provide a base current of at least

$$I_B = \frac{I_C}{\beta},$$ (2.16)

where β is the current amplification factor. For currently available devices β is approximately 50, but in the future values above 100 are predicted. With such values of β, the power consumption of the base driver is not prohibitive, not even in HVDC transmission, where auxiliary power on the submodule level is not easily available.

A comparison to the required gate power for a silicon IGCT [51] reveals that the required power consumption of the base driver of a SiC BJT is just slightly higher (depending on the switching frequency). In [46], it was found that the required base power was even lower than the required gate power of a SiC JFET at high frequencies, and in [52] it was found that the required base power was merely 0.08% of the output power for a dc-dc converter operating at 100 kHz. In the latter case, the operating conditions were favorable because the converter operated at the rated current continuously. In an MMC, where the arm current varies considerably at steady-state operating conditions, a proportional base-drive unit [53] may be preferable, because, instead of feeding the maximum base current whenever the device is in the ON-state, a proportional driver will only feed the necessary amount of base current to the device.

In contrast to a SiC JFET, however, a SiC BJT cannot conduct any significant amount of current in the reverse direction. This is the main drawback of the SiC BJT. Anti-parallel diodes will inevitably be necessary in any bridge connection. The consequences of this are additional losses and additional costs.

The third unipolar power semiconductor device based on SiC to enter the power electronics business was the SiC MOSFET. University laboratories had presented various examples of SiC MOSFETs at least a decade before the first engineering samples became available. The main reasons for this were that it was harder than expected to achieve a stable oxide layer and good mobility in the channel below the oxide.

The insufficient oxide stability manifested itself in two ways: (1) a drift of the threshold depending on the gate bias and (2) an insufficient lifetime of the oxide layer, especially at elevated temperatures. Despite the fact that the oxide layer is made of the same material as in Si MOSFETs, very disappointing results were obtained initially. The semiconductor industry was, however, determined to solve the problem and launched several parallel research and development programs, and eventually SiC power MOSFETs became available as mass-produced products. For high-temperature operation (> 175 °C), however, additional research and development is still necessary.

The main characteristics of SiC MOSFETs are very similar to those of Si MOSFETs. The device is voltage controlled, resistive in the ON-state, reverse conducting, and has an intrinsic anti-parallel body diode. The main differences compared to the Si counterpart are that the threshold voltage and maximum gate voltages are different, the ON-state resistance is very much lower, and the anti-parallel body diode has a higher voltage drop, but negligible reverse recovery. These differences make the SiC MOSFET an utterly potent power device.

The SiC MOSFET is already available as a 3.3 kV power module, and higher voltage ratings will become available over the next few years. Theoretically, it is therefore already possible to design submodules for MMCs using SiC MOSFETs today. If the aim is to beat the low conduction losses of the Si IGCT, this would require a massive amount of parallel-connected SiC MOSFET chips. At present, this would become too costly for a realistic product, but the semiconductor manufacturers unanimously declare that the cost will decrease rapidly over the next five years.

It is, therefore, time to consider the potential benefits a low-cost SiC power MOSFET will have if it is implemented in MMCs. If only the losses are considered, information is already available in [54]. It is found that efficiencies of the order of 99.9% are possible (counting only semiconductor losses). Even if this investigation of losses of MMCs with SiC power switches was performed using SiC JFETs, roughly the same results would be obtained with SiC MOSFETs in the future.

Even if the losses are important, there are also other aspects to consider when replacing one power semiconductor device with another. Reliability, cost, and secondary effects on the system design may be as important. The reliability of the SiC MOSFET has been an issue, mainly because of problems in achieving long-term stability of the gate oxide and the associated drift of the threshold voltage, as mentioned above. Lately, however, very promising test results regarding oxide-layer stability have been presented [55]. These indicate that the gate oxide may not be a serious reliability limitation in the future. It is concluded that the physical mechanisms behind the threshold voltage drift are not fully understood at present. Provided that such knowledge can be gained, however, dramatic improvements in device reliability can be achieved. Instead, issues like adverse effects on the junction terminations, caused by humidity, and cosmic ray may be of greater importance to investigate [56].

An important secondary effect is how the voltage rating of the power semiconductor devices influences the system design. Generally, higher voltage ratings are preferred because this results in a lower number of submodules. This is beneficial because some costs are related to the number of submodules and the associated complexity. These costs are not negligible. In [54] it was found that the conduction losses of the SiC-based MMC are highly dependent on the voltage rating of the semiconductor devices, and that 4.5 kV devices are very much less competitive than 3.3 kV devices from a loss perspective. With this in mind, unipolar SiC devices may be difficult to apply until devices with blocking voltages of the order of 10 kV with conduction losses similar to 3.3 kV devices become available. Such devices are possible in the future provided that the *super junction* (or *resurf*) technology has been developed in SiC. Initial experiments on this matter have been presented by several research groups, but major challenges have to be overcome before the technology can be used in mass-produced high-voltage SiC power devices.

2.2.8.2   Bipolar Devices

Bipolar devices such as IGBTs and diodes in SiC have the potential to operate at very high voltages. However, they are still in the research phase, and it is not likely that bipolar SiC devices suitable for HVDC applications will be commercialized in the next few years. There are two reasons for this. First, bipolar degradation has to be solved completely before any commercialization can take place. At some point, this will happen, but not any time soon. The second reason is that a bipolar device in SiC must not only be better than existing silicon devices. It must also be better than existing unipolar devices in SiC.

A comparison of conduction losses with silicon devices can be performed comparatively easily. The best silicon device from a conduction-loss perspective is a 4.5 kV IGCT (device A in Table 2.2) as shown in Figure 2.7. This 4.5 kV device has a voltage drop of approximately 1.7 V at 2 kA. A bipolar device in SiC can of course be designed for a much higher rated voltage, which may be very advantageous from a systems point of view because the number of submodules in the MMC can be reduced. This would in turn enable significant cost savings. Assuming that a high-voltage bipolar SiC device would have a voltage drop of approximately 3.5 V, the relative voltage drop would be the same as for the Si IGCT for a rated voltage of 9.3 kV. This means that using a 9.3 kV SiC IGBT would give no benefits in terms of conduction losses.

If a technology shift should take place, significant advantages should be foreseen because the new technology might also involve technical risks. The latter is usually a cost driver. In the case of a SiC IGBT, the reliability of the oxide layer could potentially be such a problem. A realistic choice of rated voltage of bipolar SiC device is, therefore, approximately 20 kV. For such a device, the conduction losses would be approximately 50% lower than for the 4.5 kV Si IGCT, and the number of submodules of an MMC could be reduced by a factor of more than 4. The latter would enable significant advantages on the system level, as described above.

A 20 kV device is, however not easily realized. There are two reasons for this. First, since the electrical fields inside the chip are very high in SiC, this requires delicate measures at the edges of the chip in order to prevent electric breakdown. These measures are called *junction terminations*, and the higher the voltage, the more of the chip area will be covered by these junction terminations. The only reasonable solution to this problem is to have large chips. The problem with large chips is, however, that the probability of having fatal defects on the chip increases. As a consequence, this would give a low production yield with an associated high cost for the device. Second, a 20 kV device would require a completely different module design than what is available today in order to handle the high voltage as such and the high associated electric field around the module. The development of this technology is far from easy and will be both expensive and time consuming.

Having said this, bipolar SiC devices are still worth waiting for. Half the conduction losses and four times fewer submodules in an MMC would give the system advantages that would motivate a technology shift. Until these devices become available it makes sense to follow what happens in the research laboratories.

2.2.8.3 Reliability

Before SiC power devices can be applied in commercial products, all major reliability issues must first be investigated thoroughly. In the case of HVDC transmission, the requirements of reliability are far more rigorous than in most other applications. The SiC devices must, therefore, work their way through all possible applications, starting with the simplest, next to the intermediate applications, and finally they will make their way to HVDC transmission. Several reliability issues have already been mentioned above, but the whole list of potential reliability issues is quite long.

For devices requiring a metal-oxide-semiconductor interface like the SiC MOSFET or IGBT, an obvious potential reliability issue is the stability of the oxide layer. Quite substantial work has already been carried out in this field, and dramatic improvements have been observed as a result. However, in order to reach reliability levels that are compatible with HVDC transmission, significant improvements in oxide stability must first be achieved. At this point in time, it is not easy to predict when this will be achieved.

Bipolar degradation has also been mentioned above, and this problem has been investigated for more than a decade. Still it is not fully solved, and it is likely that it will not be fully solved until the SiC carbide substrates are produced using on-axis growth. It is hard to predict when this will happen. As long as only small chips are used, there are ways to handle bipolar degradation. The first is to use a so-called buffer layer, which improves the crystal structure of the surface of the wafer. The second method is to identify the chips which are likely to degrade and sort out these chips before the final production steps.

One of the major reliability issues in the case of silicon power devices is power cycling. The devices' manufacturers unanimously declare that this should not be a major issue for SiC. It remains to be seen if this is true. As the CTE is higher for SiC than for Si, and thus closer to most other materials used in power modules, it may well be true. The Young modulus is, however, higher for SiC than for Si. This could be a disadvantage from this point of view.

In [56] several other reliability issues are raised. One potential reliability issue is the combination of humidity and high electric fields at the junction terminations. Because of the high electric fields in SiC devices, this issue is potentially far more problematic in SiC than in Si. This is currently being studied both by device manufacturers and academia. Even if this issue is still not fully investigated, there will probably be ways to handle it, for instance by trying different passivations on the junction terminations or using hermetic modules.

Another potential reliability issue mentioned in [56] is *cosmic ray* (see Section 2.2.7). A few studies have been performed, but no conclusive results can be drawn from these studies. Probably, however, SiC is at least as good as Si. To some extent this problem can also be handled by over-rating the devices voltage-wise.

2.2.8.4   Cost Competitiveness

Until recently, the cost for the SiC substrates was prohibitively high for the mass production of power devices. During recent years, however, both the cost for the substrates has reduced and the quality of them has increased. Even today, the cost continues to drop rapidly for high-quality substrates. At the same time, the device's manufacturers have received more experience in handling various manufacturing processes, which results in steadily increasing production yields. The combined effect of these two developments is a rapidly decreasing cost for SiC power device chips.

The cost for the chip is, however, only a minor part of the total cost of a power module. A result of this is that the cost of SiC power devices gets ever closer to their Si counterparts day by day. Recent predictions even indicate that the cost of a SiC device may soon be lower than that of Si devices in terms of cost per current rating.

The complete device is also only a minor part of the total system cost in almost any application. Therefore, if the system becomes more efficient with SiC devices than with Si devices, the total system cost may even be reduced by the introduction of SiC devices, because cost may be saved on the cooling system or because of size reductions. Additionally, if the costs for the losses of the equipment during a specified time are added to the initial cost of the system, SiC power devices may be motivated from a pure cost perspective in the near future.

2.3   Medium-Voltage Capacitors for Submodules

Apart from the power semiconductors, the submodule capacitors are the most important main-circuit components of an MMC. It is the voltage across these capacitors which is inserted or bypassed in order to synthesize the desired arm voltages. These capacitors also have to absorb charge pulses generated by the arm current each time a submodule is inserted. It is, therefore, extremely important that these capacitors are highly reliable. Additionally, because of the very large number of submodule capacitors required in order to build an MMC, the cost of these capacitors is one of the main contributors to the total cost of an MMC. The most suitable capacitor technology is the metallized polypropylene film capacitor [57], which is self-healing, and therefore has no catastrophic fault modes.

2.3.1   Design and Fabrication

Metallized polypropylene film capacitors are available in various values of capacitances and voltages up to 10 kV per piece. Typically, they have cuboid-shaped housings and the terminals have voltage-adapted bushings that can withstand voltages in the kilovolt range. The casing is made of aluminum or non-magnetic stainless steel and the voids between the casing and the active internal parts are filled with a polyurethane resin. A photograph of a typical submodule capacitor is shown in Figure 2.18 and in Figure 2.19 the mechanical structure holding the submodules, including the submodule capacitors, together is shown. The photograph is taken in a real MMC installation.

The active part of the capacitor consists of two polypropylene films with vacuum-deposited metal coatings of aluminum or zinc, having a thickness of a fraction of a micrometer [58]. The thickness of the plastic films depends on what voltage rating is targeted. Each of the metal layers forms an electrode. Because of the arrangement shown in Figure 2.20, one of the metal layers is contacted on one side and the other metal layer is contacted on the other side of the sheets. The contact layer on each side is formed by spraying a metal layer onto the sides of the sheets. This contact layer is also called *schoopage* after the inventor of this spraying technique, the Swiss engineer Max Schoop. A unique feature of the described capacitor layout

Figure 2.18   A 4.0 mF submodule capacitor with a voltage rating of 2.65 kV.

Figure 2.19   The mechanical structure supporting the submodules with their submodule capacitors.

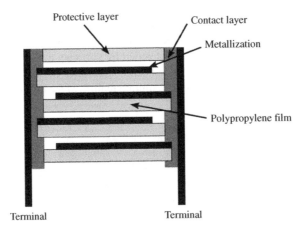

Figure 2.20   Basic layout of a polypropylene film capacitor.

is that the entire sheet is contacted on one side, which yields very low values of equivalent series resistance (ESR) and inductance (ESL).

For thin polypropylene sheets the breakdown voltage increases linearly with the thickness. However, above a certain thickness the breakdown voltage only increases approximately with the square root of the thickness of the plastic sheet. Instead of using thick sheets for high breakdown voltages, various ways of achieving intrinsic series-connections of thinner sheets are instead used. In Figure 2.21 examples of how to achieve higher voltage ratings by means of intrinsic series-connections using only partial metallizations of the polypropylene sheets are given. High voltage ratings can also be achieved by using films with metallizations on both sides combined with the method shown in Figure 2.21.

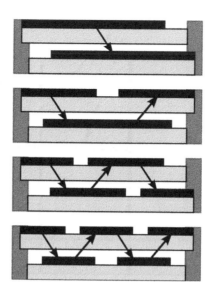

Figure 2.21 Realization of high voltage ratings by means of intrinsic series connections. The black layers are the metallizations and the light gray layers are the polypropylene films. The dark gray areas on the sides are the contact layers and the arrows indicate how the series connection is achieved.

2.3.2 Self-Healing and Reliability

The very high reliability of metallized polypropylene film capacitors is a result of two distinctly different mechanisms: self-healing and fuse segmentation. The former is an intrinsic property of metallized polypropylene, while the latter is achieved by means of slitting the metallization. Each time one of these mechanisms is activated, the capacitance of the capacitor is reduced by a small amount.

Self-healing occurs as a result of a local dielectric breakdown of the polypropylene film. A local point-defect breakdown results in an arc which vaporizes both the polypropylene film and its metallization in the immediate vicinity (typically 0.5–3.0 mm) of the point-defect [59]. The arc also causes a high pressure which removes the vapor and the arc itself. Usually, this process takes less than 10 μs, and the operation of the capacitor is unaffected. However, as some of the capacitor area has been lost, a slight reduction of the capacitance is inevitable.

The fuse segmentation is accomplished by slitting the metallization such that the metallization is segmented in smaller portions with narrow passages of interconnections between the segments; see Figure 2.22. These narrow passages act as fuses in the event of high surge currents owing to dielectric breakdown. The fuses are dimensioned such that they can withstand uncomplicated self-healing events. If more dramatic faults occur, however, the four closest fuses will burn off. When this occurs, the faulty segment is disconnected from the rest of the capacitor, and consequently a corresponding capacitance decrease is obtained.

The main factors affecting the lifetime of a metallized polypropylene film capacitor are voltage and temperature, where the latter usually is a function of the capacitor current. Accordingly, the lifetime can be increased by using a capacitor with a higher voltage rating. In the same way, the temperature may be reduced by choosing a higher value of capacitance, because the

Fuses

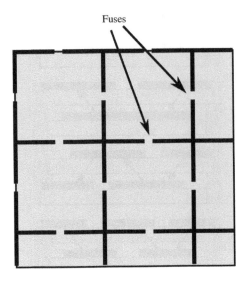

Figure 2.22   Fuse segmentation of metallized polypropylene film capacitors.

arm current will then be distributed over a larger area. In both cases the volume and cost of the capacitor will increase. This indicates that there is a delicate trade-off between reliability and low cost. In data sheets, the reliability is usually stated for specified voltages and temperatures.

The quantities used for quantifying the reliability are the FIT rate and the estimated time to failure. Typically, the FIT rate is 100 FIT and the lifetime expectancy is approximately 250,000 hours with a hot-spot temperature of 70 °C. A common criterion for determining the instant when the lifetime has elapsed is a capacitance reduction of 5%. This is a reasonable definition because each breakdown in the capacitor results in a small capacitance decrease. A consequence of this is that if the capacitor is subjected to over-voltages it is likely that a number of small breakdowns will occur and that this will result in a certain capacitance reduction. This also means that the capacitor can only withstand a low number of over-voltages of high magnitude before the capacitance has been reduced by 5%.

2.4   Arm Inductors

Arm inductors for MMCs in HVDC applications are typically dry-type air-core reactors. The three main benefits of dry-type air-core reactors over oil-immersed iron-core reactors are [60]:

- constant inductance, independent of current also at high current levels. This is especially important, for instance, during short-circuits on the dc side, when the arm inductor should limit the short-circuit current. A constant inductance may also be helpful when designing current controllers;
- absence of oil. This makes the inductor more environmentally friendly, and the risk of fire hazard is reduced;
- simple insulation to ground. Instead of a complex oil-impregnated paper insulation system, simple mechanical support insulators can be used.

Additionally, dry-type air-core reactors are maintenance free in contrast to oil-immersed iron-core reactors, which need a supervision of the quality of the oil and oil leaks. Another benefit is that the weight is lower, but this comes at the cost of a higher volume.

The winding of a dry-type air-core reactor is a solenoid with one or, typically, several concentric winding layers of film/glass-tape insulated conductors. The conductors are usually made of aluminum instead of copper because this yields lower cost and lower weight for a design with the same amount of losses [61]. The layers are connected in parallel by welding their top and bottom ends to radial metallic arms, commonly referred to as *spiders*. The configuration of the layers is such that the radial voltage stress is minimized. The turn-to-turn steady-state operating voltages are chosen to comply with limits derived from field experience. Between the layers glass-fiber sticks are inserted in order to achieve a desired radial spacing. The spacing is chosen such that sufficient air flow is allowed for natural convection air cooling.

The height of the winding is not chosen primarily to achieve maximum flux linkage. Instead, the height is chosen such that a sufficient creepage distance along the winding is obtained. Typically, this results in taller windings than if the height were chosen with respect to maximum flux linkage and minimum losses. The losses are, however, not a big concern. According to [62], losses of typically 0.02% of the rated power of the MMC are obtained. In order to achieve a good mechanical strength, the winding is impregnated and encapsulated by epoxy resin. Finally, a protective coating is applied. A photograph of two arm inductors in a real installation of an MMC is shown in Figure 2.23. From the photograph it is possible to distinguish the three layers of the winding and the glass-fiber sticks that ensure the necessary spacing between the winding layers. The three arms of the spiders are also visible.

Figure 2.23   Two arm inductors placed on top of each other.

2.5   Submodule Configurations

The dominating type of submodule for HVDC applications is the half-bridge submodule. The reason for this is that this type of submodule has the lowest losses and the lowest cost [12]. Several other types of submodules have been proposed in order to improve different properties. The most obvious alternative submodule candidate is of course the full-bridge submodule, which had been in use in cascaded-cell converters for medium-voltage drives [63] and static synchronous compensators (STATCOMs) [64] already before the invention of the MMC.

In contrast to the half-bridge submodule, the full-bridge submodule has the advantage that it can provide a counter voltage in case of dc-side short-circuits, and thereby prevent short-circuit currents flowing from the ac side to the fault on the dc side. This is, however, achieved by doubling the cost of power semiconductors and increasing the losses by approximately 80% compared to the half-bridge submodule [12].

In the literature, arms equipped with combinations of half-bridge and full-bridge submodules have also been proposed as a solution with the same dc short-circuit characteristics as the solution with only full bridges [65, 66]. This is an interesting approach because it is sufficient that only half of the submodules are full-bridges and the remaining submodules can be less costly and more efficient half-bridges [67].

This solution also allows for reactive power compensation on the ac side, which enables control of the fault current. From a mechanical and standardization point of view it would not be advantageous to have two different types of submodules in the arm. This can, however, easily be solved by having building blocks consisting of one half-bridge submodule connected in series with one full-bridge submodule. In this way, all units in the stack of the arm would be identical.

Other submodule topologies that can be used for handling dc-side faults are the *clamped single-submodule*, presented in Section 2.5.2, and the *clamped double-submodule*, presented in Section 2.5.3. Both of these solutions represent intermediate solutions between the half-bridge submodule and full-bridge submodule with respect to both total semiconductor ratings and losses. According to [67], the clamped single-submodule has a slightly lower semiconductor rating than the clamped double-submodule (with certain assumptions made), whereas the opposite is valid for the losses.

The solution with a combination of full-bridge submodules and half-bridge submodules has a higher total semiconductor rating than both the clamped single-submodule and the clamped double-submodule. The losses are, however, equivalent to the clamped double-submodule. Even if the clamped single- and double-submodules can block dc-side short-circuits, they cannot control the fault current (in contrast to the combination of full- and half-bridges), because the blocking is achieved by means of diodes. This discussion indicates that any comparison of different submodule implementations becomes complicated.

Other submodule topologies with dc-side fault current handling are the *five-level cross-connected submodule*, presented in Section 2.5.5, and the *three-level cross-connected submodule* [65], presented in Section 2.5.6, of which the latter is a simplification of the former. Also these submodules can be combined with other submodules [68].

Another derived submodule topology with dc-side short-circuit handling is the *unipolar-voltage full-bridge submodule*. This submodule is described in Section 2.5.4. As will be evident from the presentations below, the unipolar-voltage full-bridge submodule has very similar characteristics as the clamped single-submodule. The main difference is that the diode conducting during faults is connected to the positive terminal of the submodule capacitor in the unipolar-voltage full-bridge submodule, whereas the same diode is connected to the midpoint of the submodule capacitor in the clamped single-submodule. In the latter submodule, this causes an asymmetry regarding what voltage will be inserted against the fault current depending on the polarity of the fault current. In most cases, however, this difference is not very significant.

The handling of dc-side short-circuits is, however, not the only criterion for the evaluation of a new submodule topology. Other aspects, such as cost, losses, and volume, may be of similar importance. Examples like the *double submodule* and *semi-full-bridge submodule* must, therefore, also be evaluated. These are described in Sections 2.5.7 and 2.5.8. The latter submodule topology is an extension of the clamped double-submodule, with very interesting possibilities to reduce the amount of submodule capacitance. This, however, comes at the price of additional complications regarding capacitor voltage balancing.

Finally, *soft-switching submodule topologies* are discussed in Section 2.5.9. Such alternatives may become very interesting if attempts are made to use high-voltage thyristors as switching elements.

It may seem as if the intention of this section is to cover all possible submodule candidates. The reader who seeks to be convinced of the opposite will find a multitude of other options in [69], which gives an excellent overview of all the possible alternatives known today. Any reader who seeks for the ultimate solution will find that for each new submodule topology several new questions arise. The only conclusion that can be drawn today is that we have only seen the beginning of a fantastic phase of development. However, before we go into the descriptions of various alternative submodule implementations, existing half-bridge realizations must be described.

2.5.1 Existing Half-Bridge Submodule Realizations

Several manufacturers of HVDC equipment have made their own designs of MMCs. At present, all commercial installations build upon the half-bridge submodule, even though ABB refers to their topology as *cascaded two-level converter* [62]. A common feature of all commercial installations is that the half-bridge submodules employ IGBTs and anti-parallel diodes as power semiconductor elements. Another common feature is that a substantial portion of the volume of the converter is occupied by submodule capacitors. The mechanical design of the main circuit and the design choices made to handle faults are, however, far from similar. Below, therefore, three different existing MMC designs are shown and described.

2.5.1.1 Siemens Realization

Siemens was the first manufacturer to realize a commercial implementation of the MMC in 2010. It was the *Trans Bay Cable project*, which interconnects the city center of San Francisco (Potrero) and a Pacific gas and electric substation near Pittsburg, California [70, 71]. A photograph of a Siemens submodule is shown in Figure 2.24. Note that the submodule capacitor, which should be connected on the rear side of the unit shown in the figure, is missing in the photograph. Between the metal sheets in the front, the white and grooved housing of the bypass thyristor can be seen. This thyristor is triggered in order to protect one of the anti-parallel diodes in case of a dc-side short-circuit. The white structure at the bottom/front of

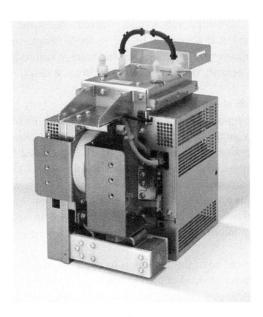

Figure 2.24 Photograph of a Siemens half-bridge submodule excluding the submodule capacitor (Reproduced with permission of Siemens AG).

the figure conceals the vacuum switch, which is used to bypass the submodule if the submodule is faulty. On the top of the submodule, pipes for de-ionized cooling water are visible. The structure to which the cooling water enters is the cooling block for the IGBT modules. The perforated sheets on the right of the figure constitute the housing for the communication and control electronics of the submodule. This also includes gate drivers for the IGBT modules.

In order to form the converter arms, the submodules must be arranged in a string in such a way that the voltage may increase by the voltage across one submodule capacitor for each subsequent submodule in the string. This means that there should be sufficient isolation both to ground and to each subsequent submodule. This mechanical arrangement of the submodules is referred to as a *converter tower*. A photograph of such a converter tower is shown in Figure 2.25. Because of the complexity of the image, the submodules are not easily distinguished. It is, however, possible to identify that the submodules are arranged in four layers, with the electrical connections to the submodules on the left and right sides, and the submodule capacitors toward the center. The arm inductors are typically located in a separate room. The metal structures on the top, bottom, and sides of the mechanical structure are there to shape the electric field around the converter tower.

2.5.1.2 ABB Realization

In the ABB design presented here, the half-bridge submodules are formed by using bridge legs where each switch consists of eight series-connected press-pack IGBTs, see Figure 2.26. These cannot be distinguished in the figure because they are hidden by the silver-colored cooling blocks to which the white pipes are attached. A photograph of such press-pack IGBTs is, however, found in Figure 2.14.

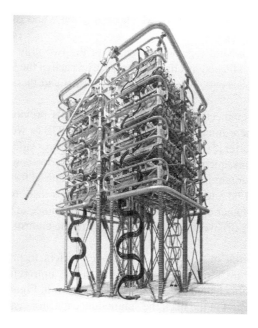

Figure 2.25 Photograph of a Siemens converter tower (Reproduced with permission of Siemens AG).

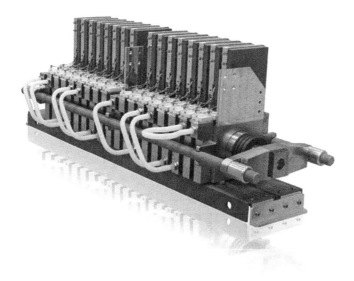

Figure 2.26 Photograph of a mechanical assembly of two stacks with eight IGBTs each, forming a high-voltage bridge leg of a half-bridge submodule (Courtesy of ABB).

Between each series-connected IGBT a cooling block is inserted. These cooling blocks are cooled by de-ionized water which flows through the white plastic pipes mentioned above. The stacked objects on top of the cooling blocks are the gate drivers for each of the series-connected IGBTs. These gate drivers are not standard off-the-shelf drivers. On the contrary, they are very advanced drivers which ensure static and dynamic voltage sharing among all IGBTs in the stack.

The stacked structure is kept together by means of the two gray reinforced plastic rods, which impose a carefully adjusted mechanical pressure on each of the IGBTs in the stack. Note that this pressure is necessary to ensure sufficient electrical and thermal conduction through the stack.

The feature of series-connection is inherited from ABB's previous design of VSCs for HVDC transmission, where hundreds of series-connected IGBTs were switched simultaneously in a two-level converter [5]. This proven technology ensures sufficient reliability through redundancy. A natural consequence of having eight IGBTs series-connected to form a switch is that a significantly lower number of submodules will be required for a certain voltage rating, and each submodule will be rated for a significantly higher voltage than if every IGBT were switched independently. Other consequences of this design choice are that the output voltage will have a lower number of levels and the complexity of the control system will be reduced because of this.

From a mechanical point of view, the ABB design differs from the others in that two half-bridge submodules are mounted together in a single unit referred to as a *valve*. A photograph of such a double half-bridge submodule is shown in Figure 2.27. In the front, the IGBT stacks can be seen and in the back the submodule capacitors can be distinguished. The metal structures on the top, bottom, and sides of the mechanical structure are intended for shaping the electric field around the valve.

In contrast to the other MMC designs presented in this subsection, the ABB design does not include any bypass valves for dc-side short-circuits. The anti-parallel diodes are designed to handle the short-circuit current for the time it takes the dc circuit breaker to interrupt the short-circuit current.

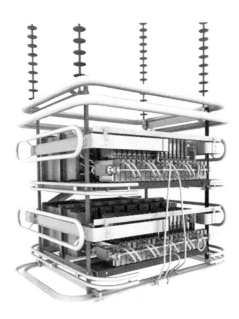

Figure 2.27   Photograph of a double half-bridge submodule (Courtesy of ABB).

2.5.1.3 Alstom Grid Realization

A photograph of the valve hall of the Alstom Grid realization is shown in Figure 2.28. At the far end of the valve hall the arm inductors can be seen. Each arm inductor is composed of six smaller arm inductors. In Figure 2.29 the valve arrangement is shown in more detail. The individual submodules can be distinguished, and the way of series-connecting sub-strings (of eight submodules each) in four layers can also be understood from the figure.

Figure 2.28 Photograph of the valve hall of an Alstom Grid realization of an MMC (Courtesy of Alstom Grid – © Alstom Grid UK Ltd).

Figure 2.29 Photograph of the valve arrangement of an Alstom Grid realization of an MMC (Courtesy of Alstom Grid – ©Alstom Grid UK Ltd).

Although not visible in the figures, Alstom makes use of industrial isolated power modules. Owing to this design choice, a bypass thyristor is necessary in order to handle dc-side short-circuits. A mechanical circuit breaker in parallel with the bypass thyristor is also necessary in order to permanently bypass faulty submodules. These devices are also hidden in the figures.

2.5.2 Clamped Single-Submodule

The clamped single-submodule, which was first presented in [72], is an extension of the half-bridge submodule that enables the blocking of dc-side short-circuits. A schematic diagram of this submodule is shown in Figure 2.30. The switches S1 and S2 along with their anti-parallel diodes, D1 and D2, form a conventional half-bridge submodule. The switch S3 and the diodes D3 and D4 have been added in order to realize the envisioned fault blocking capability.

In normal operation, S1 and S2 operate exactly as in a conventional half-bridge submodule, and S3 is always in the ON-state. A consequence of this is that the conduction losses are higher than for the half-bridge submodule. Note that S3 is connected with the opposite polarity compared to S1 and S2. This means that the arm current flows through D3 when it is positive and through S3 when it is negative.

Another characteristic property of this submodule is that the submodule capacitor is realized as a series-connection of two capacitors, each of which is charged to half of the submodule voltage. This also implies that S3, D3, and D4 are never subjected to more than half of the submodule voltage. The voltage rating of these devices can, therefore, be approximately half of that of the other devices. If the devices are rated for a lower voltage than the other devices, they may also have a slightly lower voltage drop.

In case of a short-circuit on the dc side, S1, S2, and S3 are all blocked. If the arm current is positive, the only path for the current is through D1 and D3, which means that the arm current has to flow through both capacitors of the submodule. The arm current must, therefore, flow through a stack of capacitors charged to the whole dc-side voltage when passing through one

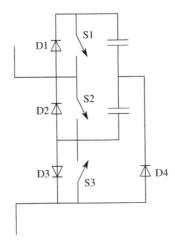

Figure 2.30   Schematic diagram of the clamped single-submodule.

arm. Consequently, in the case of a pole-to-pole fault on the dc side, two times the rated dc-side voltage is available for blocking ac-side currents flowing from one phase on the ac side through the fault on the dc side and back to another phase on the ac side.

In the case of a negative arm current, the only path for the current is through D2 and D4. Now, only one of the capacitors will be connected in the current path. As only half of the submodule capacitors in the arm are connected, each arm will provide half of the dc-side voltage against the fault current. Thus, in the event of a pole-to-pole fault on the dc side, the whole dc-side voltage is available for the blocking of fault currents flowing from one phase on the ac side through the dc-side fault and back to another phase on the ac side.

As stated in [72], the clamped single-submodule can effectively block dc-side faults without any use of circuit breakers. As a consequence, none of the semiconductor devices has to be dimensioned for handling of short-circuit currents of any significant duration. An additional feature of this submodule topology is that the switch S3 can be used to rebuild the dc-side voltage in order to restart operation immediately after a non-permanent fault.

2.5.3 Clamped Double-Submodule

This submodule topology was the first proposed alternative to the half-bridge and full-bridge submodules. It was presented by Marquardt in 2010 [73]. A follow-up with clarifications regarding the function of the submodule was presented in [74]. The main idea behind this submodule topology is to facilitate dc-side fault blocking without the high additional losses of the full-bridge submodule. A schematic diagram of this submodule topology is shown in Figure 2.31.

During normal operation, it has the same characteristics as two series-connected half-bridge submodules. These two half-bridges can easily be identified as the switches S1, S2, S3, and S4 along with their anti-parallel diodes D1, D2, D3, and D4. The series-connection is realized by means of the switch S5 and its anti-parallel diode D5, and the diodes D6 and D7 are blocking during normal operation. Owing to the additional voltage drop of the S5/D5 switch, this submodule topology has approximately 35% higher losses than the half-bridge submodule [12, 75]. This loss increase has, however, to be weighed against benefits such as the possibility of using a reduced arm inductance. This could imply a slight reduction in the converter station footprint and cost.

In case of dc-side faults all active switches, including S5, are opened, and depending on the polarity of the arm current, two different cases are obtained. If the arm current is positive, the diodes D1, D4, D6, and D7 are conducting and the arm current is divided between the two

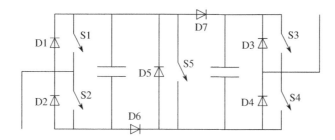

Figure 2.31   Schematic diagram of the clamped double-submodule.

capacitors. Since the two capacitors are connected in parallel, only half of the total submodule voltage will be inserted against the fault. This means that the total voltage inserted in one arm is half of the dc-side voltage. However, in case of a pole-to-pole fault on the dc side, the whole dc-side voltage is available for the blocking of fault currents flowing from one phase on the ac side through the dc-side fault and back to another phase on the ac side.

In the case of a negative arm current, the only path for the current is through D2, D3, and D5, which means that the arm current flows through both capacitors in series. In case of a pole-to-pole fault on the dc side, therefore, two times the dc-side voltage is available for the blocking of fault currents flowing from one phase on the ac side through the dc-side fault and back to another phase on the ac side.

If only the electrical characteristics of this submodule topology are evaluated, it is concluded that this solution is very competitive. However, attempts to realize the hardware of the circuit have revealed that the mechanical design is not entirely trivial. This may become an obstacle to overcome in future realizations.

2.5.4   Unipolar-Voltage Full-Bridge Submodule

This submodule topology was first proposed in [65], and is a simplification of the full-bridge submodule. From the schematic diagram in Figure 2.32, it is evident that the only difference compared to a conventional full-bridge is that the active switch in the upper position of the right bridge leg is missing. The only device left in this position is diode D3.

During normal operation S4 is in the ON-state at all times, and S1 and S2 are operated as in a half-bridge submodule. The consequence of this is that no negative voltages can be inserted (in contrast to the full-bridge submodule). As S4 is always in series with the rest of the circuit during normal operation, the conduction losses will inevitably be increased compared to the half-bridge submodule. However, no additional switching losses are associated with this switch. Based on the operation mode of the submodule, this submodule topology should have approximately the same losses as the clamped single-submodule presented in Section 2.5.2. In the latter case, however, the series-connected devices could be rated for a lower voltage, a fact which could reduce the conduction losses of these devices to some extent.

In case of a dc-side fault, S1, S2, and S4 are opened. The concept of having a switch in series with the circuit in normal operation, and then opening the switch at the inception of a fault, is the same as for the clamped single-submodule and the clamped double-submodule.

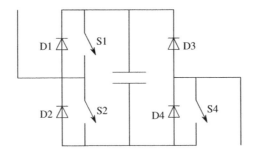

Figure 2.32   Schematic diagram of the unipolar-voltage full-bridge submodule.

The remaining devices of the unipolar-voltage full-bridge submodule during faults are the four diodes D1, D2, D3, and D4. As long as the arm current flows, two of these diodes will conduct. If the arm current is positive, D1 and D4 conduct, and if the arm current is negative, D2 and D3 conduct. This means that, regardless of the polarity of the arm current, the whole submodule voltage will be inserted against the fault current. In case of a pole-to-pole fault on the dc side, therefore, two times the dc-side voltage is available for the blocking of fault currents flowing from one phase on the ac side through the dc-side fault and back to another phase on the ac side.

2.5.5   Five-Level Cross-Connected Submodule

This submodule topology was first proposed in [68]. As with the other proposed submodules topologies, the five-level cross-connected submodule also builds upon building blocks consisting of half-bridges. A schematic diagram of the circuit is shown in Figure 2.33. Using all possible switch states five distinctive output voltages can be achieved, namely: $2U_{cap}$, U_{cap}, 0, $-U_{cap}$, and $-2U_{cap}$, where U_{cap} is the voltage across one of the submodule capacitors.

The authors of [68] conclude that it is sufficient that 25% of the submodules in one arm are five-level cross-connected submodules, and the rest of the submodules could be conventional half-bridge submodules. The negative voltage that could be inserted by the five-level cells is still sufficient to block the fault current.

A comparison performed in [68] reveals that the proposed combination of submodules has the same amount of losses as a combination of 50% half-bridge submodules and 50% full-bridges. The complication of having two different types of submodules in the arm is, however, not discussed. One possibility is of course to have building blocks consisting of three series-connected half-bridge submodules in series with one five-level cross-connected submodule.

2.5.6   Three-Level Cross-Connected Submodule

This submodule topology was first proposed in [65], and is a simplification of the five-level cross-connected submodule topology. A quick examination of the schematic diagram in Figure 2.34 and comparison with the schematic diagram of the five-level cross-connected submodule in Figure 2.33 yields that one of the active switches (S6) in the five-level cross-connected submodule has been omitted in the three-level cross-connected submodule.

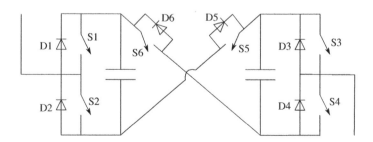

Figure 2.33   Schematic diagram of the five-level cross-connected submodule.

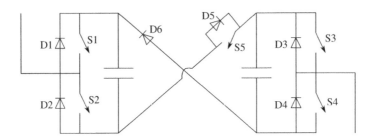

Figure 2.34   Schematic diagram of the three-level cross-connected submodule.

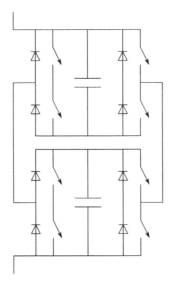

Figure 2.35   Schematic diagram of the double submodule.

This simplification has implications primarily during normal operation. During faults, however, the same negative blocking voltage can be inserted against the fault current. According to [65], this submodule technology has equivalent properties to the unipolar full-bridge submodule regarding dc-side short-circuit fault handling and losses.

2.5.7   Double Submodule

This submodule topology was first presented in [76]. A schematic diagram of the double submodule is shown in Figure 2.35. The main idea behind this submodule topology is to create a circuit that is able to connect two capacitors in series, in parallel, or fully bypass both capacitors. In the experimental results presented in [77], the elimination of low-frequency components resulted in a reduction of 18% in the peak-to-peak capacitor voltage ripple. Accordingly, the proposed submodule can reduce the capacitor voltage ripple without increasing the switching frequency.

In a practical application, this means that for a given switching frequency the size, weight, and cost of the submodule capacitors can be reduced. Or, alternatively, for a given size of the submodule capacitors, the switching frequency, and thereby the switching losses, could be reduced. Since the capacitor size is one of the dominating cost drivers in MMCs, this submodule topology deserves closer attention.

2.5.8 Semi-Full-Bridge Submodule

An important reason for using the full-bridge submodule is fault handling, since the full-bridge submodule can mitigate fault-currents originating from dc-side short-circuits. As twice the number of conducting components are connected in series during conduction compared to the half-bridge submodule, both the cost and the conduction losses of the full-bridge submodule are increased substantially. An innovative solution to this problem was proposed in [73] and [74]. This was the clamped double-submodule, which only increases the number of conducting elements by 25%. This submodule can, however, not actively insert any negative voltages, and the reason for using it is, therefore, limited to fault protection.

By replacing two of the diodes in the clamped double-submodule with active switches, the resulting submodule can act as a *semi-full-bridge* (SFB) with two positive voltage levels and one negative voltage level [78]. A schematic diagram of such a submodule is shown in Figure 2.36. The main benefit of the SFB submodule is that higher modulation indices are possible at a lower cost compared to the full-bridge submodule. Owing to the higher possible modulation indices, the energy variations in the converter arms can be reduced significantly, as shown in [79] and [80]. The energy variations in the converter arms can be minimized by increasing the modulation index to 1.41.

As can be seen from Figure 2.36, the SFB submodule employs two submodule capacitors which can be bypassed, connected in parallel, or in series. This means that one SFB submodule can replace two conventional half-bridge submodules. Therefore, when evaluating the SFB it must be compared to two series-connected half- or full-bridges. When examining the schematic diagram of the SFB submodule, it can be clearly seen that this submodule employs seven switches. It is, however, possible to control the SFB in such a way that S3 and S5 are always conducting in parallel. Consequently, S3 and S5 must only be rated for half of the arm

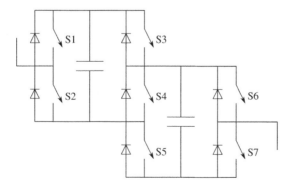

Figure 2.36   Schematic diagram for the semi-full-bridge submodule.

current. Accordingly, the combined power rating of the devices in the SFB submodule is 50% higher compared to the half-bridge submodule, and 25% lower compared to the full-bridge submodule.

When examining the different relevant switching states of the SFB submodule, it is found that the two submodule capacitors are always connected either in series, in parallel, or bypassed [78]. Even though this is the main advantage of the SFB submodule, this may potentially cause problems. When the two submodule capacitors are connected in series, they would ideally conduct the same current, and consequently the rate of change of the two capacitor voltages would ideally be identical.

However, if the capacitance values of the two submodule capacitors differ, or if the total equivalent leakage current (including components supplying various losses in the submodule) differs, the voltages across the two capacitors may diverge after some time. If the submodule capacitors are then connected in parallel, this voltage difference would cause a current spike with an associated energy loss.

A difference of 10% in the capacitor values and an increase of 2% in the capacitor voltages during the time when they were connected in series would result in that 0.01% of the energy supplied during the series connection being lost when the capacitors were connected in parallel. The details of this analysis are found in [78]. The conclusion of this is that as long as the voltage differences are kept within a few percent this is not a major problem.

Owing to the possibility of using the negative voltage levels, and that the submodule capacitors can be connected in parallel, it is possible to reduce the capacitor voltage ripple by 59%. At present, no other submodule topology can provide such capacitor voltage ripple reductions. This result is very important because it enables the use of smaller submodule capacitors and associated cost and volume reductions.

2.5.9   Soft-Switching Submodules

The development trend in power semiconductor devices has been characterized by a constant thrust toward higher power ratings in combination with high switching speeds. This is not a process which was initiated primarily by the device manufacturers. On the contrary, this development is entirely application driven by customers that develop, for instance, industrial motor drives, electric traction equipment, and renewable electric power generation systems—systems which require IGBTs having low tail currents and diodes with as little reverse recovery as possible (see Section 2.2.5).

An MMC for HVDC transmission is, however, very different from other VSC in that very low switching frequencies can be used and that the voltage drop of the device is more important than the switching properties. Most of the power semiconductor devices used in industrial isolated IGBT modules of today are, therefore, not adequately optimized for this application. Typically, too much effort has been spent on achieving fast switching transients, and too little effort is spent on achieving minimum conduction losses.

This partly explains why the IGBTs (devices D–G) in Figure 2.7 have much higher losses than the IGCTs (devices A–C). However, thyristor-based devices like the IGCT also have another advantage making them particularly suitable for MMCs in HVDC transmission applications—a more advantageous charge plasma distribution, which yields a lower ON-state voltage drop than that of the IGBT (see Section 2.2.4).

If it were possible to use *slow* thyristor-based devices with very high voltage ratings (approx. 10 kV) in MMCs for HVDC transmission, two very significant advantages could be obtained. First, by using such devices, very low relative ON-state voltage drops could be obtained. Second, by using devices with very high voltage ratings, the number of submodules, and their associated cost and complexity, could be reduced.

Thyristor-based power devices with very high voltage ratings will, however, have substantial switching losses. Without solving this issue, such a submodule design could result in increased total losses. The solution to this problem is called *soft-switching*. The typical way of achieving soft-switching is to introduce resonant elements (capacitance and inductance) to the commutation circuit.

A popular soft-switching circuit topology suitable for bridge-leg-based converters is the auxiliary resonant-commutated pole (ARCP) topology, which was first proposed by McMurray in 1989 [81]. A follow-up to this publication was published in 1990 [82]. A schematic diagram of the ARCP topology is shown in Figure 2.37.

The main differences compared to a conventional hard-switching bridge leg (or half-bridge) are that the power devices S1 and S2 have been equipped with capacitive snubbers (C_s) and that an auxiliary inductor (L_a) is connected in series with a bi-directional auxiliary switch, S3. Additionally, the main dc capacitor has been divided into two series-connected capacitors (C_{main}) in order to provide a connection point for the auxiliary switch.

The objection against this circuit is that the circuit becomes far more complex than a conventional hard-switching half-bridge, and that the additional components add cost. A closer examination of the necessary current rating of the auxiliary bi-directional switch S3 shows that this is merely a fraction of the rating of the main switches S1 and S2. In [83], it is suggested to use two comparably small anti-parallel fast thyristors for this purpose. The cost of these two thyristors will most likely not have a significant influence on the total cost of the submodule. The cost for the auxiliary inductor should not be prohibitive either. In [83], it is concluded that the size of this inductor does not have to be very much larger than the turn-ON snubber inductor of a corresponding IGCT circuit. The cost for the snubber capacitors is not significant compared to the main submodule capacitors. From a cost perspective, therefore, it is likely that the

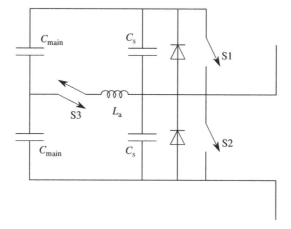

Figure 2.37 Simplified schematic diagram of a soft-switching ARCP submodule.

soft-switching submodule could be advantageous because cheaper thyristors are used instead of IGBTs, and because the number of submodules can be reduced as mentioned initially.

The objection that the losses of the ARCP submodule are increased at low output currents is not very threatening either, because HVDC links are usually operated above 50% of rated load. Hypothetical cases of HVDC links that operate at low power levels for long relative durations will never be built because the investment would not be justified.

The commutation sequence of an ARCP circuit is quite complex in comparison to a conventional hard-switching half-bridge. A detailed explanation of the commmutation from the lower diode to the upper switch of the half-bridge involves *eight* distinctive time intervals, and is definitely beyond the scope of this section. The interested reader is, instead, encouraged to study [81–83].

The findings of [83] indicate that the losses of an ARCP submodule could be 45% lower than those of an IGBT submodule. On top of that, the number of submodules could be reduced because of the increased voltage rating of the power semiconductor devices. At present no such submodule has been validated experimentally. This is, however, a very promising submodule design for the future.

2.6   Choice of Main-Circuit Parameters

The choice of main-circuit parameters is not altogether straightforward. Typically, the design starts from required power and voltage ratings, and then the initial values of the number of submodules, average voltage per submodule, and submodule capacitance are chosen. In the next stage, the arm inductance is chosen [84]. Using these initial values, various calculations and simulations are performed to fine-tune the final choice of main-circuit parameters. Because several properties are monitored to fulfill various requirements, the design process becomes iterative. This process is described briefly in this section.

2.6.1   Main Input Data

Any HVDC transmission project starts with a need for a certain amount of power transmission. Projects that involve MMCs usually require a high-voltage extruded cross-bound polyethylene (XLPE) cable for the power transmission. Since these cables are available at standardized voltage levels, these levels typically dictate the voltage rating on the dc side of the MMC. At present three voltage levels are available: 80, 150, and 300 kV. The latter class is sometimes used up to 320 kV. In 2003, 500 kV XLPE HVDC cables were already available on a laboratory scale [85]. The present technology seems, however, to be limited to a maximum voltage of 600 kV because of risk for thermal runaway. ABB has developed a new technology for HVDC XLPE cables for 525 kV, and claims that this technology could be scaled up for even higher voltages without the risk of thermal runaway. According to [86], these 525 kV XLPE HVDC cables have recently become commercially available. For HVDC grids, voltage levels of 800 and 1100 kV are discussed, but such cables are currently unavailable.

The cables are offered at a multitude of current ratings, corresponding to various conductor areas and choice of conductor material (copper or aluminum). At present, conductor areas from

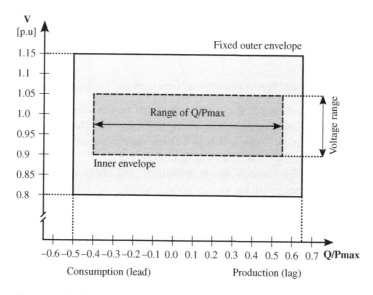

Figure 2.38 Requirements on reactive support proposed by ENTSO-E.

Table 2.5 Parameters for the inner envelope of Figure 2.38.

Synchronous Area	Maximum Range of Q/Pmax	Maximum Range of Voltage in p.u.
Continental Europe	0.95	0.225
Nordic States	0.95	0.150
Great Britain	0.95	0.100
Ireland	1.08	0.218
Baltic States	1.00	0.220

95 to 3000 square millimeters are available. In the latter case current ratings of up to 3 kA are possible in moderate climates. The largest cables have outer diameters of 150 mm.

From the transmission needs, the average values of the direct current and direct voltage are known. On the ac side, however, not only rated voltages and currents are necessary. Also ratings for reactive currents at various voltages are specified.

According to recent publications of committees of the European Network of Transmission System Operators for Electricity (ENTSO-E), the reactive power rating of an MMC should fulfill the specification in Figure 2.38. The outer envelope is fixed, and the inner envelope is specified depending on the location of the connection point. The proposed values for the inner envelope are given in Table 2.5. Note that the inner envelope may very well be smaller than the specified one. It may also have another shape than the rectangular envelope shown in Figure 2.38. It is, however, important that the inner envelope does not exceed the specified one in any point in the diagram.

Apart from these requirements several other requirements must also be fulfilled. The most important ones regarding the choice of main-circuit parameters are connected to fault cases.

2.6.2 Choice of Power Semiconductor Devices

As already discussed in Section 2.2, the optimal choice of power semiconductor devices in an MMC differs substantially from other VSCs. Owing to the low switching frequency per device, slow devices with low voltage drops and high voltage ratings are preferable. At present, the best device is probably a 4.5 kV IGCT, as shown in Figure 2.7.

Depending on if the submodules are half-bridges, full-bridges, or any other type of submodule, the current ratings of the power semiconductor devices must be chosen based on two different requirements. First, normal operation at rated current must be considered. All operating points dictated by the specification as shown in Figure 2.38 must be covered. The current that should be handled is the peak value of the arm current at rated operation as defined by

$$I_n = \frac{I_{d,n}}{3} + \frac{\hat{I}_{s,n}}{2},$$
(2.17)

where $I_{d,n}$ is the rated current on the dc side (absolute value) and $\hat{I}_{s,n}$ is the peak value of the ac-side current at the rated operation.

The choice of power semiconductor devices, however, also depends on how dc-side short-circuits should be handled and what type of module technology should be used. In the case of half-bridge submodules the dc-side short-circuit current flows through one of the antiparallel diodes of the submodules, unless special measures to prevent this are taken. For point-to-point HVDC transmission links, the short-circuit current is interrupted by means of the ac-side circuit breakers, and therefore a path for the dc-side fault current must be provided by the converter until the ac-side circuit breakers have interrupted the current.

As described in Section 2.5.1, Siemens and Alstom use industrial isolated power modules, whereas ABB uses press-pack devices. In the former case dc-side short-circuits are handled by a large thyristor, which is triggered immediately when a dc-side short-circuit is detected. The antiparallel diode of the switch which would otherwise conduct the short-circuit current, therefore, does not have to be dimensioned for the fault case. As an alternative to the bypass thyristor, a separate diode module could be connected in parallel with the diode that would conduct the fault current. In this way a sufficient fault current capability can be achieved.

In the case of press-pack power modules, bypass thyristors are not necessary because the antiparallel diode which should conduct the fault current is dimensioned to endure the fault current. This is achieved by equipping the module with additional diode chips and by choosing an appropriate value of arm inductance.

If the MMC operates in a dc grid, it may be equipped with a dc-side circuit breaker. In that case, over-dimensioned diodes or bypass thyristors may not be necessary. This, however, depends on the clearing time of the dc-side circuit breaker.

In the case of full-bridge submodules, the dc-side fault current can be controlled. Even if full controllability is available, this may be the dimensioning case for the power semiconductor devices in the submodules. This current will, however, be significantly lower than is the case with half-bridge submodules.

2.6.3   Choice of the Number of Submodules

The choice of the number of submodules per arm may seem to be trivial as soon as the voltage rating of the power semiconductor devices has been chosen. On the contrary, this choice involves at least three design processes, each of which may involve several investigations. The procedure is not entirely straightforward because several of the investigations require initial data, which may not be available, and therefore have to be determined either from previous experience or from numerous time-domain circuit simulations.

At this stage, for instance, the submodule capacitance is still unknown. A low value would imply a high submodule capacitor voltage ripple. This would increase the minimum number of submodules per arm in order to produce a sufficiently high output alternating voltage. Conversely, a high value of submodule capacitance would generate an unnecessarily costly converter. The solution to this is to use an initial value of submodule capacitance which yields a total energy storage of the converter of the order of 30–40 kJ/MVA [62].

Below, one possible way of performing these design processes is described. Several other procedures may exist, of which some may be more efficient than the one described. The three design processes mentioned above should find the following three quantities:

- the average sum capacitor voltage per arm, $\overline{v_{cu,l}^{\Sigma}}$;
- the average submodule capacitor voltage, $V_{c,n}$;
- the number of redundant submodules, N_{red}.

A reasonable choice of $\overline{v_{cu,l}^{\Sigma}}$ would be to let

$$\overline{v_{cu,l}^{\Sigma}} = V_{d,n}, \tag{2.18}$$

where $V_{d,n}$ is the rated dc-side pole-to-pole voltage. This is also a common choice, but it is not the only option. If the current rating of the power semiconductor devices is not sufficient, $\overline{v_{cu,l}^{\Sigma}}$ can be chosen slightly higher than $V_{d,n}$, and by choosing a suitable turns ratio of the transformer, a sufficient current rating on the ac side can be achieved without violating the current rating of the power semiconductor devices. A higher value of $\overline{v_{cu,l}^{\Sigma}}$, however, implies a modulation index in excess of one, which can only be realized if at least a fraction of the submodules can insert a negative voltage. Below, however, it is assumed that the choice of $\overline{v_{cu,l}^{\Sigma}}$ is made according to Equation (2.18).

The next step is to choose the average submodule capacitor voltage, $V_{c,n}$. The five main quantities that have to be considered when choosing $V_{c,n}$ are:

- the average submodule capacitor voltage with respect to reliability due to cosmic ray;
- the submodule capacitor voltage ripple in the operation point where this ripple is maximum;
- the over-voltage due to the stray inductance of the submodule when switching the maximum arm current;
- the maximum submodule capacitor voltage during single phase ac-side faults;
- the maximum submodule capacitor voltage during dc-side faults.

The average submodule capacitor voltage with respect to reliability due to cosmic ray is typically chosen based on experience and in order to achieve a desired reliability on the system level. As already pointed out in Section 2.2.7, a suitable choice for a 4.5 kV device may be approximately 2.8 kV or slightly lower.

The submodule capacitor voltage ripple in the operation point where this ripple is at its greatest is not straightforward to determine because the submodule capacitance has not yet been determined. Two possible ways to acquire a preliminary value would be to either assume a ripple of, for instance, ±10%, or to assume a submodule capacitor value corresponding to 30–40 kJ/MVA for the total converter and perform time-domain simulations to determine the submodule capacitor ripple.

On top of this voltage, the over-voltage due to the stray inductance of the submodule when switching the maximum arm current is added. At this stage it is necessary to know in what case the maximum arm current is obtained (probably a fault case), and the value of this current. It is also necessary to know the stray inductance of the commutation path of the submodule and the rate at which the semiconductor devices turn off the current.

The value of the stray inductance can be found in two ways. Switching waveforms from previous designs of submodules, or an early prototype, may give a reasonable value of the stray inductance. Another alternative to finding the stray inductance might be to establish this value by means of numerical methods, such as partial element equivalent circuit (PEEC) modeling. This, however, requires a detailed *paper design*. The current derivative at turn-OFF is often chosen based on experience from previous designs. If, however, another value of the current derivative is chosen, the switching losses, the turn-ON derivative, and the associated diode over-voltages due to the diode turn-OFF (*reverse recovery*, as explained in Section 2.2.5) have to be investigated.

Based on the obtained value of over-voltage, the average submodule capacitor voltage may be chosen equal to or slightly lower than the value obtained with respect to reliability due to cosmic ray. Using this preliminary choice of $V_{c,n}$, a first value for the number of submodules per arm is found from

$$N = \frac{\overline{v_{cu,l}^{\Sigma}}}{V_{c,n}}. \tag{2.19}$$

This preliminary value of N must now be evaluated with respect to single-phase ac-side faults and dc-side faults. These fault cases must be evaluated by means of time-domain simulations using the implemented fault-handling strategies implemented in the control system.

In [87] it is shown that a single-phase ac-side fault causes the submodule capacitor voltage in one of the six arms to increase to approximately 1.3 times $V_{c,n}$ in the case of half-bridge submodules, and approximately 1.9 times $V_{c,n}$ for full-bridge submodules. In the former case the initial value of $V_{c,n}$ may very well be kept, whereas in the latter case it is very likely that $V_{c,n}$ must be reduced compared to the initial value.

In the case of full-bridge submodules also dc-side short-circuits must be investigated. In [88] it is found that one consequence of a dc-side short-circuit is that the submodule capacitor voltages of one phase may start to diverge from those in another phase unless the power transfer across the dc-link is interrupted. It is also stated that it is important to control down the dc-side voltage as soon as the short-circuit is detected. Provided that these measures are taken, the submodule capacitor voltages may increase to higher values than during normal operation, but not very much higher. This is, however, strongly dependent on the chosen fault-handling strategy and on the value of the submodule capacitance. As the latter value is still not chosen, a preliminary value has to be used in the time-domain simulations. A reasonable preliminary value of total submodule capacitance is approximately 30–40 kJ/MVA considering all sub-modules of the converter and the total power rating of the converter. Depending on the results from this simulation, the value of $V_{c,n}$ may or may not have to be reduced again. Most likely, however, it will remain unchanged compared to the value obtained from the simulation of the ac-side fault.

At this stage a reasonable initial value of N can be established using Equation (2.19). This value will then be used in the following steps of the design procedure.

Finally, the number of redundant submodules per arm, N_{red}, has to be chosen. This quantity can, however, be chosen independently, and based entirely on estimations of the reliability on the system level. In most other investigations, except losses, it is assumed that all redundant submodules are bypassed, as this is usually the worst case. In Section 2.7, the handling of faulty and redundant submodules is discussed.

2.6.4   Choice of Submodule Capacitance

The value of the submodule capacitance must be chosen very carefully. An unnecessarily high value of submodule capacitance has a drastic effect on the total cost of the MMC, whereas an insufficient value causes a high voltage ripple. The latter effect may not be acceptable with respect to the voltage rating of the power semiconductors, or even the submodule capacitor itself. Typically, the submodule capacitance is chosen such that the total stored energy in all submodule capacitors of the converter is approximately 30–40 kJ/MVA [62]. An initial value is chosen within this interval. The final value is, however, found by means of an optimization where the cost for the submodule capacitors is weighed against the cost for the power semi-conductor devices. At this stage the expressions for the capacitor voltage ripple in [84] may be helpful. Finally, a number of time-domain or real-time simulations are performed, where vari-ous quantities are checked with respect to limitations as described in the previous subsections.

2.6.5   Choice of Arm Inductance

As already described in Section 2.4, the arm inductors (or valve reactors [84]) of an MMC are typically dry-type air-core reactors. The dimensioning of the arm inductance can be made in three different ways depending on $\underline{\text{how}}$ dc-side short-circuits are handled, and depending on whether the design option increase $\overline{v_{cu,l}^{\Sigma}}$ beyond $V_{d,n}$ is utilized (see Section 2.6.3).

Where the submodules are equipped with bypass thyristors or separate additional power modules entirely equipped with diodes, the current fed from the ac side into the fault on the dc side does not have to be limited to protect the anti-parallel diodes of the submodules. The purpose of the arm inductance is merely to limit high-frequency components in the circulating current, and to enable a smooth circulating current control. This dimensioning typically yields an inductance value of the order of 0.05 p.u.

In the second case, where the submodules are not equipped with bypass thyristors or separate additional power modules entirely equipped with diodes, the anti-parallel diodes of the sub-modules must be protected from the high currents that are fed from the ac side into the fault on the dc side. As this fault current is interrupted by slow-acting ac circuit breakers, the magnitude of the current must be limited by the arm inductors in order to protect the anti-parallel diodes that conduct the fault current. The value is typically determined through real-time simulations of dc-side faults, considering detection delays, circuit breaker interruption times, and fault handling strategies implemented in the control system. This dimensioning typically yields an inductance value of approximately 0.10–0.15 p.u.

The third design option is used to increase $v_{cu,l}^{\Sigma}$ beyond $V_{d,n}$ in order to increase the current capability on the ac side or to minimize the losses. In order to achieve this, the arm inductance has to be increased in comparison to the first design case. The necessary increase cannot, however, be determined in a straightforward way, because part of the necessary additional inductance may originate from an augmented transformer leakage inductance.

2.7 Handling of Redundant and Faulty Submodules

Several faults can occur internally within a submodule, and in order to be able to continue the operation of the MMC without interruption, any faulty submodule must be short-circuited such that the arm current can continue its flow through the series-connected submodules in the arm. The arm must, therefore, be equipped with an additional number of submodules, such that when a maximum number of submodules are short-circuited, the operation of the arm can continue with a tolerable voltage stress on each remaining operating submodule in the arm. The additional number of submodules are usually referred to as *redundant submodules*, and the number of redundant submodules is denoted as N_{red} below. When a faulty submodule is short-circuited, it will stay bypassed until the next planned maintenance occasion. This long-term bypass function is typically realized by a mechanical vacuum switch which is connected in parallel to the power semiconductor switch which is used as the bypass switch during normal operation. In parallel to this mechanical vacuum switch there is also a thyristor, which may be triggered simultaneously as the vacuum switch in order to facilitate the bypass function faster than what would be possible if only the vacuum switch would have been triggered. This bypass function is implemented in both the Siemens and Alstom realizations described in Section 2.5.1.

The use of redundant submodules is described in [89–92]. The presentation below, however, is based primarily on the latter work. Two different methods of handling redundant and faulty submodules will be presented and compared. Method 1 keeps the average sum capacitor voltage constant regardless of the number of faulty submodules, N_f, whereas Method 2 keeps the average submodule capacitor voltage of each submodule constant. These two methods are compared in Section 2.7.3 and, finally, in Section 2.7.4, some comments are made about how redundancy can be handled if the MMC is implemented using stacks of IGBTs as in the ABB realization described in Section 2.5.1.

2.7.1 Method 1

By this method the average sum capacitor voltage per arm, $\overline{v_{cu,l}^{\Sigma}}$, is kept constant, and for simplicity it is assumed to be equal to the rated value of the dc-side voltage $V_{d,n}$. Consequently, the individual average submodule capacitor voltage is given by

$$V_c = \frac{\overline{v_{cu,l}^{\Sigma}}}{N + N_{red} - N_f} = \frac{V_{d,n}}{N + N_{red} - N_f}. \tag{2.20}$$

Obviously, as long as N_f is lower than N_{red}, V_c will be lower compared to a reference case, where the total number of submodules in operation per arm is N. However, for each failing (and thus bypassed) submodule, the value of V_c increases as $\overline{v^\Sigma_{cu,1}}$ remains constant. This method produces an $N + N_{red} - N_f + 1$-level arm voltage, which is a higher number of levels than for an MMC without redundant submodules. The average energy stored in each arm of the converter equals

$$E_{arm} = \frac{1}{2}(N + N_{red} - N_f)C\left(\frac{V_{d,n}}{N + N_{red} - N_f}\right)^2. \tag{2.21}$$

As long as N_f is lower than N_{red}, this energy will be lower compared to the case of an MMC without redundant submodules. However, for each failing (and thus bypassed) submodule this energy will increase. For this method the operation of the MMC can be summarized as:

- The average sum capacitor voltage per arm, $\overline{v^\Sigma_{cu,1}}$, remains unchanged regardless of N_{red} and N_f.
- The actual number of healthy submodules $N + N_{red} - N_f$ must be detected during operation. Typically, this information is provided by each submodule to the central control unit of the MMC.
- The control system in each arm will be given a reference for E_{arm}, which changes depending on N_f because the resulting arm capacitance changes each time a submodule fails.
- The insertion index is unaffected by N_{red} and N_f.
- The modulator must know N_f for each arm at all times in order to be able to decide how many submodules to insert and bypass depending on the desired insertion index.

2.7.2   Method 2

By this method the individual average submodule capacitor voltages, V_c, in the arm are kept constant [90] regardless of the values of N_{red} and N_f. The average sum capacitor voltage of the arm, $\overline{v^\Sigma_{cu,1}}$, is therefore given by

$$\overline{v^\Sigma_{cu,1}} = (N + N_{red} - N_f)\frac{V_{d,n}}{N}. \tag{2.22}$$

Clearly, $\overline{v^\Sigma_{cu,1}}$ is greater than $V_{d,n}$ as long as N_{red} is greater than N_f. However, it decreases each time a submodule fails (and thus gets bypassed). In this case the average energy stored in each arm of the MMC is given by

$$E_{arm} = \frac{1}{2}(N + N_{red} - N_f)C\left(\frac{V_{d,n}}{N}\right)^2. \tag{2.23}$$

Obviously, E_{arm} is increased compared to a case without redundant submodules as long as N_{red} is greater than N_f. However, this energy decreases for each failing (and thus bypassed) submodule. In contrast to Method 1, it is necessary to reduce the insertion index depending on the value of N_f when using Method 2. The reason is that a higher average sum capacitor voltage is available, as given by Equation (2.22). Among the $N + N_{red} - N_f$ submodules in the arm, the submodules are selected in such a way that an $N + 1$-level arm voltage is generated.

Consequently, the number of levels in the arm voltage remains constant irrespective of N_{red} and N_f. Moreover, all submodules in the arm are treated in the same way regarding balancing control of the individual submodule capacitor voltages. For this method the operation of the MMC can be summarized as:

- The reference voltage for each submodule capacitor remains unchanged.
- The actual number of healthy submodules $N + N_{red} - N_f$ must be detected. Depending on the number of healthy submodules in the arm, $v_{cu,1}^{\Sigma}$ is increased, as given by Equation (2.22).
- The control system is given a reference for the total energy in the submodule capacitors which changes depending on N_f because the resulting arm capacitance changes each time a submodule fails.
- The insertion index must be reduced because $\overline{v_{cu,1}^{\Sigma}}$ is greater than $V_{d,n}$ as long as N_{red} is greater than N_f.
- The modulator must know N_f for each arm at all times in order to be able to decide how many submodules to insert and bypass depending on the desired insertion index.

2.7.3 Comparison of Method 1 and Method 2

The investigation in [92] covers both simulations using a detailed equivalent model, and experimental verifications of events when submodules are bypassed. Especially, the transients following the event when a submodule is bypassed are studied.

For both methods, *the ac-side output voltage is unaffected* by the bypassing event of a faulty submodule. This also holds for the arm currents and the circulating current for Method 2. For Method 1, however, transients in the arm currents and in the circulating current are inevitable, because each of the healthy submodules must be charged to a slightly higher voltage. These transients are not severe, and cannot be used as the reason for choosing one method or the other.

Regarding the sum capacitor voltages, transients are observed for both methods. For Method 1, on the one hand, the transients are constituted by distorted waveforms with a slight over-voltage in the arm that did not experience a fault. This over-voltage is still well below the sum capacitor voltage for normal operation when using Method 2. For Method 2, on the other hand, the transients are mainly constituted by a reduction of the average sum capacitor voltage for the arm that experienced the fault. However, also for Method 2 there is a slight over voltage in the arm that did not experience a fault.

Finally, the individual submodule capacitor voltages are examined. For Method 2, the individual capacitor voltages of all healthy submodules are unaffected by the bypassing event. For Method 1, however, the main transient observed was an increase in the average value of the submodule capacitor voltage. Just a slight over-voltage (almost negligible) is observed. The most striking result is that the submodule capacitor voltages are always lower when using Method 1 than when using Method 2, even during the transients. This is probably the single most important finding of the investigation. As the submodule capacitor voltages are lower when using Method 1 than when using Method 2, it is likely that the reliability is improved when using Method 1. Both the power semiconductor devices and the submodule capacitors benefit from being subjected to lower voltages. These are the two most important components of an MMC, and if the reliability of these components is improved, it is wise to choose this method. The conclusion is, therefore, that Method 1, which uses a constant average sum capacitor voltage, may be a better choice than Method 2, which uses constant individual average submodule capacitor voltages.

2.7.4 *Handling of Redundancy Using IGBT Stacks*

In the ABB realization described in Section 2.5.1, redundancy is achieved in a different way than for other realizations. When an IGBT in the stack breaks down, it will always behave as a short-circuit. For this mechanism to be successful in case more than one of the eight IGBTs are faulty, it is the opinion of the author that the average submodule capacitor voltage has to be reduced each time a faulty IGBT is permanently short-circuited. When this occurs, it is reasonable to assume that the average capacitor voltage of the other submodules has to increase slightly such that all IGBTs in the whole arm have approximately the same voltage stress. After a certain duration of operation, therefore, it is likely that the submodules have different average submodule capacitor voltages even if the individual IGBTs operate under similar conditions.

2.8   Auxiliary Power Supplies for Submodules

Each submodule of an MMC contains at least two gate-drive units, i.e. one for each active power semiconductor device. Each of these gate drivers requires a certain amount of power both during conduction and switching operations of the active device, and when the device, for some reason, is idle. In the case where IGBTs are used as switching elements in the submodule, the necessary average value of auxiliary power is of the order of 1 W, whereas in the case of IGCTs this value may be approximately 50 W (for two IGCTs). Depending on whether the gate signals are directly provided from a central control unit (CCU), or if the submodule itself contains a distributed control unit (DCU), there may also be a need for auxiliary power supply (APS) for this purpose. A reasonable estimation of the required power consumption of such a distributed control unit would be of the order of 10 W. A simplified schematic diagram of such a system for a full-bridge submodule with a DCU is shown in Figure 2.39.

There are several alternative methods for supplying this power. Supplying the power from a power source that has the same ground reference as that of the entire MMC is, however, not practical [84] because of the need for excessive insulating systems. If all possible alternatives of achieving the required auxiliary power are evaluated it is usually found that the best option is to take the power from the submodule capacitor.

2.8.1 *Using the Submodule Capacitor as Power Source*

In the case when only 1 W of auxiliary power is required this is comparably straightforward. If the APS is assumed to be lossless, and if the input voltage of the APS is assumed to be approximately 100 V, the input current will be approximately 10 mA. If such a current is fed to the APS by means of ordinary voltage division through a resistor, a significant power dissipation will be obtained. Such a circuit is shown in Figure 2.40. The resistor R_{eq} represents the input quantities (100 V, 10 mA) of the APS, and the resistor R_{hv} is used as the channel of power to the APS. As the voltage across R_{hv} is approximately equal to the submodule capacitor voltage,

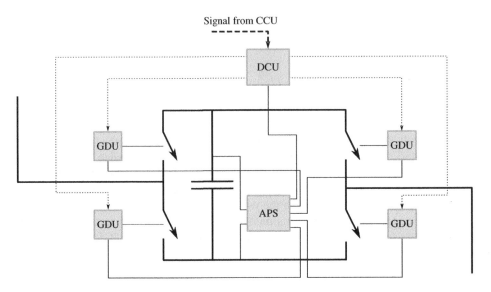

Figure 2.39 Auxiliary power supply (APS) for a full-bridge submodule. The APS receives its power from the submodule capacitor. Each gate-drive unit (GDU) receives power from a galvanically isolated power supply channel from the APS. An optional distributed control unit (DCU) also receives power from the APS. The DCU, on the other hand, distributes control signals to each GDU. These control channels are shown as dashed lines. The DCU receives its control commands from a central control unit (CCU, not shown).

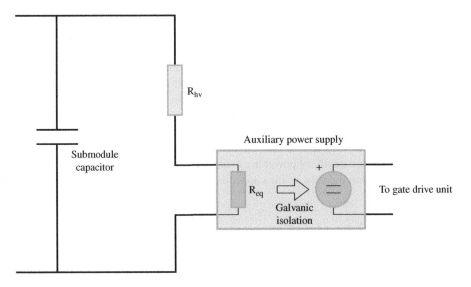

Figure 2.40 Resistive voltage divider for achieving a low-voltage input to an auxiliary power supply.

i.e. 3 kV, the power dissipation in R_{hv} will be approximately 3000 V times 0.01 A, i.e. 30 W. Conversely, if the input voltage of the APS is 20 V, the power dissipation in R_{hv} will be 150 W. At first sight, the latter number seems very high, but after a comparison with the conduction losses of a 4.5 kV IGBT, it may still be acceptable. Assuming that the average current during a fundamental cycle is 700 A and that this current flows through either an IGBT or a diode with the voltage drop 2.2 V, this would yield a value of conduction losses of approximately 1.5 kW. On top of this, there will also be switching losses. With this in mind, losses of the order of 150–200 W may be acceptable from an energy perspective.

When several tens to hundreds of watts of auxiliary power are required, a voltage divider would be far too lossy to be acceptable. Here, a switch-mode power converter with a reasonable efficiency is required. The problem can be subdivided into two steps, one step where the high-voltage submodule capacitor voltage is converted into a low voltage referred to, for instance, the minus terminal of the submodule capacitor, and another step where this low voltage is converted into galvanically isolated controlled low-voltage outputs; see Figure 2.41.

The first step is not easily realized, for several reasons, but the second conversion step is simpler to realize and such converters are available as commercial products. Such a converter with a 18 kV isolation voltage and four galvanically isolated low-voltage outputs is shown in Figure 2.42.

2.8.2 Power Supplies with High-Voltage Inputs

To realize the conversion of the submodule capacitor voltage into a low direct voltage referred to the minus terminal of the submodule capacitor is difficult because suitable high-voltage, i.e. 4.5 kV, low-current power transistors are not available [4, 51]. To use a 1 kA, 4.5 kV device for this purpose is not a realistic option. A more realistic option is to series-connect 900 or 1200 V devices with appropriate current ratings. A problem with series connection of switch-mode power devices is that each of the series-connected elements must be provided with a gate signal for turn-ON and turn-OFF. This usually requires both galvanically isolated signal transmission and galvanically isolated power transmission to the gate-drive units of each of the series-connected switch-mode power devices. Both of these tasks are challenging.

The first publication addressing this issue was [93], where a 10 W APS for MMC submodules was presented. In that case, however, a nominal submodule capacitor voltage of 850 V was used. This is a significant simplification because several options for the choice of switch-mode power devices capable of handling the whole voltage were available. The problem of providing

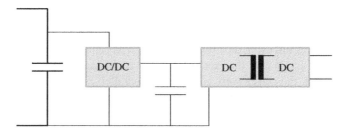

Figure 2.41 Two-stage auxiliary power supply with an input stage that converts the high-voltage submodule capacitor voltage into a low-voltage output, and a galvanically isolated low-voltage dc/dc converter.

Figure 2.42   Auxiliary power supply with low-voltage input and four galvanically isolated low-voltage outputs. The galvanic isolation is guaranteed up to 18 kV.

galvanically isolated gate signals and auxiliary power supplies to the gate-drive units of the series-connected devices did, therefore, not have to be solved. An interesting feature of this APS, however, was that it could start the operation already at 100 V, a feature which may be beneficial during the start-up process.

In [94], the problem of handling a high input voltage was addressed. In this work two different options of 100 W, 1500 V APSs were considered. The first one was an neutral-point-clamp-derived flyback converter and the second option was a flying-capacitor-derived flyback converter. Both options are realistic multilevel converter approaches. The common problem for both approaches is, however, that the power-semiconductor switches need galvanically isolated signal transmission and power supplies as described above.

In [95], a four-stage cascaded flyback converter aiming for APSs for 4.5 kV IGBTs with a submodule capacitor voltage of 3 kV was presented. The circuit was, however, only tested at 1400 V and 15 W, but an efficiency of almost 80% was recorded. As for the previous work, the multi-stage cascaded flyback converter needs galvanically isolated signal transmission and power supplies. This is a considerable complication.

A variation of this circuit, using only three stages, was presented [96]. In this case, however, the power rating was 100 W. On the input side the flyback stages were connected in series, whereas on the output, the stages were connected in parallel. This converter was referred to as an input-series output-parallel (ISOP) topology which should be particularly suitable for the application. Also in this circuit solution, each power semiconductor device needs both galvanically isolated signal transmission and power supply to the individual gate-drive units. As for the previous design options, this is a significant drawback for the implementation.

The most promising solution is a tapped-inductor buck converter [51]. The advantages of this solution is that the series-connected high-voltage switch is autonomous without the need for galvanically isolated signal transmission, and that no galvanically isolated auxiliary power has to be supplied to the series-connected power semiconductor devices. Additionally, the circuit is soft switching, a fact which reduces both switching losses and electromagnetic interference.

2.8.3   The Tapped-Inductor Buck Converter

The tapped-inductor buck (TIB) converter was first described in [97]. An advantageous feature of the TIB converter is that the voltage conversion factor not only depends on the duty ratio of the switch but also on the turns ratio of the tapped inductor [98–100]. This allows for an extension of the voltage conversion factor while maintaining a comparably high duty ratio. A consequence of this is that high efficiencies are possible also at high voltage conversion ratios [101].

In order to achieve zero-voltage switching (ZVS) of the high-voltage power switch, however, the concept of *current injection* was used [102]. In contrast to [102], however, the TIB converter in [51] employs a high-side switch consisting of a number of series-connected super-junction MOSFETs. A simplified schematic diagram of this circuit is shown in Figure 2.43.

The ZVS feature is beneficial not only because it reduces the switching losses. It also facilitates good voltage sharing between the series-connected super-junction MOSFETs of the high-side switch. A third advantage of the implementation presented in [51] is autonomous operation of the high-side switch, which needs no turn-ON or turn-OFF signals to operate. A photograph of the TIB converter is shown in Figure 2.44.

The converter has been tested using input voltages up to 3 kV and output powers up to 70 W. The efficiency at 3 kV and 70 W output power was found to be approximately 82%.

A high reliability can be achieved because of the redundancy of the series-connected transistors of the high-side switch. Several converters can also be operated in parallel.

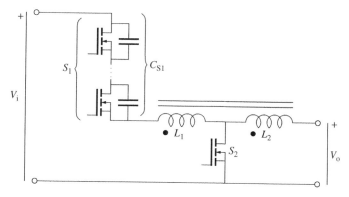

Figure 2.43   Simplified schematic diagram of a TIB converter employing a high-side switch consisting of a number of super-junction MOSFETs (S_1).

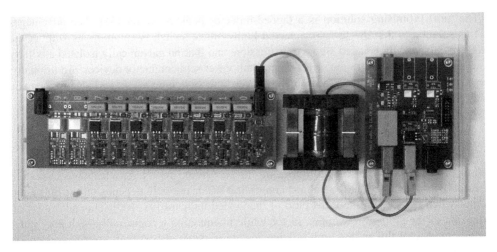

Figure 2.44 Photograph of the TIB converter presented in [51]. On the left-hand side, the high-side switch with the series-connected super-junction MOSFETs can be seen. At the center, the tapped inductor can be seen, and on the right-hand side the low-voltage side with the control unit can be seen.

2.9 Start-Up Procedures

A totally de-enegized MMC is started by means of a two-stage process. The first stage starts by the closing of the ac-side circuit breaker, such that the diodes of all half-bridges can charge the dc side of the converter. However, in series with the ac-side circuit breaker, there is also a set of charging resistors (one in each phase) which are intended to limit the inirush current. At a certain point, the voltage on the dc side is approximately equal to the peak value of the line-to-line voltage times the turns ratio of the transformer. At this point, when the dc side is charged, the submodule capacitors in all arms are charged to approximately half of their rated values. Simultaneously, the charging resistors can be short-circuited by means of a mechanical circuit breaker. In the case of point-to-point HVDC transmission both converters have been partly energized. The same holds in the case of an HVDC grid. Typically, however, the energizing of the dc side can be made by closing the dc-side circuit breakers such that the un-energized grid part gets charged. Also in this case charging resistors are necessary.

In stage two one of the submodules starts to operate in pulse width modulation (PWM) mode [103]. By swapping the submodules operating with PWM and by subsequently having ever fewer submodules inserted, the submodule capacitor voltages can finally be charged to their rated values. Similar ideas are found in [104], but here the charging current is constant, and in [105] this is done by means of a closed-loop control scheme.

2.10 Summary

The chapter starts with an introduction to power semiconductor devices used for VSC-HVDC transmission and possible future alternatives. It is concluded that for MMCs switch-mode power semiconductor devices optimized for low conduction losses should be used because of the low switching frequency per device. At present, the best choice from a loss perspective is probably a 4.5 kV IGCT. This device is also very robust against surge currents and has excellent reliability.

An introduction to silicon carbide power semiconductor devices is also given. The conclusion is that 3.3 kV silicon carbide MOSFETs could probably be used already in a few years. This is, however, unlikely because the cost would probably still be too high, and because a sufficient reliability could probably not be guaranteed.

Submodule capacitors and arm inductors are treated next. The main design features of metallized polypropylene film capacitors and dry-type air-core reactors are presented. Both technologies are excellent regarding reliability and losses.

Different submodule configurations are described regarding losses, submodule capacitor sizing, and fault handling. A brief description of practical realizations of MMCs by Siemens, ABB, and Alstom is also presented.

Next, an overview of how main-circuit parameters are chosen is given. From the presentation it is clear that the design procedure is far from straightforward, and that several steps may be iterative.

A key feature of the MMC is its redundancy. A comparison of two possible methods of handling redundant and faulty submodules is, therefore, given. It is concluded that the method resulting in a lower average submodule capacitor voltage is probably superior, because a lower voltage stress may increase the reliability.

As already mentioned, IGCTs may be the best choice of power semiconductors for MMCs. The gate-drive units of these devices, however, require a significantly higher power supply. Several possible solutions are presented, and a version using a TIB converter is identified as especially interesting.

The chapter ends with a very brief description of start-up procedures.

References

[1] U. Lamm, "Mercury-arc valves for high voltage dc transmission," *IEE Proceedings*, vol. 3, no. 10, pp. 1747–1753, 1964.

[2] S. Hofstein and F. Heiman, "The silicon insulated-gate field-effect transistor," *Proceedings of the IEEE*, vol. 51, no. 9, pp. 1190–1202, Sep. 1963.

[3] H. Becke and C. Wheatley Jr,, "Power MOSFET with an anode region," United States of America Patent 4,364,073, Dec. 14, 1982.

[4] T. Modeer, S. Norrga, and H.-P. Nee, "High-voltage tapped-inductor buck converter auxiliary power supply for cascaded converter submodules," *Proceedings of the 2012 IEEE Energy Conversion Congress and Exposition (ECCE)*, pp. 19–25, Sep. 2012.

[5] G. Asplund, K. Eriksson, and K. Svensson, "DC transmission based on voltage source converters," *CIGRÉ SC14 Colloquium in South Africa 1997*, 1997.

[6] M. Glinka and R. Marquardt, "A new AC/AC-multilevel converter family applied to a single-phase converter," *Proceedings of the Fifth International Conference on Power Electronics and Drive Systems*, vol. 1, pp. 16–23, 2003.

[7] A. Lesnicar and R. Marquardt, "An innovative modular multilevel converter topology suitable for a wide power range," *2003 IEEE Bologna Power Tech Conference Proceedings*, vol. 3, 2003.

[8] D. Blackburn, "Power MOSFET failure revisited," *Proceedings of the 19th Annual IEEE Power Electronics Specialists Conference*, pp. 681–688, 1988.

[9] M. Rahimo, A. Kopta, S. Eicher, U. Schlapbach, and S. Linder, "Switching-self-clamping-mode 'SSCM', a breakthrough in SOA performance for high voltage IGBTs and diodes," *Proceedings of the 16th International Symposium on Power Semiconductor Devices and ICs*, pp. 437–440, 2004.

[10] J. Lutz, H. Schlangenotto, U. Scheuermann, and R. De Doncker, *Semiconductor Power Devices: Physics, Characteristics, Reliability*. Springer Verlag, 2011.

[11] J. Linnros, R. Revsäter, L. Heijkenskjöld, and P. Norlin, "Correlation between forward voltage drop and local carrier lifetime for a large area segmented thyristor," *IEEE Transactions on Electron Devices*, vol. 42, no. 6, pp. 1174–1179, Jun. 1995.

[12] T. Modeer, H.-P. Nee, and S. Norrga, "Loss comparison of different sub-module implementations for modular multilevel converters in HVDC applications," *EPE Journal*, vol. 22, no. 3, pp. 32–38, Sep. 2012.

[13] P. Steimer, H. Grüning, J. Werninger, E. Carroll, S. Klaka, and S. Linder, "IGCT: A new emerging technology for high power, low cost inverters," *Conference Record of the 1997 IEEE Industry Applications Conference, 1997. Thirty-Second IAS Annual Meeting*, vol. 2, pp. 1592–1599, 1997.

[14] U. Vemulapati, M. Arnold, M. Rahimo, T. Stiasny, and J. Vobecky, "3.3 kV RC-IGCTs optimized for multi-level topologies," *Proceedings of International Exhibition and Conference for Power Electronics, Intelligent Motion, Renewable Energy and Energy Management*, pp. 1–8, 2014.

[15] M. Kitagawa, I. Omura, S. Hasegawa, T. Inoue, and A. Nakagawa, "A 4500 V injection enhanced insulated gate bipolar transistor (IEGT) operating in a mode similar to a thyristor," *International Electron Devices Meeting*, pp. 679–682, 1993.

[16] K. Fujii, P. Koellensperger, and R. De Doncker, "Characterization and comparison of high blocking voltage IGBTs and IEGTs under hard- and soft-switching conditions," *IEEE Transactions on Power Electronics*, vol. 23, no. 1, pp. 172–179, Jan. 2008.

[17] I. Omura, T. Ogura, K. Sugiyama, and H. Ohashi, "Carrier injection enhancement effect of high voltage MOS devices: Device physics and design concept," *IEEE International Symposium on Power Semiconductor Devices and IC's*, pp. 217–220, 1997.

[18] P. Ranstad and H.-P. Nee, "On dynamic effects influencing IGBT losses in soft-switching converters," *IEEE Transactions on Power Electronics*, vol. 26, no. 1, pp. 260–271, Jan. 2011.

[19] S. Ochi and N. Zommer, "Driving and protecting the latest high voltage and current power MOS and IGBTs," *Conference Record of the 1992 IEEE Industry Applications Society Annual Meeting*, pp. 1196–1203, 1992.

[20] V. John, B.-S. Suh, and T. Lipo, "Fast-clamped short-circuit protection of IGBTs," *IEEE Transactions on Industry Applications*, vol. 35, no. 2, pp. 477–486, Mar. 1999.

[21] M. Nguyen, R. Cassel, J. Delamare, and G. Pappas, "Gate drive for high-speed high-power IGBTs," *2001 IEEE Conference Record on Pulsed Power Plasma Science*, pp. 288–291, Jun. 2001.

[22] S. Gekenidis, E. Ramezani, and H. Zeller, "Explosion tests on IGBT high voltage modules," *Proceedings of the 11th International Symposium on Power Semiconductor Devices and ICs*, pp. 129–132, 1999.

[23] Y. Nishimura, K. Kido, F. Momose, and T. Goto, "Development of ultrasonic welding for IGBT module structure," *Proceedings of the 2010 22nd International Symposium on Power Semiconductor Devices and ICs*, pp. 293–296, 2010.

[24] L. Feller, S. Hartmann, D. Schneider, D. Granata, and B. Behzadi, "Evaluation of insulation material in advanced high power IGBT modules with extended operation temperature," *Proceedings of the 2010 6th International Conference on Integrated Power Electronics Systems (CIPS)*, pp. 1–6, 2010.

[25] C. Bayer, E. Baer, U. Waltrich, D. Malipaard, and A. Schletz, "Simulation of the electric field strength in the vicinity of metallization edges on dielectric substrates," *IEEE Transactions on Dielectrics and Electrical Insulation*, vol. 22, no. 1, pp. 257–265, Feb. 2015.

[26] F. Richardeau, Z. Dou, J.-M. Blaquiere, E. Sarraute, D. Flumian, and F. Mosser, "Complete short-circuit failure mode properties and comparison based on IGBT standard packaging: Application to new fault-tolerant inverter and interleaved chopper with reduced parts count," *Proceedings of the 2011-14th European Conference on Power Electronics and Applications*, pp. 1–9, 2011.

[27] M. Billmann, D. Malipaard, and H. Gambach, "Explosion proof housings for IGBT module based high power inverters in HVDC transmission application," *Proceedings of PCIM Europe 2009 Conference*, pp. 352–357, 2009.

[28] S. Eicher, M. Rahimo, E. Tsyplakov, D. Schneider, A. Kopta, U. Schlapbach, and E. Carroll, "4.5 kV press pack IGBT designed for ruggedness and reliability," *Conference Record of the 2004 IEEE Industry Applications Conference, 39th IAS Annual Meeting*, vol. 3, pp. 1534–1539, 2004.

[29] S. Gunturi, J. Assal, D. Schneider, and S. Eicher, "Innovative metal system for IGBT press pack modules," *Proceedings of the IEEE 15th International Symposium on Power Semiconductor Devices and ICs*, pp. 110–113, 2003.

[30] C. Toechterle, F. Pfirsch, C. Sandow, and G. Wachutka, "Evolution of current filaments limiting the safe-operating area of high-voltage trench-IGBTs," *Proceedings of the 2014 IEEE 26th International Symposium on Power Semiconductor Devices & IC's*, pp. 135–138, 2014.

[31] H. Zeller, "Cosmic ray induced breakdown in high voltage semiconductor devices, microscopic model and phenomenological lifetime prediction," *Proceedings of the 6th International Symposium on Power Semiconductor Devices and ICs*, pp. 339–340, 1994.

[32] P. Steimer, O. Apeldoorn, and E. Carroll, "IGCT devices-applications and future opportunities," *Proceedings of the IEEE Power Engineering Society Summer Meeting*, vol. 2, pp. 1223–1228, 2000.

[33] H. Kabza, H.-J. Schulze, Y. Gerstenmaier, P. Voss, and F. Pfirsch, "Cosmic radiation as a cause for power device failure and possible countermeasures," *Proceedings of the 6th International Symposium on Power Semiconductor Devices and ICs*, pp. 9–12, 1994.

[34] H.-J. Schulze, F.-J. Niedernostheide, F. Pfirsch, and R. Baburske, "Limiting factors of the safe operating area for power devices," *IEEE Transactions on Electron Devices*, vol. 60, no. 2, pp. 551–562, Feb. 2013.

[35] C. Zorn and N. Kaminski, "Acceleration of temperature humidity bias (THB) testing on IGBT modules by high bias levels," *Proceedings of the 2015 IEEE 27th International Symposium on Power Semiconductor Devices & IC's*, pp. 385–388, 2015.

[36] H. Lendenmann, A. Mukhitdinov, F. Dahlqvist, H. Bleichner, M. Irwin, and R. Söderholm, "4.5 kV 4H-SiC diodes with ideal forward characteristic," *Proceedings of the 13th International Symposium on Power Semiconductor Devices and ICs, 2001*, pp. 31–34, 2001.

[37] W. Vetter, J. Liu, M. Dudley, M. Skowronski, H. Lendenmann, and C. Hallin, "Dislocation loops formed during the degradation of forward-biased 4H–SiC p-n junctions," *Materials Science and Engineering*, vol. B98, pp. 220–224, 2003.

[38] M. Bhatnagar, P. McLarty, and B. Baliga, "Silicon-carbide high-voltage (400 V) schottky barrier diodes," *IEEE Electron Device Letters*, vol. 13, no. 10, pp. 501–503, Oct. 1992.

[39] B. Baliga, "Trends in power semiconductor devices," *IEEE Transactions on Electron Devices*, vol. 43, no. 10, pp. 1717–1731, Oct. 1996.

[40] B. Baliga, "Semiconductors for high-voltage, vertical channel field-effect transistors," *Journal of Applied Physics*, vol. 53, no. 3, pp. 1759–1764, 1982.

[41] B. Ållebrand and H.-P. Nee, "On the possibility to use SiC JFETs in power electronic circuits," *Proceedings of the 2001 European Conference on Power Electronics and Applications (EPE 2011)*, pp. 1–9, 2001.

[42] B. Ållebrand and H.-P. Nee, "On the choice of blanking times at turn-on and turn-off for the diode-less SiC JFET inverter bridge," *Proceedings of the 2001 European Conference on Power Electronics and Applications*, pp. 1–9, 2001.

[43] H. Mitlehner, W. Bartsch, K. Dohnke, P. Friedrichs, R. Kaltschmidt, U. Weinert, B. Weis, and D. Stephani, "Dynamic characteristics of high voltage 4H-SiC vertical JFETs," *Proceedings of the 11th International Symposium on Power Semiconductor Devices and ICs, 1999*, pp. 339–342, 1999.

[44] P. Friedrichs, H. Mitlehner, R. Schörner, K.-O. Dohnke, R. Elpelt, and D. Stephani, "The vertical silicon carbide JFET: A fast and low loss solid state power switching device," *Proceedings of the 9th European Conference on Power Electronics and Applications, Graz*, pp. 1–6, 2001.

[45] S. Round, M. Heldwein, J. Kolar, I. Hofsajer, and P. Friedrichs, "A SiC JFET driver for a 5 kW, 150 kHz three-phase PWM converter," *Conference Record of the 2005 IEEE Industry Applications Society Conference*, 2005, pp. 410–416.

[46] D. Peftitsis, J. Rabkowski, G. Tolstoy, and H.-P. Nee, "Experimental comparison of DC-DC boost converters with SiC JFETs and SiC bipolar transistors," *Proceedings of the 2011 14th European Conference on Power Electronics and Applications*, pp. 1–9, 2001.

[47] D. Peftitsis, J. Rabkowski, and H.-P. Nee, "Self-powered gate driver for normally on silicon carbide junction field-effect transistors without external power supply," *IEEE Transactions on Power Electronics*, vol. 28, no. 3, pp. 1488–1501, Mar. 2013.

[48] D. Peftitsis, J. Rabkowski, and H.-P. Nee, "Self-powered gate driver for normally-on SiC JFETs: Design considerations and system limitations," *IEEE Transactions on Power Electronics*, vol. 29, no. 10, pp. 5129–5135, Oct. 2014.

[49] A. Lindgren and M. Domeij, "1200 V 6 A SiC BJTs with very low VCESAT and fast switching," *Proceedings of the 2010 6th International Conference on Integrated Power Electronics Systems*, pp. 1–5, 2010.

[50] A. Lindgren and M. Domeij, "Degradation free fast switching 1200 V 50 A silicon carbide BJT's," *Proceedings of the 2011 Twenty-Sixth Annual IEEE Applied Power Electronics Conference and Exposition*, pp. 1064–1070, 2011.

[51] T. Modeer, S. Norrga, and H.-P. Nee, "High-voltage tapped-inductor buck converter utilizing an autonomous high-side switch," *IEEE Transactions on Industrial Electronics*, vol. 62, no. 5, pp. 2868–2878, May 2015.

[52] J. Rabkowski, G. Tolstoy, D. Peftitsis, and H.-P. Nee, "Low-loss high-performance base-drive unit for SiC BJTs," *IEEE Transactions on Power Electronics*, May 2012.

[53] G. Tolstoy, D. Peftitsis, J. Rabkowski, P. Palmer, and H.-P. Nee, "A discretized proportional base driver for silicon carbide bipolar junction transistors," *IEEE Transactions on Power Electronics*, May 2014.

[54] D. Peftitsis, G. Tolstoy, A. Antonopoulos, J. Rabkowski, J.-K. Lim, M. Bakowski, L. Ängquist, and H.-P. Nee, "High-power modular multilevel converters with SiC JFETs," *IEEE Transactions on Power Electronics*, vol. 27, no. 1, pp. 28–36, Jan. 2012.

[55] A. Lelis, R. Green, D. Habersat, and M. El, "Basic mechanisms of threshold-voltage instability and implications for reliability testing of SiC MOSFETs," *IEEE Transactions on Electron Devices*, vol. 62, no. 2, pp. 316–323, Feb. 2015.

[56] J. Lutz and R. Baburske, "Some aspects on ruggedness of SiC power devices," *Microelectronics Reliability*, vol. 54, no. 54, pp. 49–56, 2014.

[57] V. Najmi, J. Wang, R. Burgos, and D. Boroyevich, "High reliability capacitor bank design for modular multilevel converter in MV applications," *Proceedings of 2014 IEEE Energy Conversion Congress and Exposition*, pp. 1051–1058, 2014.

[58] R. Anderson, "Select the right plastic film capacitor for your power electronic applications," *Conference Record of the 1996 IEEE Thirty-First IAS Annual Meeting*, vol. 3, pp. 1327–1330, 1996.

[59] M. Schneider, J. Macdonald, M. Schalnat, and J. Ennis, "Electrical breakdown in capacitor dielectric films: Scaling laws and the role of self-healing," *Proceedings of the 2012 IEEE International Power Modulator and High Voltage Conference*, pp. 284–287, 2012.

[60] K. Papp, G. Christineer, H. Popelka, and M. Schwan, "High voltage series reactors for load flow control," *Proceedings of CIGRÉ 2004*, vol. C2-206, pp. 1–8, 2004.

[61] K. Papp, M. Sharp, and D. Peelo, "High voltage dry-type air-core shunt reactors," *Elektrotechnik & Informationstechnik*, vol. 131, no. 8, pp. 349–354, 2014.

[62] B. Jacobson, P. Karlsson, G. Asplund, L. Harnefors, and T. Jonsson, "VSC-HVDC transmission with cascaded two-level converters," *Proceedings of CIGRÉ 2010*, vol. B4-110, 2010.

[63] P. Hammond, "A new approach to enhance power quality for medium voltage AC drives," *Transactions on Industry Applications*, vol. 33, no. 1, pp. 202–208, Jan./Feb. 1997.

[64] J. Ainsworth, M. Davies, P. Fitz, K. Owen, and D. Trainer, "Static VAr compensator (STATCOM) based on single-phase chain circuit converters," *IEE Proceedings: Generation, Transmission and Distribution*, vol. 145, no. 4, pp. 381–386, Jul. 1998.

[65] J. Qin, M. Saeedifard, A. Rockhill, and R. Zhou, "Hybrid design of modular multilevel converters for HVDC systems based on various submodule circuits," *IEEE Transactions on Power Delivery*, vol. 30, no. 1, pp. 385–394, Feb. 2015.

[66] R. Zeng, L. Xu, L. Yao, and B. Williams, "Design and operation of a hybrid modular multilevel converter," *IEEE Transactions on Power Electronics*, vol. 30, no. 3, pp. 1137–1146, Mar. 2015.

[67] S. Norrga, X. Li, and L. Ängquist, "Converter topologies for HVDC grids," *Proceedings of the 2014 IEEE International Energy Conference*, pp. 1554–1561, 2014.

[68] A. Nami, L. Wang, F. Dijkhuizen, and A. Shukla, "Five level cross connected cell for cascaded converters," *Proceedings of the 2013 15th European Conference on Power Electronics and Applications*, pp. 1–9, 2013.

[69] A. Nami, J. Liang, F. Dijkhuizen, and G. Demetriades, "Modular multilevel converters for HVDC applications: Review on converter cells and functionalities," *IEEE Transactions on Power Electronics*, vol. 30, no. 1, pp. 18–36, Jan. 2015.

[70] S. Teeuwsen, "Modeling the trans bay cable project as voltage-sourced converter with modular multilevel converter design," *Proceedings of the 2011 IEEE Power and Energy Society General Meeting*, pp. 1–8, 2011.

[71] H.-J. Knaak, "Modular multilevel converters and HVDC/FACTS: A success story," *Proceedings of the 2011 14th European Conference on Power Electronics and Applications*, pp. 1–6, 2011.

[72] X. Li, W. Liu, Q. Song, H. Rao, and S. Xu, "An enhanced mmc topology with DC fault ride-through capability," *Proceedings of the 39th Annual Conference of the IEEE Industrial Electronics Society*, pp. 6182–6188, 2013.

[73] R. Marquardt, "Modular multilevel converter: An universal concept for HVDC-networks and extended DC-bus-applications," *Proceedings of the 2010 International Power Electronics Conference*, pp. 502–507, 2010.

[74] R. Marquardt, "Modular multilevel converter topologies with DC-short circuit current limitation," *Proceedings of the 2011 IEEE 8th International Conference on Power Electronics and ECCE Asia*, pp. 1425–1431, 2011.

[75] T. Modeer, S. Norrga, and H.-P. Nee, "Loss comparison of different sub-module implementations for modular multilevel converters in HVDC applications," *Proceedings of the 2011-14th European Conference on Power Electronics and Applications*, pp. 1–7, 2011.

[76] K. Ilves, F. Taffner, S. Norrga, A. Antonopoulos, L. Harnefors, and H.-P. Nee, "A submodule implementation for parallel connection of capacitors in modular multilevel converters," *Proceedings of the 2013 15th European Conference on Power Electronics and Applications*, pp. 1–10, 2013.

[77] K. Ilves, F. Taffner, S. Norrga, A. Antonopoulos, L. Harnefors, and H.-P. Nee, "A submodule implementation for parallel connection of capacitors in modular multilevel converters," *IEEE Transactions on Power Electronics*, vol. 30, no. 7, pp. 3518–3527, Jul. 2015.

[78] K. Ilves, L. Bessegato, L. Harnefors, S. Norrga, and H.-P. Nee, "Semi-full-bridge submodule for modular multilevel converters," *Proceedings of the 2015 IEEE ECCE ASIA*, Seoul, pp. 1–8, Jun. 2015.

[79] L. Baruschka and A. Mertens, "Comparison of cascaded H-bridge and modular multilevel converters for BESS application," *Proceedings of the IEEE Energy Conversion Congress and Exposition 2011*, pp. 909–916, 2011.

[80] K. Ilves, S. Norrga, and H.-P. Nee, "On energy variations in modular multilevel converters with full-bridge submodules for AC-DC and AC-AC applications," *Proceedings of the 2013 15th European Conference on Power Electronics and Applications*, pp. 1–10, 2013.

[81] W. McMurray, "Resonant snubbers with auxiliary switches," *Conference Record of the 1989 IEEE Industry Applications Society Annual Meeting*, vol. 1, pp. 829–834, 1989.

[82] R. De Doncker and J. Lyons, "The auxiliary resonant commutated pole converter," *Conference Record of the 1990 IEEE Industry Applications Society Annual Meeting*, vol. 2, pp. 1228–1235, 1990.

[83] M. Heuvelmans, T. Modeer, and S. Norrga, "Soft-switching cells for high-power converters," *Proceedings of the 40th Annual Conference of the IEEE Industrial Electronics Society*, pp. 1806–1812, 2014.

[84] C. Oates, "Modular multilevel converter design for VSC HVDC applications," *IEEE Journal of Emerging and Selected Topics in Power Electronics*, vol. 3, no. 2, pp. 505–515, Jun. 2015.

[85] T. Yamanaka, S. Maruyama, and T. Tanaka, "The development of DC +/− 500 kV XLPE cable in consideration of the space charge accumulation," *Proceedings of the 7th International Conference on Properties and Applications of Dielectric Materials*, vol. 2, pp. 689–694, 2003.

[86] G. Chen, M. Hao, Z. Xu, A. Vaughan, J. Cao, and H. Wang, "Review of high voltage direct current cables," *CSEE Journal of Power and Energy Systems*, vol. 1, no. 2, pp. 9–21, Jun. 2015.

[87] T. Jonsson, P. Lundberg, S. Maiti, and Y. Jiang-Häfner, "Converter technologies and functional requirements for reliable and economical HVDC grid design," *Proceedings of the 2013 CIGRÉ Canada Conference*, Calgary, Sep. 2013.

[88] D. Soto-Sanchez and T. Green, "Control of a modular multilevel converter-based HVDC transmission system," *Proceedings of the 2011 14th European Conference on Power Electronics and Applications*, pp. 1–10, 2011.

[89] G. Konstantinou, M. Ciobotaru, and V. Agelidis, "Effect of redundant sub-module utilization on modular multilevel converters," *Proceedings of the IEEE International Conference on Industrial Technology*, pp. 1–6, Mar. 2012.

[90] G. Konstantinou, J. Pou, S. Ceballos, and V. Agelidis, "Active redundant submodule configuration in modular multilevel converters," *IEEE Transactions on Power Delivery*, vol. 28, no. 4, pp. 2333–2341, Oct. 2013.

[91] G. Son, H.-J. Lee, T. Nam, Y.-H. Chung, U.-H. Lee, S.-T. Baek, K. Hur, and J.-W. Park, "Design and control of a modular multilevel HVDC converter with redundant power modules for noninterruptible energy transfer," *IEEE Transactions on Power Delivery*, vol. 27, no. 3, pp. 1611–1619, Jul. 2012.

[92] N. Ahmed, L. Ängquist, A. Antonopoulos, L. Harnefors, S. Norrga, and H.-P. Nee, "Performance of the modular multilevel converter with redundant submodules," *Proceedings of 41st Annual Conference of the Industrial Electronics Society*, Dec. 2015.

[93] M. Glinka, "Prototype of multiphase modular-multilevel-converter with 2 MW power rating and 17-level-output-voltage," *Proceedings of the 2004 IEEE 35th Annual Power Electronics Specialists Conference*, vol. 4, pp. 2572–2576, 2004.

[94] H. Torresan, D. Holmes, and I. Shraga, "Auxiliary power supplies for high voltage converter systems," *Proceedings of the 2004 IEEE 35th Annual Power Electronics Specialists Conference*, vol. 1, pp. 645–651, 2004.

[95] T. Barth, S. Semmler, M. Buschendorf, R. Alvarez, and S. Bernet, "Gate drive unit DC-DC power supply for multi-level converters or series connection of IGBTs with high voltage insulation," *Proceedings of the 2014 11th International Multi-Conference on Systems, Signals & Devices*, pp. 1–5, 2014.

[96] O. Senturk, T. Maerki, P. Steimer, and S. McLaughlin, "High voltage cell power supply for modular multilevel converters," *Proceedings of the 2014 IEEE Energy Conversion Congress and Exposition*, pp. 4416–4420, 2014.

[97] M. Rico, J. Uceda, J. Sebastian, and F. Aldana, "Static and dynamic modeling of tapped-inductor DC-to-DC converters," *Proceedings of the 1987 IEEE Power Electronics Specialists Conference*, pp. 281–288, 1987.

[98] D. Edry, M. Hadar, O. Mor, and S. Ben-Yaakov, "A Spice compatible model of tapped-inductor PWM converters," *Proceedings of the IEEE Applied Power Electronics Conference and Exposition*, vol. 2, pp. 1021–1027, Feb. 1994.

[99] D. Grant, Y. Darroman, and J. Suter, "Synthesis of tapped-inductor switched-mode converters," *IEEE Transactions on Power Electronics*, vol. 22, no. 5, pp. 1964–1969, Sep. 2007.

[100] B. Williams, "Unified synthesis of tapped inductor DC-to-DC converters," *IEEE Transactions on Power Electronics*, vol. 29, no. 10, pp. 5370–5383, Oct. 2014.

[101] J.-H. Park and B.-H. Cho, "The zero voltage switching (ZVS) critical conduction mode (CRM) buck converter with tapped-inductor," *IEEE Transactions on Power Electronics*, vol. 20, no. 4, pp. 762–774, Jul. 2005.

[102] J.-H. Park and B. Cho, "Nonisolation soft-switching buck converter with tapped-inductor for wide-input extreme step-down applications," *IEEE Transactions on Circuits and Systems I: Regular Papers*, vol. 54, no. 8, pp. 1809–1818, Aug. 2007.

[103] K. Shi, F. Shen, D. Lv, P. Lin, M. Chen, and D. Xu, "A novel start-up scheme for modular multilevel converter," *Proceedings of the 2012 IEEE Energy Conversion Congress and Exposition*, pp. 4180–4187, Se. 2012.

[104] B. Li, Y. Zhang, D. Xu, and R. Yang, "Start-up control with constant precharge current for the modular multilevel converter," *Proceedings of the 2014 IEEE 23rd International Symposium on Industrial Electronics*, pp. 673–676, 2014.

[105] B. Li, D. Xu, Y. Zhang, R. Yang, G. Wang, W. Wang, and D. Xu, "Closed-loop precharge control of modular multilevel converters during start-up processes," *IEEE Transactions on Power Electronics*, vol. 30, no. 2, pp. 524–531, Feb. 2015.

3

Dynamics and Control

3.1 Introduction

In this chapter, fundamental aspects of modular multilevel converter (MMC) dynamics and control are considered. Unlike many other converter topologies, the MMC has relatively complex internal dynamics. For this reason, the chapter starts with a description of the converter topology for the purpose of dynamic modeling. This is made in Section 3.2. An averaged dynamic model is then derived in Section 3.3. This model constitutes the basis for the control-system design.

Figure 3.1 shows an overall block diagram of a typical MMC control system. The control loop for the output current has the highest bandwidth of all nested control loops. It can be regarded as the "heart" of the converter control system. Its design on a per-phase basis is considered in Section 3.4. The output-voltage reference, which is determined by the output-current controller, can in the control system for, for example, a two-level voltage source converter (VSC) be forwarded directly to the pulsewidth modulator, which produces the gate signals (switching pulses). The higher complexity of the MMC requires yet another stage, for arm-balancing control. Considered in Section 3.5, this control stage includes control of the circulating current and of the sum capacitor voltages. Its output signals are the insertion indices, which go to the modulation and submodule balancing stage. Here, gate signals for all submodules are generated, as shown in Chapter 5.

Up to and including Section 3.5, modeling and control are made on a per-phase basis. The theory is thus general and applicable to MMCs with any number of phases. Virtually all MMCs are three-phase converters, though. General theory for three-phase systems and space vectors is considered in Section 3.6. Output-current control is then revisited in Section 3.7, but using space vectors in a synchronously rotating reference frame rather than on a per-phase basis. The first block in the chain shown in Figure 3.1 is considered in Section 3.8: the higher-level control. This includes the phase-locked loop (PLL), which is used for grid synchronization, active- and reactive-power control, dc-bus-voltage control, and the principle of power-synchronization control. The chapter is finished by an overview of control architectures in Section 3.9.

Design, Control, and Application of Modular Multilevel Converters for HVDC Transmission Systems, First Edition.
Kamran Sharifabadi, Lennart Harnefors, Hans-Peter Nee, Staffan Norrga, and Remus Teodorescu.
© 2016 John Wiley & Sons, Ltd. Published 2016 by John Wiley & Sons, Ltd.
Companion Website URL: www.wiley.com/go/Sharifabadi/ModularConverters

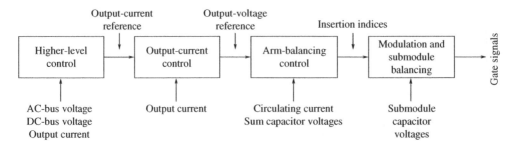

Figure 3.1   Overview of a typical MMC control system.

3.2   Fundamentals

Figure 3.2(a) shows a per-phase schematic of the MMC. This figure will serve as a reference throughout the chapter. Although three-phase MMCs are of primary interest, a general number of phases, M, is assumed in the converter modeling. That is, the model to be developed is valid

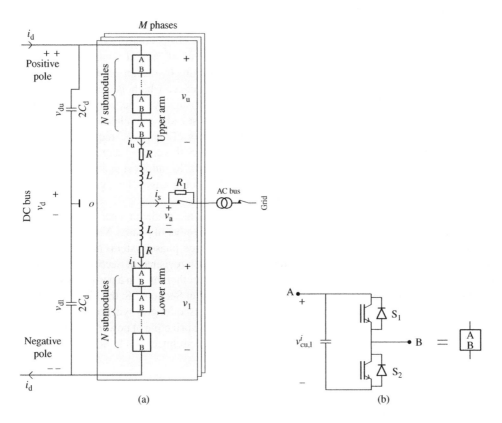

Figure 3.2   (a) Circuit diagram and (b) submodule of an MMC.

also for single-phase converters and for converters with more phases than three. The converter dynamics can be adequately modeled on a phase-by-phase basis. Explicit phase notation is not made in expressions that are formally identical for all phases. A list of symbols can be found in the Nomenclature at the beginning of this book.

3.2.1 Arms

Each phase leg of the MMC comprises two arms: the upper and the lower. Each arm is equipped with N series-connected submodules (also called *cells*). The submodules can be considered as variable voltage sources, which, when their output voltages are added together, give the inserted voltages $v_{u,l}$. To prevent unreasonably large switching harmonics in the arm currents, it is necessary to equip each arm with an inductor, with inductance L. A parasitic resistance, R, results from losses in the arm inductors and in the submodules. R—which to some extent is varying with the operating conditions—is normally small.

3.2.2 Submodules

The standard submodule topology is depicted in Figure 3.2(b). It consists of a half-bridge circuit employing two switches: S_1 and S_2. Each one of the switches consists of a controllable semiconductor device—generally a power transistor, e.g. an insulated-gate bipolar transistor (IGBT)—with an anti-parallel-connected diode. Each submodule is equipped with a capacitor, which is connected across both switches. The submodules can be switched in three different ways.

- **Inserted:** S_2 is turned on and S_1 is turned off. This inserts the submodule capacitor in the arm circuit, giving $n_{u,l}^i = 1$, allowing the capacitor to charge or discharge, depending on the direction of the arm current.
- **Bypassed:** S_1 is turned on and S_2 is turned off. The capacitor becomes bypassed, giving $n_{u,l}^i = 0$, and its voltage remains constant regardless of the arm current.
- **Blocked:** S_1 and S_2 are both turned off. In this operating mode the capacitor may charge through the diode of S_2, but it cannot discharge. This mode is used for energization of the converter. It may also be used during brief time periods—typically tens of milliseconds or less, during grid faults—to protect the transistor from overcurrent. By turning both switches off, the current pulse is shunted through the anti-parallel-connected diode, which often is rated to withstand a larger current than the transistor. An alternative is to install a thyristor in parallel with S_1, through which the surge current can be shunted.

By choosing at each time instant the appropriate numbers of submodules to insert, the inserted voltages can be varied as desired between zero and the sum capacitor voltage of the arm, i.e.

$$0 \le v_{u,l} \le v_{cu,l}^{\Sigma}, \tag{3.1}$$

where $v_{u,l} = 0$ when all submodules are bypassed and $v_{u,l} = v_{cu,l}^{\Sigma}$ when all submodules are inserted.

3.2.3   AC Bus

A grid-connected converter is considered. In some cases it may be possible to give the converter a voltage rating that matches the grid voltage. But often the converter voltage rating is different from the grid voltage, usually lower. A transformer is therefore generally used to step the voltage down (or up) to a level that matches the converter voltage rating. This transformer also provides galvanic isolation and a tap changer is generally used to allow adjustment of the turns ratio with variations in the grid voltage. In the three-phase case, the star point on the converter side of the transformer is often left ungrounded, which gives a high zero-sequence impedance; see Section 3.6. A main ac breaker is placed at the grid interface, also known as the point of common coupling (PCC). In many cases, a pre-insertion resistor with a parallel bypass breaker is used to prevent overvoltages in the converter when energization is made from the grid by closing the main breaker.

3.2.4   DC Bus

For the purpose of dynamic modeling, the dc bus can generally be considered to have pure capacitive characteristics, with a capacitance $2C_d$ from the neutral to the positive and negative poles, respectively, i.e. a pole-to-pole capacitance C_d. These capacitances represent a lumped model of the pole-to-neutral capacitances of the positive- and negative-pole dc cables interconnecting two MMCs in an high-voltage direct current (HVDC) transmission, plus the capacitance of the pole-to-neutral capacitor in the converter station if such is installed. The dc bus is assumed to be balanced, i.e. except where noted,

$$v_{du} = v_{dl} = \frac{v_d}{2}. \tag{3.2}$$

In a so-called symmetric monopole configuration the dc-bus midpoint is grounded, as shown in Figure 3.2(a).

3.2.5   Currents

By definition from the Nomenclature we have $i_s = i_u - i_l$ [this also follows directly from Figure 3.2(a)] and $i_c = (i_u + i_l)/2$. Combining these two relations allows the arm currents to be expressed in the output and circulating currents as

$$i_u = \frac{i_s}{2} + i_c \qquad i_l = -\frac{i_s}{2} + i_c. \tag{3.3}$$

The M upper arms all connect at the positive pole, whereas the M lower arms all connect at the negative pole. The mean values of the arm currents must add up to the dc-bus current in order for the dc-bus voltage to remain constant, i.e.

$$\sum_{k=1}^{M} \overline{i_{u,l}^k} = i_d. \tag{3.4}$$

This implies that, unless $i_d = 0$, there will be a dc component in the arm currents. Under balanced ac-side conditions, i.e. when the phase voltages and currents have identical but time-shifted waveforms, this dc component is given by

$$\bar{i}_u = \bar{i}_l = \bar{i}_c = \frac{i_d}{M}. \tag{3.5}$$

To keep the converter losses and the rms arm currents at a minimum, it is desirable that the circulating current i_c should be pure dc, i.e. $i_c = i_d/M$.

3.3 Converter Operating Principle and Averaged Dynamic Model

We shall in this section review the operating principle and derive an averaged dynamic model of the MMC. This model is the foundation on which the control designs of the subsequent sections are built.

3.3.1 Dynamic Relations for the Currents

Let us direct our attention to the circuits formed by the arms. Assuming that the dc bus is balanced, so that Equation (3.2) holds, an inspection of Figure 3.2(a) gives

$$\frac{v_d}{2} - v_u - Ri_u - L\frac{di_u}{dt} = v_a \qquad -\frac{v_d}{2} + v_l + Ri_l + L\frac{di_l}{dt} = v_a. \tag{3.6}$$

Respectively adding and subtracting these two relations, and introducing the output and circulating currents according to Equation (3.3), results in

$$\frac{L}{2}\frac{di_s}{dt} = \underbrace{\frac{-v_u + v_l}{2}}_{v_s} - v_a - \frac{R}{2}i_s \qquad L\frac{di_c}{dt} = \frac{v_d}{2} - \underbrace{\frac{v_u + v_l}{2}}_{v_c} - Ri_c \tag{3.7}$$

where the underbraced identities follow from the definitions made in the Nomenclature. The rationale for naming v_s as the output voltage now becomes obvious: v_s drives the output current i_s, as shown by the first relation in Equation (3.7). In a similar fashion, the internal voltage v_c drives the circulating current i_c (but with opposite sign), as shown by the second relation in Equation (3.7). For i_c to be pure dc—as desired—the converter has to be controlled such that $v_c = v_d/2 - Ri_c \approx v_d/2$, since the right-hand side of the second relation in Equation (3.7) then vanishes.

It may be noted in the first relation in Equation (3.7) that the effective inductance and resistance seen by the output voltage and current respectively are $L/2$ and $R/2$. This is because the two arms appear parallel connected as observed from the ac bus.

3.3.2 Selection of the Mean Sum Capacitor Voltages

Consider the definitions of v_s and v_c in Equation (3.7). The maximum output voltage is obviously obtained by bypassing all submodules in the upper arm and inserting all submodules in the lower arm, giving

$$v_u = 0, \quad v_l = v_{cl}^\Sigma \qquad \Rightarrow \qquad v_s^{max} = \frac{v_{cl}^\Sigma}{2}. \tag{3.8}$$

The minimum output voltage is obtained by the opposite operation, i.e. by inserting all submodules in the upper arm and bypassing all submodules in the lower arm, giving

$$v_u = v_{cu}^\Sigma, \quad v_l = 0 \qquad \Rightarrow \qquad v_s^{min} = -\frac{v_{cu}^\Sigma}{2}. \tag{3.9}$$

To get $v_s^{min} = -v_s^{max}$, $v_{cu}^\Sigma = v_{cl}^\Sigma$ is required. That is, there should—at least ideally—be a charge balance between the upper and lower arms. How should then the ideal common value $v_{cu}^\Sigma = v_{cl}^\Sigma$ of the sum capacitor voltages be selected? As noted, to keep the circulating current at a constant value, $v_c \approx v_d/2$ is required. Let us inspect the second relation in Equation (3.7). To get $v_c = v_d/2$ for the maximum value of the output voltage, i.e. according to Equation (3.8) for $v_u = 0$ and $v_l = v_{cl}^\Sigma$, we need to have $v_{cl}^\Sigma = v_d$. To get $v_c = v_d/2$ for the minimum value of the output voltage, i.e. according to Equation (3.9) for $v_u = v_{cu}^\Sigma$ and $v_l = 0$, we need to have $v_{cu}^\Sigma = v_d$. The ideal common value of the sum capacitor voltages should thus be selected as

$$v_{cu,l}^\Sigma = v_d. \tag{3.10}$$

It is duly noted that this is the ideal value. Since each submodule capacitor—when inserted—will be charged with the current in the arm where it is placed, a ripple will appear on each capacitor voltage. The capacitor-voltage ripples add up constructively to a sum-capacitor-voltage ripple. For this reason, only the mean values of the sum capacitor voltages can equal v_d

$$\overline{v_{cu,l}^\Sigma} = v_d. \tag{3.11}$$

If the sum-capacitor-voltage ripples for the time being yet are neglected, it follows from Equations (3.8) and (3.9) that

$$v_s^{max} = -v_s^{min} = \frac{v_d}{2}. \tag{3.12}$$

A balanced distribution of the sum capacitor voltage over the submodules is generally desirable, i.e. each submodule capacitor is generally charged such that its mean voltage is given by

$$\overline{v_{cu,l}^i} = \frac{\overline{v_{cu,l}^\Sigma}}{N} = \frac{v_d}{N}. \tag{3.13}$$

Example 3.1 *Equation (3.13) can be used to determine the number of submodules that are required in a certain situation. Suppose, for example, that the semiconductor devices of each submodule are rated such that $\overline{v_{cu,l}^i} = 2$ kV is a suitable mean capacitor voltage. The converter is to be connected without a transformer to a 330 kV (rms line-to-line voltage) three-phase grid. The nominal phase-voltage peak value is thus $330\sqrt{2/3} = 270$ kV. From Equation (3.12) it follows that at least $v_d = 540$ kV is required. Equation (3.13) then gives that the required minimum number of submodules is $N = 540/2 = 270$.*

3.3.3 Averaging Principle

The key to obtaining a dynamic model of relative simplicity for the MMC lies in averaging. Let us begin by noting that the submodule insertion indices can attain only two values: $n_{u,l}^i = 0$

implies that the ith capacitor in the upper/lower arm is bypassed, whereas $n^i_{u,l} = 1$ implies that the capacitor is inserted. Consequently, the inserted voltages can be expressed as

$$v_{u,l} = \sum_{i=1}^{N} n^i_{u,l} v^i_{cu,l}. \tag{3.14}$$

It is usually desired that all capacitor voltages should be controlled to the same mean value, according to Equation (3.13). The submodule balancing can often be made accurate enough to allow the individual differences between the capacitor voltages to be neglected in the dynamic model of the converter. The individual submodule capacitor voltages in Equation (3.14) thus can all be approximated by the averaged value $v^\Sigma_{cu,l}/N$ with good accuracy [1]. This reduces the summation in Equation (3.14) to

$$v_{u,l} = \sum_{i=1}^{N} n^i_{u,l} v^i_{cu,l} \approx \sum_{i=1}^{N} n^i_{u,l} \frac{v^\Sigma_{cu,l}}{N} = \frac{v^\Sigma_{cu,l}}{N} \sum_{i=1}^{N} n^i_{u,l}. \tag{3.15}$$

The remaining summation in Equation (3.15) can be removed as well by introducing the per-arm insertion indices (which henceforth will be referred to as *insertion indices*, for simplicity). They are the averages of the submodule insertion indices, given as

$$n_{u,l} = \frac{1}{N} \sum_{i=1}^{N} n^i_{u,l}. \tag{3.16}$$

The insertion indices obviously can attain $N + 1$ discrete values: $0, 1/N, 2/N, \ldots, 1$, where $n_{u,l} = 0$ indicates that all submodules in the arm are bypassed, and $n_{u,l} = 1$ indicates that all submodules are inserted. Yet, it is assumed that the number of submodules, N, is large enough to allow approximating the insertion indices as continuous on $[0, 1]$. This assumption together with the approximation in Equation (3.15) form the basis for the averaged model of the MMC. The assumptions allow Equation (3.14) to be simplified as the product of two continuous variables:

$$v_{u,l} = n_{u,l} v^\Sigma_{cu,l}. \tag{3.17}$$

Substitution of Equation (3.17) in the expressions for v_s and v_c in Equation (3.7) yields the following averaged dynamic model for the currents:

$$\frac{L}{2} \frac{di_s}{dt} = \underbrace{\frac{-n_u v^\Sigma_{cu} + n_l v^\Sigma_{cl}}{2}}_{v_s} - v_a - \frac{R}{2} i_s \qquad L \frac{di_c}{dt} = \frac{v_d}{2} - \underbrace{\frac{n_u v^\Sigma_{cu} + n_l v^\Sigma_{cl}}{2}}_{v_c} - R i_c. \tag{3.18}$$

The averaging principle implies that the switched-mode operation of the MMC is disregarded. This is similar to the often made consideration of a two-level VSC as a controllable voltage source, where the discrete effects of switching are disregarded.

The output and circulating currents are two state variables in the MMC per-phase dynamic model. In addition, the submodule capacitors add $2N$ more state variables per phase, namely the capacitor voltages v^i_{cu} and v^i_{cl} for $i = 1, 2, \ldots, N$. This brings the total system order to $2(N + 1)$ per phase. As there may be hundreds of submodules per arm, the complexity of a dynamic model where the submodule capacitor voltages are explicitly included would be immense.

The complexity can be reduced to a manageable level by extending the averaging principle to the capacitor voltages. This can be done without making any other approximation than regarding $n_{u,l}$ as continuous. Each capacitor voltage is governed by

$$C\frac{dv^i_{cu,l}}{dt} = n^i_{u,l}i_{u,l}, \quad i = 1, 2, \ldots, N. \tag{3.19}$$

Summation over all N capacitors in the arm yields

$$C\underbrace{\sum^N_{i=1}\frac{dv^i_{cu,l}}{dt}}_{dv^\Sigma_{cu,l}/dt} = \sum^N_{i=1}n^i_{u,l}i_{u,l} = i_{u,l}\underbrace{\sum^N_{i=1}n^i_{u,l}}_{Nn_{u,l}} \tag{3.20}$$

which can be simplified to

$$\frac{C}{N}\frac{dv^\Sigma_{cu,l}}{dt} = n_{u,l}i_{u,l}. \tag{3.21}$$

Expressing $i_{u,l}$ in the output and circulating currents according to Equation (3.3), the relations

$$\frac{C}{N}\frac{dv^\Sigma_{cu}}{dt} = n_u\left(\frac{i_s}{2} + i_c\right) \qquad \frac{C}{N}\frac{dv^\Sigma_{cl}}{dt} = n_l\left(-\frac{i_s}{2} + i_c\right) \tag{3.22}$$

are obtained from Equation (3.21). It may be noted that the effective arm capacitance is C/N. This is logical, since the N submodules in each arm are series-connected.

Equations (3.18) and (3.22) comprise the averaged dynamic model of the MMC. The averaged model is merely of order four per phase, with the state variables i_s, i_c, v^Σ_{cu}, and v^Σ_{cl}, regardless of the number of submodules.

3.3.4 Ideal Selection of the Insertion Indices

The averaged model given by Equations (3.18) and (3.22) is useful for analysis as well as for controller design. One fundamental step in the design of the control system for an MMC is to find appropriate selections of the insertion indices n_u and n_l. Suppose that references $v^\star_s$ and $v^\star_c$ respectively for the output and internal voltages are available. These references are generally set by the output- and circulating-current controllers, as shown in Sections 3.4, 3.5, and 3.7. Given these references, we wish to find appropriate selections of n_u and n_l. This can be accomplished by considering the underbraced relations in Equation (3.18), i.e.

$$v_s = \frac{-n_u v^\Sigma_{cu} + n_l v^\Sigma_{cl}}{2} \qquad v_c = \frac{n_u v^\Sigma_{cu} + n_l v^\Sigma_{cl}}{2}. \tag{3.23}$$

By respectively adding and subtracting these two relations, n_u and n_l can be solved

$$n_u = \frac{v_c - v_s}{v^\Sigma_{cu}} \qquad n_l = \frac{v_c + v_s}{v^\Sigma_{cl}}. \tag{3.24}$$

The appropriate selection formulas for the insertion indices are then obtained by substituting v_s and v_c with their respective references, as

$$n_u = \frac{v_c^\star - v_s^\star}{v_{cu}^\Sigma} \qquad n_l = \frac{v_c^\star + v_s^\star}{v_{cl}^\Sigma}. \qquad (3.25)$$

These relations form the ideal basis for how the insertion indices should be selected, given $v_s^\star$ and $v_c^\star$. Equation (3.25) can either be used as it stands or be slightly modified. See Section 3.5 for details.

3.3.5 Sum-Capacitor-Voltage Ripples

Consider the dynamic relation for the sum capacitor voltages given by Equation (3.21); the right-hand side is given as the product of the respective insertion index and arm current as $n_{u,l} i_{u,l}$. Both these quantities are periodic with the fundamental frequency. This has the effect of inducing a ripple in the sum capacitor voltages, which can be quantified simply by integrating Equation (3.21), scaled by N/C. The ripple amplitudes consequently become inversely proportional to C. Theoretically, they could be reduced to a minimum by selecting C as a very large capacitance, but in practice this would be prohibitively costly. Capacitance C should not be made larger than necessary for obtaining tolerable ripple amplitudes. Typically, these are in the range of 10% of the mean sum capacitor voltage.

Analytic determination of the ripple amplitudes is useful, not only for dynamic-analysis and control purposes, but also in the converter design stage, particularly the selection of C. By using the derived averaged model, the steady-state sum-capacitor-voltage ripples can be calculated analytically. To do so, Equation (3.25) is substituted in Equation (3.22), giving

$$\frac{C}{N}\frac{dv_{cu}^\Sigma}{dt} = \frac{v_c^\star - v_s^\star}{v_{cu}^\Sigma}\left(\frac{i_s}{2} + i_c\right) \qquad \frac{C}{N}\frac{dv_{cl}^\Sigma}{dt} = \frac{v_c^\star + v_s^\star}{v_{cl}^\Sigma}\left(-\frac{i_s}{2} + i_c\right). \qquad (3.26)$$

We then multiply both sides of these two relations respectively by v_{cu}^Σ and v_{cl}^Σ:

$$v_{cu}^\Sigma\frac{C}{N}\frac{dv_{cu}^\Sigma}{dt} = (v_c^\star - v_s^\star)\left(\frac{i_s}{2} + i_c\right) \qquad v_{cl}^\Sigma\frac{C}{N}\frac{dv_{cl}^\Sigma}{dt} = (v_c^\star + v_s^\star)\left(-\frac{i_s}{2} + i_c\right). \qquad (3.27)$$

Noting that

$$\frac{1}{2}\frac{d(v_{cu,l}^\Sigma)^2}{dt} = v_{cu,l}^\Sigma\frac{dv_{cu,l}^\Sigma}{dt} \qquad (3.28)$$

allows Equation (3.27) to be rearranged as

$$\underbrace{\frac{C}{2N}\frac{d(v_{cu}^\Sigma)^2}{dt}}_{dW_u/dt} = (v_c^\star - v_s^\star)\left(\frac{i_s}{2} + i_c\right) \qquad \underbrace{\frac{C}{2N}\frac{d(v_{cl}^\Sigma)^2}{dt}}_{dW_l/dt} = (v_c^\star + v_s^\star)\left(-\frac{i_s}{2} + i_c\right) \qquad (3.29)$$

where the arm energies $W_{u,l} = C(v_{cl}^\Sigma)^2/(2N)$ have been introduced in the underbraced expressions. Calculating the ripples becomes easier by, in addition, introducing the per-phase and imbalance energies $W_\Sigma = W_u + W_l$ and $W_\Delta = W_u - W_l$. By respectively adding and subtracting the relations in Equation (3.29), the following simplification is obtained:

$$\frac{dW_\Sigma}{dt} = 2v_c^\star i_c - v_s^\star i_s \qquad \frac{dW_\Delta}{dt} = v_c^\star i_s - 2v_s^\star i_c. \qquad (3.30)$$

Let us quantify the right-hand sides of the two relations of Equation (3.30). It was previously concluded that $v_c \approx v_d/2$, so we let $v_c^\star = v_d/2$. Furthermore, the circulating current i_c can be assumed to be pure dc, whereas $v_s^\star$ and i_s are assumed to be pure sinusoids. With the output voltage taken as phase reference, we can write

$$v_s^\star = \hat{V}_s \cos \omega_1 t \qquad i_s = \hat{I}_s \cos(\omega_1 t - \varphi). \qquad (3.31)$$

Substitution of Equation (3.31) in Equation (3.30) yields

$$\frac{dW_\Sigma}{dt} = v_d i_c - \frac{\hat{V}_s \hat{I}_s}{2} \cos \varphi - \frac{\hat{V}_s \hat{I}_s}{2} \cos(2\omega_1 t - \varphi) \qquad (3.32)$$

$$\frac{dW_\Delta}{dt} = \frac{v_d \hat{I}_s}{2} \cos(\omega_1 t - \varphi) - 2\hat{V}_s i_c \cos \omega_1 t. \qquad (3.33)$$

It may be noted that the term $-(\hat{V}_s \hat{I}_s/2) \cos \varphi$ on the right-hand side of Equation (3.32) equals the mean active input power per phase, i.e. $-P/M$. In order for the mean value of W_Σ to be constant in the steady state, this term and the first term on the right-hand side of Equation (3.32), i.e. $v_d i_c$, must sum up to zero. The following identity is thus obtained:

$$v_d i_c = \frac{P}{M} \Rightarrow i_c = \frac{P}{M v_d}. \qquad (3.34)$$

This is an obvious result; it shows that the ac-bus output power must be equal to the dc-bus input power if the converter losses are neglected. All remaining terms on the right-hand sides of Equations (3.32) and (3.33) are of zero mean. Integration yields

$$W_\Sigma = W_{\Sigma 0} \underbrace{- \frac{\hat{V}_s \hat{I}_s}{4\omega_1} \sin(2\omega_1 t - \varphi)}_{\Delta W_\Sigma} \qquad (3.35)$$

$$W_\Delta = W_{\Delta 0} + \underbrace{\frac{v_d \hat{I}_s}{2\omega_1} \sin(\omega_1 t - \varphi) - \frac{2\hat{V}_s i_c}{\omega_1} \sin \omega_1 t}_{\Delta W_\Delta} \qquad (3.36)$$

where the integration constants $W_{\Sigma 0}$ and $W_{\Delta 0}$ are the mean values. It can be noted that the total-energy ripple is of twice the fundamental frequency, whereas the imbalance-energy ripple is of the fundamental frequency. As normally $\overline{v_{u,l}} = v_d$ is desired and the effective per-arm capacitance is C/N, the stored energy per arm is normally $Cv_d^2/(2N)$. The total arm energy is twice this value and it is balanced among the arms, giving

$$W_{\Sigma 0} = \frac{Cv_d^2}{N} \qquad W_{\Delta 0} = 0. \qquad (3.37)$$

Now, $W_{u,l}$ can be solved from the identities $W_\Sigma = W_u + W_l$ and $W_\Delta = W_u - W_l$ as $W_u = (W_\Sigma + W_\Delta)/2$ and $W_l = (W_\Sigma - W_\Delta)/2$. Expressing W_Σ and W_Δ as mean value plus ripple, according to Equations (3.35) and (3.36), we get

$$W_u = \frac{W_{\Sigma 0} + \Delta W_\Sigma + \Delta W_\Delta}{2} \qquad W_l = \frac{W_{\Sigma 0} + \Delta W_\Sigma - \Delta W_\Delta}{2}. \tag{3.38}$$

The sum capacitor voltages are then obtained as

$$v_{cu}^\Sigma = \sqrt{\frac{2N}{C} W_u} = \sqrt{\frac{N}{C}(W_{\Sigma 0} + \Delta W_\Sigma + \Delta W_\Delta)} = \sqrt{v_d^2 + \frac{N}{C}(\Delta W_\Sigma + \Delta W_\Delta)} \tag{3.39}$$

$$v_{cl}^\Sigma = \sqrt{\frac{2N}{C} W_l} = \sqrt{\frac{N}{C}(W_{\Sigma 0} + \Delta W_\Sigma - \Delta W_\Delta)} = \sqrt{v_d^2 + \frac{N}{C}(\Delta W_\Sigma - \Delta W_\Delta)}. \tag{3.40}$$

Since the ripples are much smaller than the mean value v_d, the following approximations can be made:

$$v_{cu}^\Sigma = v_d \sqrt{1 + \frac{N}{C v_d^2}(\Delta W_\Sigma + \Delta W_\Delta)} \approx v_d + \underbrace{\frac{N}{2 C v_d}(\Delta W_\Sigma + \Delta W_\Delta)}_{\Delta v_{cu}^\Sigma} \tag{3.41}$$

$$v_{cl}^\Sigma = v_d \sqrt{1 + \frac{N}{C v_d^2}(\Delta W_\Sigma - \Delta W_\Delta)} \approx v_d + \underbrace{\frac{N}{2 C v_d}(\Delta W_\Sigma - \Delta W_\Delta)}_{\Delta v_{cl}^\Sigma}. \tag{3.42}$$

Some observations can be made from these equations.

- As predicted, the amplitudes of the sum-capacitor-voltage ripples are inversely proportional to C.
- In both arms, the ripples consist of two components: one component of the fundamental frequency [resulting from W_Δ; see Equation (3.36)] and one component of twice the fundamental frequency [resulting from W_Σ; see Equation (3.35)].
- The twice-the-fundamental-frequency components of the upper and lower arms are in phase, whereas the fundamental-frequency components are 180° phase shifted. Respectively adding and subtracting Equations (3.41) and (3.42) we obtain

$$v_c^\Sigma = v_{cu}^\Sigma + v_{cl}^\Sigma \approx 2v_d + \frac{N}{C v_d} \Delta W_\Sigma \tag{3.43}$$

$$v_c^\Delta = v_{cu}^\Sigma - v_{cl}^\Sigma \approx \frac{N}{C v_d} \Delta W_\Delta \tag{3.44}$$

and with the expressions for ΔW_Σ and ΔW_Δ from Equations (3.41) and (3.42) shown explicitly

$$v_c^\Sigma \approx 2v_d - \frac{N}{C v_d} \frac{\hat{V}_s \hat{I}_s}{4\omega_1} \sin(2\omega_1 t - \varphi) \tag{3.45}$$

$$v_c^\Delta \approx \frac{N}{C v_d} \left[\frac{v_d \hat{I}_s}{2\omega_1} \sin(\omega_1 t - \varphi) - \frac{2\hat{V}_s i_c}{\omega_1} \sin \omega_1 t \right]. \tag{3.46}$$

These relations can be verified to accurately predict the sum-capacitor-voltage ripples [1].

3.3.6   Maximum Output Voltage

In Equation (3.12) the instantaneous minimum and maximum values of the output voltage are shown to be $-v_d/2$ and $v_d/2$, respectively, assuming that the sum-capacitor-voltage ripples can be neglected. The maximum attainable output-voltage peak value $\hat{V}_{max}$ is then consequently limited as

$$\hat{V}_{max} \leq \frac{v_d}{2}. \tag{3.47}$$

Neglecting the sum-capacitor-voltage ripples is possible only if C is unrealistically large; Equation (3.47) represents an idealized scenario. The analytic ripple expressions, given by Equations (3.41) and (3.42) respectively for the upper and lower arms, allow Equation (3.47) to be generalized. To this end, we substitute Equations (3.41) and (3.42) in the expressions for the insertion indices given by Equation (3.25). This allows $\hat{V}_{max}$ to be determined from the constraint $0 \leq n_{u,l} \leq 1$. Assuming balance between the arms, n_u and n_l have identical waveforms, but they are $180°$ phase shifted over the fundamental period. It is therefore sufficient to consider either one of the insertion indices. Let us pick n_l. With $v_s^\star$ given by Equation (3.31) and $v_c^\star = v_d/2$, we have

$$n_l = \frac{\frac{v_d}{2} + \hat{V}_s \cos \omega_1 t}{v_d + \Delta v_{cl}^\Sigma},$$

$$\Delta v_{cl}^\Sigma = \frac{N}{2Cv_d}\left[-\frac{\hat{V}_s\hat{I}_s}{4\omega_1}\sin(2\omega_1 t - \varphi) - \frac{v_d\hat{I}_s}{2\omega_1}\sin(\omega_1 t - \varphi) + \frac{2\hat{V}_s i_c}{\omega_1}\sin \omega_1 t\right]. \tag{3.48}$$

The largest value of n_l is obtained for $t = 0$, since then $\cos \omega_1 t = 1$

$$n_l|_{max} = \frac{\frac{v_d}{2} + \hat{V}_s}{v_d + \frac{N\hat{I}_s}{4\omega_1 C}\left(\frac{\hat{V}_s}{2v_d} + 1\right)\sin \varphi}. \tag{3.49}$$

Setting $n_l|_{max} = 1$ and solving for $\hat{V}_s$ yields

$$\hat{V}_s = \frac{\frac{v_d}{2} - \frac{N\hat{I}_s \sin \varphi}{4\omega_1 C}}{1 + \frac{N\hat{I}_s \sin \varphi}{8\omega_1 Cv_d}}. \tag{3.50}$$

This expression gives the maximum output-voltage peak value as a function not only of v_d but also of some parameters as well. Equation (3.50) can be employed to generalize Equation (3.47) as

$$\hat{V}_{max} \leq \frac{\frac{v_d}{2} - \frac{NQ}{2M\omega_1 C\hat{V}_s}}{1 + \frac{NQ}{4M\omega_1 C\hat{V}_s v_d}}. \tag{3.51}$$

where the identity $(\hat{I}_s/2)\sin \varphi = Q/(M\hat{V}_s)$ has been applied. The following conclusions can be drawn from Equation (3.51).

- $\hat{V}_{max}$ is dependent on the ac-side operating point, i.e. $\hat{V}_s$ and Q.
- As $C \to \infty$, the right-hand side of Equation (3.51) tends to $v_d/2$ irrespective of the ac-side operating point, i.e. Equation (3.51) approaches Equation (3.47) as C increases.

- The largest value (which is larger than the nominal $\hat{V}_{max} \leq v_d/2$) is obtained for the maximum reactive-power production, i.e. $\varphi = 90°$, giving $Q > 0$.
- The smallest value is obtained for the maximum reactive-power consumption, i.e. $\varphi = -90°$, giving $Q < 0$.
- For unity power factor, $\varphi = 0 \Rightarrow Q = 0$, the nominal $\hat{V}_{max} \leq v_d/2$ as given by Equation (3.47) is obtained.

The found influence of Q on $\hat{V}_{max}$ is not necessarily a drawback. Injection of reactive power into the grid often requires a high output voltage, whereas when drawing reactive power from the grid a lower voltage is often adequate.

It should finally be noted that Equation (3.51) is derived assuming a sinusoidal output voltage. As shown in Section 3.7.4, for a three-phase converter it is possible to increase $\hat{V}_{max}$ as given by Equation (3.51) roughly by 15%, through the addition of zero-sequence components to the output-voltage references.

Example 3.2 *Let us introduce a Test Converter, with data as shown in Table 3.1, that will be used in several examples in the following. The equations for the Test Converter, with $C = 4$ mF, $\hat{V}_s = 100$ kV, and $\hat{I}_s = 0.5$ kA (i.e. half the rated current), are utilized to produce traces of the insertion indices and the sum-capacitor-voltage ripples. Three operating points are considered: successively $\varphi = 0$, $\varphi = 90°$, and $\varphi = -90°$, each for two fundamental periods. The results are shown in Figure 3.3. The following can be noted.*

- *The maximum values of the insertion indices agree with the theoretical results: $n_{u,l}^{max} = 1$ for $\varphi = 0$, $n_{u,l}^{max} < 1$ for $\varphi = 90°$, and $n_{u,l}^{max} > 1$ for $\varphi = -90°$. (The occurrence of $n_{u,l}^{max} > 1$ would in practice be prevented by hard limiting the insertion indices to 1, resulting in a lower output voltage than desired as well as a distorted voltage waveform.)*
- *It is evident that there are components at ω_1 as well as $2\omega_1$ in the sum-capacitor-voltage ripples.*
- *The sum-capacitor-voltage ripples are larger for operation with a low power factor ($\varphi = \pm90°$ yields zero power factor) than with a high power factor. There is then also a significant asymmetry among the positive and negative half periods.*

Table 3.1   Data of the Test Converter.

M	3
v_d	200 kV
$\hat{V}_{max}$	100 kV (line-to-neutral, at $\varphi = 0$)
$\hat{I}_{max}$	1 kA
rated power	$M\hat{V}_{max}\hat{I}_{max}/2 = 150$ MVA
ω_1	$2\pi \cdot 50$ rad/s
N	100
C	16 mF, 8 mF, or 4 mF as noted
C_d	100 μF
L	50 mH
R	0.1 Ω
T_d	Negligible

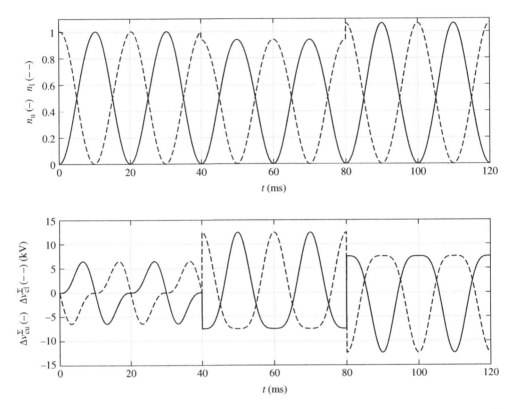

Figure 3.3 Insertion indices and sum-capacitor-voltage ripples for $\varphi = 0$ (0 ms $\leq t < 40$ ms), $\varphi = 90°$ (40 ms $\leq t < 80$ ms), and $\varphi = -90°$ (80 ms $\leq t < 120$ ms).

3.3.7 DC-Bus Dynamics

Let us reiterate that, in order for the mean value of the dc-bus voltage to stay constant, the mean values of the arm currents must add up to the dc-bus current in the steady state, according to Equation (3.5). Yet, ac components and transient imbalances between the dc-bus current and the arm currents may incur ripple and transients in the dc-bus voltage. This effect can be quantified. The transient imbalance between the dc-bus current and the arm currents of all phases at each pole—see Figure 3.2(a)—charges the corresponding dc-bus capacitors according to

$$2C_d \frac{dv_{du,l}}{dt} = i_d - \sum_{k=1}^{M} i_{u,l}^k. \tag{3.52}$$

A better understanding of the dc-bus dynamics is gained by introducing the pole-to-pole dc-bus voltage $v_d = v_{du} + v_{dl}$ and the imbalance dc-bus voltage $v_d^\Delta = v_{du} - v_{dl}$. By respectively adding and subtracting the expressions for the upper and lower arms given by Equation (3.52), and using Equation (3.3), the following dynamic relations are obtained:

$$C_d \frac{dv_d}{dt} = i_d - \sum_{k=1}^{M} i_c^k \qquad C_d \frac{dv_d^\Delta}{dt} = -\frac{1}{2} \sum_{k=1}^{M} i_s^k. \tag{3.53}$$

The circulating current can be controlled to a (near-)constant value irrespective of the number of phases M. Consequently, from the first relation in Equation (3.53) it is found that it is possible to maintain a (near-) constant pole-to-pole dc-bus voltage v_d. This holds irrespective of M, i.e. also for a single-phase MMC. This is a different—and beneficial—property as compared to a two-level single-phase VSC, where pulsations of twice the fundamental frequency appear on the dc bus. For an MMC, these pulsations instead appear as sum-capacitor-voltage ripples.

The second relation in Equation (3.53) shows that no imbalance dc-bus voltage appears as long as the phase output currents add up to zero. That is, imbalances among the phase output currents will not give rise to a dc-bus imbalance as long as there is no zero-sequence current component (see also Section 3.6 and Chapter 4). This motivates the assumption $v_d^\Delta = 0$, implying $v_{du} = -v_{dl} = v_d/2$, which is made throughout the chapter.

3.3.7.1 Effective DC-Bus Dynamics Including Submodule Capacitors

The circulating currents of all phases add at the positive and negative poles; see Figure 3.2(a). In equation form, they sum up on the right-hand side of the first relation in Equation (3.53). The circulating current thus transfers energy between the submodule capacitors and the dc-bus capacitors. A simplified dynamic model for this interaction can be obtained by adding the two relations in Equation (3.22)

$$\frac{C}{N}\frac{d(v_{cu}^\Sigma + v_{cl}^\Sigma)}{dt} = n_u\left(\frac{i_s}{2} + i_c\right) + n_l\left(-\frac{i_s}{2} + i_c\right).\tag{3.54}$$

Introducing $v_c^\Sigma = v_{cu}^\Sigma + v_{cl}^\Sigma$ yields the simplification

$$\frac{C}{N}\frac{dv_c^\Sigma}{dt} = (n_u - n_l)\frac{i_s}{2} + (n_u + n_l)i_c.\tag{3.55}$$

The insertion indices—see Equation (3.25)—can be approximated by letting $v_c^\star = v_d/2$ in the numerators and neglecting the sum-capacitor-voltage ripples in the denominators, i.e.

$$n_u \approx \frac{v_d/2 - v_s^\star}{v_d} \qquad n_l \approx \frac{v_d/2 + v_s^\star}{v_d}.\tag{3.56}$$

This allows Equation (3.55) to be further simplified as

$$\frac{C}{N}\frac{dv_c^\Sigma}{dt} = -\frac{v_s^\star i_s}{v_d} + i_c.\tag{3.57}$$

The sum-capacitor-voltage ripples are then disregarded by averaging. Assuming i_c to be pure dc, we get

$$\frac{C}{N}\frac{d\overline{v_c^\Sigma}}{dt} = -\underbrace{\frac{\overline{v_s^\star i_s}}{v_d}}_{P/(Mv_d)} + i_c.\tag{3.58}$$

Normally, the mean sum capacitor voltages are controlled to v_d. The dynamics of this control can often be assumed to be much faster than the time scale on which v_d evolves [1]. This motivates substituting $\overline{v_c^{\Sigma}} \to 2v_d$ in Equation (3.58), giving

$$\frac{2C}{N}\frac{dv_d}{dt} = -\frac{P}{Mv_d} + i_c. \tag{3.59}$$

As the final steps, balance among the phase legs is assumed, i.e. $\sum_{k=1}^{M} i_c^k = Mi_c$ in Equation (3.53)

$$C_d \frac{dv_d}{dt} = i_d - Mi_c. \tag{3.60}$$

Eliminating i_c among Equations (3.59) and (3.60) results in

$$\underbrace{\left(C_d + \frac{2MC}{N} \right)}_{C_d'}\frac{dv_d}{dt} = i_d - \frac{P}{v_d}. \tag{3.61}$$

This relation determines the effective dc-bus dynamics in the sense that the submodule capacitors are included. The effective dc-bus capacitance C_d' is larger than C_d, owing to the contribution from the submodule capacitors.

3.3.8 Time Delays

The control system of an MMC is invariably implemented digitally, generally using digital signal processors (DSPs), often combined with field-programmable gate arrays (FPGAs). In such a control system there are some time delays.

- Computation in DSPs and FPGAs together with communication delays and delays in sensors, anti-aliasing filters, and in the analog-to-digital conversion process give the computational time delay T_c. It is typically in the same order of magnitude as the sampling period T_s, and possibly equal to T_s.
- If the pulse frequency (i.e. the effective switching frequency taken over all submodules; see Chapter 5) is equal to half the sampling frequency or higher, then pulse-width modulation (PWM) results in a switching time delay of $0.5T_s$ [2].

The total time delay, T_d, is the sum of these two components, i.e. $T_d = T_c + 0.5T_s$.

3.4 Per-Phase Output-Current Control

Having considered the dynamics of the MMC and developed a per-phase averaged model, it is now time to consider the control of the output current. The control loop for the output current can be considered as the "heart" of the converter control system. It should be designed with care, as otherwise the performance of the entire converter may deteriorate.

In this section, the fundamentals of current control are first reviewed. This is made on a per-phase basis. The methods developed are consequently applicable to MMCs with an arbitrary number of phases, including single-phase MMCs. (In fact, the methods for output-current control considered in this chapter are, after minor modifications, applicable to VSCs of other topologies than the MMC as well.) Output-current control of a three-phase converter is normally made using the concept of space vectors, i.e. in a corresponding two-phase system. Section 3.7 is devoted to vector current control, where more details specific to the three-phase case can be found.

3.4.1 Tracking of a Sinusoidal Reference Using a PI Controller

The proportional–integral–derivative (PID) controller is by far the most common controller. For example, it is used in more than 90% of all control loops in the process industry. In most cases, the D part is inhibited or not implemented, sometimes simply because it is considered difficult to tune (even though its inclusion might improve performance) [3]. In many cases, though, a D part is not needed; good enough performance can be achieved by a PI controller [4]. This is the case particularly for controlled processes that have first-order dynamics or can be approximated as such.

Suppose that the current i through an inductor (with inductance L and with negligible inner resistance) is to be controlled. This is a first-order process, so a PI controller would seem to be appropriate. The controller output signal is the reference for the voltage v across the inductor, which in turn is created by a converter. Assuming that the total time delay can be neglected, so that v equals its reference, a block diagram for the closed-loop system is illustrated in Figure 3.4. The open-loop transfer function $G_k(s)$ is given by the product of the controller and inductor-admittance transfer functions as[1]

$$G_k(s) = \left(K_p + \frac{K_i}{s} \right) \frac{1}{sL} = \frac{K_p s + K_i}{s^2 L} \tag{3.62}$$

and the closed-loop transfer function from reference $i^\star$ to i is given by

$$G_c(s) = \frac{G_k(s)}{1 + G_k(s)} = \frac{K_p s + K_i}{s^2 L + K_p s + K_i}. \tag{3.63}$$

The I part of the PI controller is included to ensure that the output signal (here i) tracks its reference (here $i^\star$) without a static error. For a constant reference, zero static error is obtained if the closed-loop system has unity static gain, $G_c(0) = 1$, as then $i = i^\star$ in the steady state. In

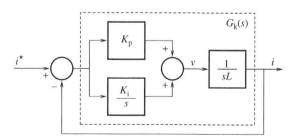

Figure 3.4   Control of current i through an inductor using a PI controller.

---

[1] Observe that s shall, where appropriate, be interpreted either as the complex Laplace variable or as the differential operator $s = d/dt$.

Equation (3.63) it can be observed that, even with pure P control, $K_i = 0$, we get $G_c(0) = 1$, because the controlled process—i.e. the inductor—has integrator characteristics (if its internal resistance can be neglected).

However, $i^\star$ is normally not constant, but a sinusoid. Do we then also obtain zero static error? The static gain can in this case be calculated by setting $s = j\omega$, which yields

$$G_c(j\omega) = \frac{K_i + j\omega K_p}{K_i - \omega^2 L + j\omega K_p}. \tag{3.64}$$

Only for $\omega = 0$ do we get $G_c(j\omega) = 1$, whereas for $\omega > 0$, $|G_c(j\omega)| \neq 1$ and $\arg G_c(j\omega) \neq 0$. This leads us to conclude that the PI controller is *inherently incapable* of tracking a sinusoidal reference without incurring amplitude and phase errors! This property of the PI controller was found experimentally for variable-speed drives, i.e. converters feeding ac motors, already in the 1970s. Somewhat surprisingly—since the above analysis is straightforward—the origin of the phenomenon was debated for several years. The correct explanation was given in [5]. To quote [5, p. 679]:

> The fact that this error exists in a current regulated $R - L$ load suggests that the error observed in ac drives at higher frequencies previously attributed to so-called back EMF or "running up against the bus" is neither, but rather a consequence of the stationary regulator itself. Therefore, whether the load is static or dynamic is immaterial from the standpoint of phase shift and magnitude error present in the steady state currents.

3.4.2 Resonant Filters and Generalized Integrators

As the first step to remedy the failure of the PI controller to accurately track a sinusoidal reference, a versatile controller building block is introduced: the resonant filter. As will repeatedly be demonstrated, many filters and controllers—most used in this chapter, in fact—can be realized using variants of the resonant filter. Its transfer function is denoted by the subscript h, which indicates multiples of the fundamental frequency [6]

$$H_h(s) = \frac{K_h(s \cos\phi_h - h\omega_1 \sin\phi_h)}{s^2 + \alpha_h s + (h\omega_1)^2} \tag{3.65}$$

where K_h is the gain, α_h is the bandwidth, and ϕ_h is the so-called compensation angle. The resonant filter acts as a band-pass filter (BPF) about the angular frequency $h\omega_1$, which is its angular resonant frequency. We have

$$H_h(jh\omega_1) = \frac{K_h(jh\omega_1 \cos\phi_h - h\omega_1 \sin\phi_h)}{jh\omega_1 \alpha_h} = \frac{K_h(\cos\phi_h + j\sin\phi_h)}{\alpha_h} = \frac{K_h e^{j\phi_h}}{\alpha_h}. \tag{3.66}$$

So, if $K_h = \alpha_h$ and $\phi_h = 0$, then $H_h(jh\omega_1) = 1$, i.e. a frequency component at $h\omega_1$ is admitted without amplitude or phase shift. A positive ϕ_h gives phase lead, which is useful for the compensation of a phase lag that results from a time delay (this is exemplified later in the text). Bandwidth α_h is that of the passband about $h\omega_1$. If $\alpha_h \ll h\omega_1$, then, with $K_h = \alpha_h$,

$$|H_h[j(h\omega_1 \pm \alpha_h/2)]| \approx \frac{1}{\sqrt{2}} \tag{3.67}$$

with good accuracy. Some additional interesting properties of the resonant filter are as follows.

- For $h = 0$, Equation (3.65) reduces to a low-pass filter

$$H_0(s) = \frac{K_0 \cos \phi_0}{s + \alpha_0} \tag{3.68}$$

where letting $K_0 = \alpha_0$ and $\phi_0 = 0$ (a different selection of ϕ_0 is meaningless) yields unity static gain: $H_0(0) = 1$.
- Zero bandwidth, $\alpha_h = 0$

$$H_h(s) = \frac{K_h(s \cos \phi_h - h\omega_1 \sin \phi_h)}{s^2 + (h\omega_1)^2} \tag{3.69}$$

gives infinite gain at the resonant frequency. The poles of the filter are then located at $s = \pm jh\omega_1$, i.e. on the imaginary axis. A zero-bandwidth resonant filter is used as the resonant (R) part in a so-called proportional–resonant (PR) controller [2, 7, 8], which we shall consider momentarily. Another name is second-order generalized integrator (SOGI) [9]. This is because the SOGI has infinite gain for $s = jh\omega_1$, whereas an integrator has infinite gain for $s = 0$. Not surprisingly, the SOGI reduces to a pure integrator for $h = 0$ (again with $\phi_0 = 0$):

$$H_0(s) = \frac{K_0}{s}. \tag{3.70}$$

3.4.2.1 Discrete-Time Realization

For discrete-time realization of a resonant filter, often implemented as code on a DSP, the continuous-time transfer function given by Equation (3.65) needs to be discretized. Care is needed in this process.

- Discretization must be made such that the left half of the s plane is mapped onto the unit disc of the z plane in order to preserve the stability properties of $H_h(s)$.
- The resonant frequency should not be shifted because of the discretization. The prewarped Tustin method, which involves the substitution [7, 10]

$$s \rightarrow \frac{h\omega_1}{\tan\left(\frac{h\omega_1 T_s}{2}\right)} \frac{z - 1}{z + 1} \tag{3.71}$$

has the desired mapping and frequency properties. This can be verified as follows. The analytical relation between the Laplace and z-transform variables is $z = e^{sT_s}$. Putting $s = jh\omega_1 \Rightarrow z = e^{jh\omega_1 T_s}$, both sides of Equation (3.71) equal $jh\omega_1$.
- Realization should preferably be made using a structure that has good numerical properties. The "direct form II transposed" (DFIIt) structure in the so-called delta operator

$$\delta = z - 1 \tag{3.72}$$

has excellent numerical properties [10].

Applying Equations (3.71) and (3.72) to Equation (3.65) yields

$$H(z) = \frac{b_0 \delta^2 + b_1 \delta + b_2}{\delta^2 + a_1 \delta + a_2} \tag{3.73}$$

where straightforward but tedious algebraic manipulations result in the following expressions for the coefficients:

$$r_h = \frac{1}{1 + \frac{\alpha_h}{2h\omega_1} \sin(h\omega_1 T_s)} \tag{3.74}$$

$$K'_h = \frac{K_h r_h}{2h\omega_1} \tag{3.75}$$

$$b_0 = K'_h[\sin(h\omega_1 T_s + \phi_h) - \sin \phi_h] \tag{3.76}$$

$$b_1 = K'_h[3 \sin(h\omega_1 T_s + \phi_h) - 4 \sin \phi_h - \sin(h\omega_1 T_s - \phi_h)] \tag{3.77}$$

$$b_2 = K'_h[2 \sin(h\omega_1 T_s + \phi_h) - 4 \sin \phi_h - 2 \sin(h\omega_1 T_s - \phi_h)] \tag{3.78}$$

$$a_1 = 2[1 - r_h \cos(h\omega_1 T_s)] \tag{3.79}$$

$$a_2 = 2r_h[1 - \cos(h\omega_1 T_s)]. \tag{3.80}$$

At a glance, Equations (3.74)–(3.80) may look complicated, but this is deceiving. There are just five trigonometric operations, four of which are added with different weights in Equations (3.76)–(3.80). For $\phi_h = 0$ there are just two trigonometric operations—$\cos(h\omega_1 T_s)$ and $\sin(h\omega_1 T_s)$—since Equations (3.76)–(3.78) then reduce to

$$b_0 = K'_h \sin(h\omega_1 T_s) \qquad b_1 = 2b_0 \qquad b_2 = 0. \tag{3.81}$$

A signal-flow graph for the DFIIt realization of Equation (3.73) is shown in Figure 3.5, where u and y are generic input and output signals and x_1 and x_2 are state variables. The realization is obtained in the form of two delayed numerical integrators $\delta^{-1} = 1/(z - 1)$. It can equivalently be described in equation form as

$$y(n) = b_0 u(n) + x_1(n) \tag{3.82}$$

$$x_1(n + 1) = x_1(n) + x_2(n) + b_1 u(n) - a_1 y(n) \tag{3.83}$$

$$x_2(n + 1) = x_2(n) + b_2 u(n) - a_2 y(n) \tag{3.84}$$

where n is the sample number (corresponding to the time instant $t = nT_s$).

3.4.3   Tracking of a Sinusoidal Reference Using a PR Controller

We now replace the I part in Figure 3.4 with an R part that consists of a SOGI—Equation (3.69)—with $\phi_h = 0$ for simplicity. This is shown in Figure 3.6. The open-loop transfer function

$$G_k(s) = \frac{K_p[s^2 + (h\omega_1)^2] + K_h s}{[s^2 + (h\omega_1)^2]sL} \tag{3.85}$$

is obtained and the closed-loop transfer function becomes

$$G_c(s) = \frac{G_k(s)}{1 + G_k(s)} = \frac{K_p[s^2 + (h\omega_1)^2] + K_h s}{(sL + K_p)[s^2 + (h\omega_1)^2] + K_h s}. \tag{3.86}$$

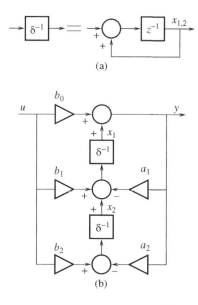

(a)

(b)

Figure 3.5 (a) Realization of the delayed numerical integrator δ^{-1}. (b) DFIIt structure for resonant-filter realization.

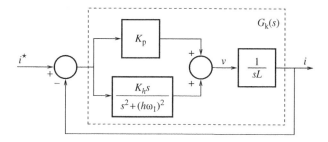

Figure 3.6 Control of current i through an inductor using a PR controller.

This time—unlike for the PI controller—accurate reference tracking is obtained, but only for a sinusoid of the angular frequency $h\omega_1$, since $G_c(jh\omega_1) = 1$, whereas $G_c(j\omega) \neq 1$ for $\omega \neq h\omega_1$. In order to track a reference that has multiple frequency components, R parts that are tuned to each of these frequencies are needed in the controller.

3.4.4 Parameter Selection for a PR Current Controller

3.4.4.1 Selection of the P Gain

Let us consider the closed-loop system given by Equation (3.86) and assume for the moment that pure P control is used, i.e. $K_h = 0$. This gives

$$G_c(s) = \frac{K_p}{sL + K_p} = \frac{K_p/L}{s + K_p/L}. \tag{3.87}$$

This is a first-order system with bandwidth K_p/L, i.e. $|G_c(jK_p/L)| = 1/\sqrt{2}$ and $G_c(0) = 1$. The closed-loop system bandwidth is an important parameter, as it determines the exponential convergence rate for closed-loop-system transients. The output signal—i.e. the inductor current in our case—responds to a reference step I_0 as $I_0(1 - e^{-K_p t/L})$. The time constant of this exponential response is L/K_p, i.e. the inverse of the bandwidth. A useful parametrization of the P gain is thus [11]

$$K_p = \alpha_c L \tag{3.88}$$

where α_c is the desired closed-loop-system bandwidth, giving

$$G_c(s) = \frac{\alpha_c}{s + \alpha_c}. \tag{3.89}$$

3.4.4.2 Selection of the Closed-Loop-System Bandwidth

Because the total time delay, T_d, is not generally negligible, there is an upper limit for α_c that must be observed for the closed-loop system to remain stable with large enough margins. A useful rule of thumb is [11]

$$\alpha_c \leq \frac{\omega_s}{10} \tag{3.90}$$

where ω_s is the angular sampling frequency

$$\omega_s = \frac{2\pi}{T_s}. \tag{3.91}$$

With the total time delay taken into account, we obtain, with K_p given by Equation (3.88) and for pure P control,

$$G_k(s) = \frac{K_p e^{-sT_d}}{sL} = \frac{\alpha_c e^{-sT_d}}{s}. \tag{3.92}$$

The crossover frequency, i.e. the value of ω for which $|G_k(j\omega)| = 1$, is obviously α_c. This yields the phase margin

$$\phi_m = \pi - \arg G_k(j\alpha_c) = \frac{\pi}{2} - \alpha_c T_d. \tag{3.93}$$

If the upper limit of Equation (3.90) is applied, then

$$\phi_m = \frac{\pi}{2} - \frac{\omega_s T_d}{10}. \tag{3.94}$$

Suppose that the total time delay is given as $T_d = 1.5 T_s$ (see Section 3.3.8). Then we obtain $\omega_s T_d = 1.5 \cdot 2\pi$ and

$$\phi_m = 0.2\pi \text{ rad} = 36°. \tag{3.95}$$

This is a somewhat small, but yet often acceptable, phase margin. The recommendation given in Equation (3.90) typically yields an α_c in the range of kiloradians per second. For example, a sampling frequency of 10 kHz gives $\alpha_c \leq 6.3$ krad/s.

3.4.4.3 Selection of the R Gain

A simple inspection shows that the dimension of K_h is angular frequency times impedance, or equivalently, angular frequency squared times inductance. The following parametrization gives the correct dimension:

$$K_h = 2\alpha_h \alpha_c L = 2\alpha_h K_p. \tag{3.96}$$

What do we gain from this parametrization? Substituting Equations (3.88) and (3.96) in Equation (3.86) yields

$$G_c(s) = \frac{\alpha_c[s^2 + (h\omega_1)^2] + 2\alpha_h \alpha_c s}{(s + \alpha_c)[s^2 + (h\omega_1)^2] + 2\alpha_h \alpha_c s}. \tag{3.97}$$

This may equivalently be expressed as

$$G_c(s) = \frac{\alpha_c[s^2 + 2\alpha_h s + (h\omega_1)^2]}{(s + \alpha_c)[s^2 + 2\alpha_h s + (h\omega_1)^2] - 2\alpha_h s^2} \tag{3.98}$$

which can be approximated as

$$G_c(s) \approx \frac{\alpha_c[s^2 + 2\alpha_h s + (h\omega_1)^2]}{(s + \alpha_c)[s^2 + 2\alpha_h s + (h\omega_1)^2]} = \frac{\alpha_c}{s + \alpha_c} \tag{3.99}$$

provided that the last term in the denominator of $G_c(s)$ in Equation (3.98) can be neglected. That is, Equation (3.99) simplifies to the closed-loop transfer function shown in Equation (3.89), which is obtained for pure P control. The approximation holds if

$$\alpha_h \ll \alpha_c \tag{3.100}$$

because if the transfer-function denominators of Equations (3.97) and (3.99) are expanded, the second-degree terms are respectively $\alpha_c s^2$ and $(\alpha_c + 2\alpha_h)s^2 \approx \alpha_c s^2$, whereas all other terms are identical. This translates to a selection of α_h in the range of hundreds of radians per second, typically such that

$$\alpha_h < \omega_1. \tag{3.101}$$

The conclusion of this is that, if the R gain is selected according to Equations (3.96) and (3.100), then the closed-loop-system dynamics will be dominated by a pole at

$$s \approx -\alpha_c. \tag{3.102}$$

In addition, there will be a pole pair at

$$s \approx -\alpha_h \pm \sqrt{\alpha_h^2 - (h\omega_1)^2} = -\alpha_h \pm j\sqrt{(h\omega_1)^2 - \alpha_h^2} \tag{3.103}$$

but this pole pair tends to cancel with the zero pair of $G_c(s)$ in Equation (3.97), which is located exactly as given by Equation (3.103). Parameter α_h, which may be called the *resonant bandwidth*, determines the exponential convergence rate of the adjustment made by the R part to obtain accurate reference tracking. This is illustrated in the following example.

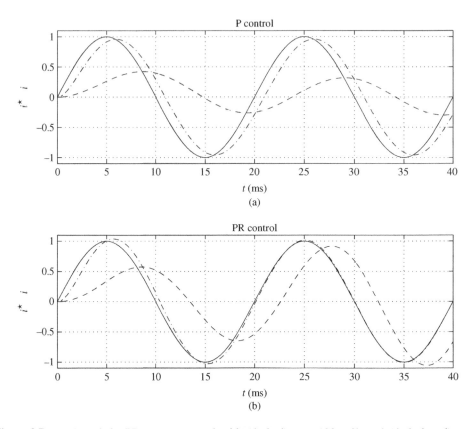

Figure 3.7 (a) P and (b) PR current control with (dashed) $\alpha_c = 100$ rad/s and (dash-dotted) $\alpha_c = 1000$ rad/s. In the PR controller, $\alpha_1 = 100$ rad/s. Solid curves show the reference.

Example 3.3 *Pure P control, as well as PR control, of an inductor current are simulated without time delay. The R part is placed at $h = 1$, for a fundamental frequency of 50 Hz. Two closed-loop system bandwidths are tested: $\alpha_c = 100$ rad/s and $\alpha_c = 1000$ rad/s. These correspond to the time constants 10 ms and 1 ms, respectively. In the PR controller, $\alpha_1 = 100$ rad/s is used (hence, $\alpha_1 \ll \alpha_c$ only for the case $\alpha_c = 1000$ rad/s). Simulation results (with the reference peak value normalized to 1) are shown in Figure 3.7. The following observations can be made.*

- *P control with $\alpha_c = 100$ rad/s not only gives a significant tracking error, both of the amplitude and of the phase angle, but also an apparent initial transient of the current. (Note that the first peak of i is larger than the second.) This is not surprising, since the closed-loop time constant is $1/\alpha_c = 10$ ms, whereas the fundamental period is 20 ms.*
- *For P control with $\alpha_c = 1000$ rad/s, the tracking error is much reduced and it is hard to discern the initial transient.*
- *PR control with $\alpha_c = 100$ rad/s gives very slow convergence. Even at the end of the displayed time interval, the tracking error is significant. (In fact, this parameter setting is so poor that is takes about 500 ms until good tracking is obtained.)*

- *PR control with $\alpha_c = 1000$ rad/s gives a quick initial response—much like that obtained for pure P control with $\alpha_c = 1000$ rad/s—followed by a slower settling, which corresponds to the time constant $1/\alpha_1 = 10$ ms. The latter is induced by the R part. At the end of the displayed time interval, very good tracking can be verified. This performance is often quite acceptable, whereas those shown by the other curves are not.*

3.4.5 Output-Current Controller Design

It is now time to apply the results gathered so far to the task of controlling the output current of the MMC. The output current evolves according to the first relation in Equation (3.18):

$$\frac{L}{2}\frac{di_s}{dt} = v_s - v_a - \frac{R}{2}i_s \Rightarrow i_s = \frac{2}{sL+R}(v_s - v_a). \tag{3.104}$$

Since $i_s^\star$ is a fundamental-frequency sinusoid, a PR controller with an R part at $h = 1$ is an appropriate choice.

In Equation (3.104), the ac-bus voltage v_a subtracts from v_s; the output current is driven by the difference voltage $v_s - v_a$. In control-engineering terminology v_a is said to be a *load disturbance*. The ac-bus voltage is a fundamental-frequency sinusoid, whose amplitude and phase angle ideally should be constant. However, the smaller the short-circuit ratio (SCR) is, the more v_a will vary as a function of the output current. This is because the grid impedance as seen from the PCC is inversely proportional to the SCR, which is defined as

$$\text{SCR} = \frac{\text{short-circuit power of the grid at the PCC (MVA)}}{\text{converter power rating (MVA)}}. \tag{3.105}$$

In addition, when faults occur in the grid at locations close to the converter, transients in v_a—phase-angle jumps, sags, and (less frequently) swells [12]—appear, typically with a duration of a few hundred milliseconds. Since v_a is measurable, the dynamic performance can be improved by feedforward. That is, a term is included in the control law, which acts to (partially) cancel v_a. In order to avoid that, high-frequency disturbances (harmonics) are fed forward, v_a should be passed through a BPF centred at ω_1. This BPF can be realized as a resonant filter $H_h(s)$, given by Equation (3.65) for $h = 1$ and $K_h = \alpha_h = \alpha_f$, i.e.

$$H_1(s) = \frac{\alpha_f(s\cos\phi_1 - \omega_1\sin\phi_1)}{s^2 + \alpha_f s + \omega_1^2} \tag{3.106}$$

where $\phi_1 = \omega_1 T_d$ to compensate for the total time delay [9]. This yields the control law

$$v_s^{\star 0} = K_p e + \underbrace{\frac{K_1 s}{s^2 + \omega_1^2}e}_{v_R} + \underbrace{\frac{\alpha_f(s\cos\phi_1 - \omega_1\sin\phi_1)}{s^2 + \alpha_f s + \omega_1^2}v_a}_{v_a^f} \qquad e = i_s^\star - i_s \tag{3.107}$$

where e is the control error and where the superscript ^{0} indicates that the voltage reference computed by Equation (3.107) is an ideal value that may need to be limited, as will be shown momentarily. Since the parasitic arm resistance R typically is small, the parameter selections of

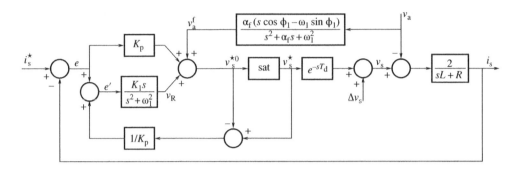

Figure 3.8 Output-current control loop.

Equations (3.88) and (3.96)—derived for current control through a pure inductor—are appropriate, though $L \to L/2$ should be substituted, i.e.

$$K_p = \frac{\alpha_c L}{2} \qquad K_1 = 2\alpha_1 K_p = \alpha_1 \alpha_c L \qquad \alpha_1 \ll \alpha_c. \qquad (3.108)$$

Before the voltage reference is forwarded to the arm-balancing control stage for computation of the insertion indices—either as given by the ideal selection in Equation (3.25) or by an alternative selection as is discussed in Section 3.5—it has to be limited to avoid that $n_{u,l}$ exceed the permitted interval $[0, 1]$. The simplest way to perform this limitation is a saturation

$$v_s^\star = \mathrm{sat}(v_s^{\star 0}) = \begin{cases} v_s^{\star 0}, & |v_s^{\star 0}| \le \hat{V}_{max} \\ \hat{V}_{max} \mathrm{sgn}(v_s^{\star 0}), & |v_s^{\star 0}| > \hat{V}_{max} \end{cases} \qquad (3.109)$$

where $\mathrm{sgn}(\cdot)$ indicates the signum function. ($\hat{V}_{max}$ is a function of the operating point, as shown in Section 3.3.6.) The relation between $v_s^\star$ and v_s can be modeled simply as the total time delay and the addition of a parasitic component Δv_s:

$$v_s = e^{-sT_d} v_s^\star + \Delta v_s. \qquad (3.110)$$

The parasitic component represents switching harmonics. In addition, for some arm-balancing control schemes Δv_s will contain a parasitic fundamental-frequency component, as we shall see in Section 3.5. Its impact on the output current is suppressed by the R part. The control-system block diagram that results from this control law is depicted in Figure 3.8, with the exception that the feedback loop with gain $1/K_p$ is not yet included, i.e. $e' = e$ in the control law given by Equation (3.107).

Example 3.4 *For the Test Converter—see Table 3.1—the nominal peak value of v_a is set to 98 kV, i.e. slightly below $\hat{V}_{max}$. Reference $i_s^\star$ is set to a sinusoid with peak value 1 kA and in phase with v_a.*

We first investigate the response to a two-period sag in the ac-bus voltage (down to the peak value 80 kV), at $t = 20$ ms. The bandwidths are selected as $\alpha_c = 1000$ rad/s and $\alpha_1 = 100$ rad/s. Two values of the feedforward-filter bandwidth are evaluated: $\alpha_f = 0$ and

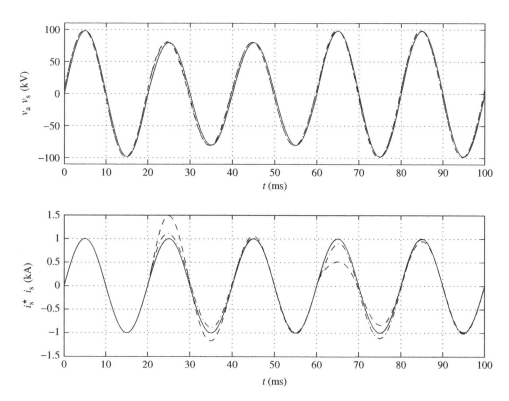

Figure 3.9   Responses to a two-period voltage sag for PR output-current control. Solid curves show v_a and $i_s^\star$, dashed curves show the responses in v_s and i_s with $\alpha_f = 0$, whereas dash-dotted curves show the responses in v_s and i_s with $\alpha_f = 1000$ rad/s.

$\alpha_f = 1000$ rad/s. The dashed curve for i_s in Figure 3.9 shows that, if ac-bus-voltage feedforward is not used, i.e. $\alpha_f = 0$, the voltage sag results in a 50% transient overcurrent. At the end of the sag there is a 50% transient undercurrent. Usage of feedforward with $\alpha_f = 1000$ rad/s reduces the over- and undercurrents to merely 10%. Obviously, feedforward can be highly recommended.

Example 3.5 Next, the response to a two-period swell in the ac-bus voltage (up to the peak value 110 kV), at $t = 20$ ms, is investigated. The bandwidths are the same as before, and feedforward with $\alpha_f = 1000$ rad/s is used. The results are shown by the dashed curves in Figure 3.10. The following observations can be made.

- When the swell occurs, $v_s^{\star 0}$ exceeds $\hat{V}_{\max}$ twice per period and $v_s^\star$ goes into saturation.
- Saturation according to Equation (3.109) gives a waveform distortion. As a result, the output current becomes distorted as well.
- At the end of the voltage swell, the saturation of $v_s^\star$ remains for three more periods. It then ends, but with a 40% transient overcurrent as a very undesirable side effect.

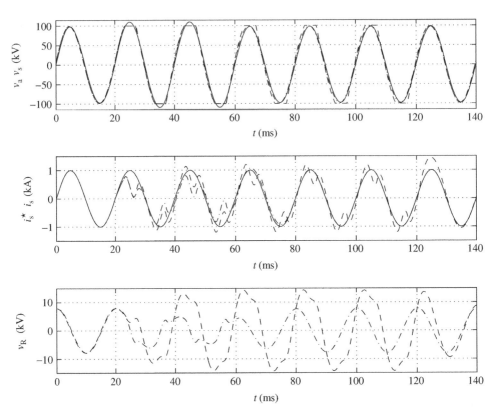

Figure 3.10 Responses to a two-period voltage swell for PR output-current control. Solid curves show v_a and $i_s^\star$, dashed curves show the responses in v_s, i_s, and v_R without anti-windup, whereas dash-dotted curves show the responses in v_s, i_s, and v_R with anti-windup.

3.4.5.1 Anti-Windup

Despite the usage of voltage feedforward, the performance shown in Example 3.5 is unacceptable. Particularly bad is the prolonged saturated operation after the end of the voltage swell and the associated transient overcurrent. The reason is so-called windup of the R part. Windup occurs since the saturation of $v_s^\star$ introduces a nonlinearity in the control loop that makes the control error larger than in the unsaturated situation. The excess control error accumulates in the R part during saturated operation, and the excess has to be worked down before unsaturated operation returns.

There are several possible ways to prevent windup. One straightforward method is known as *back calculation* [13]. Consider the control law given by Equation (3.107)

$$v_s^{\star 0} = K_p e + v_R + v_a^f \qquad e = i_s^\star - i_s. \tag{3.111}$$

Suppose that a modified reference, $i_s^{\star\prime}$, were used, such that the saturated reference $v_s^\star$ would be obtained without actually going into saturation, i.e.

$$v_s^\star = K_p e' + v_R + v_a^f \qquad e' = i_s^{\star\prime} - i_s. \tag{3.112}$$

That is, the control system would balance on the brink of saturated operation. The modified reference results in a modified control error, e', which can be calculated simply by subtracting Equation (3.111) from Equation (3.112) and solving for e':

$$e' = e + \frac{1}{K_p}(v_s^\star - v_s^{\star 0}). \tag{3.113}$$

The modified control error is used as the input signal to the R part, as shown in Figure 3.8. This way, the output signal v_R of the R part will evolve as if the reference were $i_s^{\star\prime}$ instead of $i_s^\star$, effectively preventing windup. (Note that e shall still be used in the P part and that $i_s^{\star\prime}$ is not actually computed!)

Example 3.6 *Anti-windup is added to the controller in Example 3.5. The results are illustrated by the dash-dotted curves in Figure 3.10. Waveform distortion during the swell is not prevented, but at the end of the swell unsaturated operation returns immediately. The impact of anti-windup can be clearly observed by comparing the dashed and dash-dotted curves for v_R. In the latter case, the amplitude reduces rather than increases during the swell.*

3.5 Arm-Balancing (Internal) Control

Having considered in the previous section the output-current control, three state variables (per phase) of the averaged dynamic model remain to be controlled: the circulating current and the two sum capacitor voltages. These state variable are governed the second relation in Equation (3.18) and both relations in Equation (3.22), repeated here for convenience:

$$L\frac{di_c}{dt} = \underbrace{\frac{v_d}{2} - \frac{n_u v_{cu}^\Sigma + n_l v_{cl}^\Sigma}{2} - Ri_c}_{v_c} \tag{3.114}$$

$$\frac{C}{N}\frac{dv_{cu}^\Sigma}{dt} = n_u\left(\frac{i_s}{2} + i_c\right) \qquad \frac{C}{N}\frac{dv_{cl}^\Sigma}{dt} = n_l\left(-\frac{i_s}{2} + i_c\right). \tag{3.115}$$

As the output current is not considered, Equations (3.114) and (3.115) may be called the internal dynamics of the MMC [1]. The control objectives are that i_c should converge to (ideally) a dc component $P/(Mv_d)$ according to Equation (3.34), whereas the mean values of v_{cu}^Σ and v_{cl}^Σ both should converge to v_d. This we call *arm-balancing* (or internal) control.

An inspection of Equations (3.114) and (3.115) shows that arm-balancing control is a nontrivial task. The input variables that are available for manipulation are the insertion indices n_u and n_l. They enter the internal dynamics in a nonlinear fashion, multiplying with the sum capacitor voltages in Equation (3.114) and with the output and circulating currents in Equation (3.115). In addition, there are only two input variables, whereas three state variables are to be controlled. The task may thus appear to be unsolvable at a first glance.

The output current shall be controlled so as to track its reference regardless of the state of the internal dynamics. Therefore, the output current is not available for manipulation in the arm-balancing control. The vehicle for transferring energy into and out of the arms has to be the circulating current. Charging and discharging of the submodule capacitors can be controlled by varying i_c in a suitable way. For this sake we consider the control of the circulating current in Section 3.5.1.

The remainder of the arm-balancing comprises the control of the sum capacitor voltages, here called *voltage control*, for short. The voltage control must ensure that the sum capacitor voltages converge to their desired common mean value, v_d. The core of the voltage control is the selection of the insertion indices. The ideal selection in Equation (3.25) is one of several feasible choices. Four alternative schemes are presented, respectively in Sections 3.5.2–3.5.5. The schemes have different pros and cons; which scheme to choose is essentially a matter of preference. The four alternative schemes are:

- **Direct voltage control** (also known as *direct modulation*): $v_{cu,l}^\Sigma$ in the denominators of Equation (3.25) are substituted with their common mean value, v_d, as given by Equation (3.11):

$$n_u = \frac{v_c^\star - v_s^\star}{v_d} \qquad n_l = \frac{v_c^\star + v_s^\star}{v_d}. \tag{3.116}$$

Direct voltage control inherently gives an asymptotically stable system [1] and has low computational complexity. On the other hand, parasitic components appear in the output and internal voltages, but the effects of these can easily be suppressed by the output- and circulating-current controllers, as discussed in Sections 3.4.5 and 3.5.1.
- **Closed-loop voltage control** [14]: The relations in the ideal insertion-index selection, Equation (3.25), are used as they stand, i.e.

$$n_u = \frac{v_c^\star - v_s^\star}{v_{cu}^\Sigma} \qquad n_l = \frac{v_c^\star + v_s^\star}{v_{cl}^\Sigma}. \tag{3.117}$$

Parasitic voltage components do not appear, but two arm-energy controllers are needed to obtain an asymptotically stable system.
- **Open-loop voltage control** [15]: $v_{cu,l}^\Sigma$ in the denominators of Equation (3.25) are substituted with their respective references $v_{cu,l}^{\Sigma\star}$:

$$n_u = \frac{v_c^\star - v_s^\star}{v_{cu}^{\Sigma\star}} \qquad n_l = \frac{v_c^\star + v_s^\star}{v_{cl}^{\Sigma\star}}. \tag{3.118}$$

The references include estimates of the sum-capacitor-voltage ripples. As for direct voltage control, inherent asymptotic stability is obtained [16, 17]. None, or just small, parasitic voltage components appear.
- **Hybrid voltage control**: $v_{cu,l}^\Sigma$ in the denominators of Equation (3.25) are band-pass filtered, such that their mean values vanish but the ripple components remain. The desired mean value v_d is then added:

$$n_u = \frac{v_c^\star - v_s^\star}{v_d + \text{BPF}\{v_{cu}^\Sigma\}} \qquad n_l = \frac{v_c^\star + v_s^\star}{v_d + \text{BPF}\{v_{cl}^\Sigma\}}. \tag{3.119}$$

The system becomes inherently asymptotically stable.

3.5.1 Circulating-Current Control

The circulating-current control is intimately linked to the voltage control. The circulating-current controller therefore to some extent must be adapted to suit the voltage control scheme that is chosen (out of the four variants that are presented in Sections 3.5.2–3.5.5). These adaptations are discussed in the following. The circulating-current dynamics are governed by Equation (3.114):

$$L\frac{di_c}{dt} = \frac{v_d}{2} - v_c - Ri_c \Rightarrow i_c = \frac{1}{sL + R}\left(\frac{v_d}{2} - v_c\right). \tag{3.120}$$

Similarly to the output voltage v_s, see Equation (3.110), v_c equals its reference, delayed T_d, plus a parasitic component Δv_c:

$$v_c = e^{-sT_d}v_c^\star + \Delta v_c. \tag{3.121}$$

This parasitic component represents switching harmonics. For direct voltage control a component of twice the fundamental frequency adds to Δv_c. An R part for $h = 2$ may therefore be incorporated in the circulating-current controller. In addition, a feedforward term, $v_d/2 - Ri_c^\star$ (where, in practice, an estimate must be used in place of R), can be added to compensate for the resistive voltage drop and term $v_d/2$ in Equation (3.120). Since the dc-bus voltage generally has a lower amount of harmonics (and other disturbances) than the ac-bus voltage, unlike Equation (3.107), filtering of v_d is often not required. We obtain the control law

$$v_c^\star = \frac{v_d}{2} - Ri_c^\star - R_a\left(1 + \frac{2\alpha_2 s}{s^2 + (2\omega_1)^2}\right)(i_c^\star - i_c) \qquad i_c^\star = \frac{i_d}{M} + \Delta i_c^\star \tag{3.122}$$

where R_a is the P gain. Since v_c enters Equation (3.120) with a negative sign, the term proportional to the control error $i_c^\star - i_c$ must have a negative sign in Equation (3.122). The reference is set to the desired dc value $i_d/M = P/(Mv_d)$, plus an optional increment, $\Delta i_c^\star$. Except when closed-loop voltage control is used-see Section 3.5.3—$\Delta i_c^\star = 0$. The R gain is selected similarly to Equation (3.96).

The P gain effectively increases the resistance from R to $R + R_a$; it can be considered as an "active resistance" [18] (also called a *virtual resistance*), hence, the subscript $_a$. The circulating-current controller should be designed for good rejection of the parasitic component Δv_c and for increasing the damping [from R/L to $(R + R_a)/L$], but it need not be as aggressive as the output-current controller. The following selection is therefore appropriate:

$$R \ll R_a \le K_p \tag{3.123}$$

where K_p is the P gain of the output-current controller. R-part bandwidth α_2 can be selected according to Equation (3.101), i.e. $\alpha_2 < \omega_1$. Figure 3.11 shows the block diagram of the circulating-current control loop. Since $v_c \approx v_d/2$, i.e. the feedforward term dominates and the contribution from the PR controller is small, saturation and anti-windup schemes need not be applied in this loop.

3.5.2 Direct Voltage Control

A controller block diagram illustrating this scheme is shown in Figure 3.12.

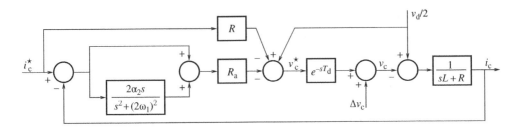

Figure 3.11   Circulating-current control loop.

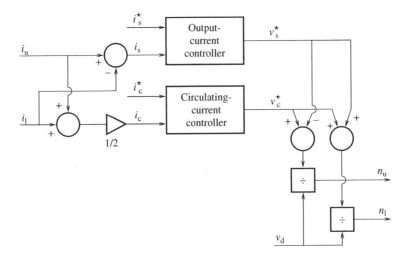

Figure 3.12   Direct voltage control.

Example 3.7 *The alteration of the ideal insertion-index selection of Equation (3.25) to that shown in Equation (3.116) gives as a result that the output and internal voltages do not perfectly track their respective references. Parasitic components Δv_s and Δv_c appear. As the sum-capacitor-voltage ripples decrease with C—compare Equations (3.41) and (3.42)—so do the parasitic components. Let us verify this for the Test Converter, for $\hat{I}_s = 1$ kA. The output current is controlled to be in phase with the output voltage, i.e. the MMC is operated as an inverter with the output power $P/M = \hat{V}_s \hat{I}_s/2 = 50$ MW per phase. Neglecting losses, the mean value of the circulating current is 250 A.*

Simulations using the averaged model derived in Section 3.3.3 are made, which result in the waveforms depicted in Figure 3.13. For the larger $C = 16$ mF there is only a small error between the output voltage and its reference, whereas for the smaller $C = 4$ mF there are noticeable amplitude and phase errors. The output voltage is still a fundamental-frequency sinusoid, which shows that the parasitic component Δv_s is of the fundamental frequency. Consequently, it will be suppressed by the PR output-current controller, such that the output current tracks its reference without a static error.

In Figure 3.14 the sum capacitor voltages and the circulating current are depicted for three different values of C. Initially, $R_a = 0$ in the circulating-current controller. At $t = 0$,

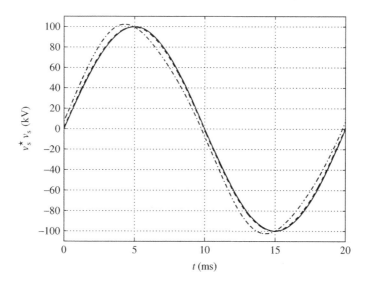

Figure 3.13 One period of the output voltage for direct voltage control, showing (solid) $v_s^\star$ and v_s for (dashed) $C = 16$ mF and (dash-dotted) $C = 4$ mF.

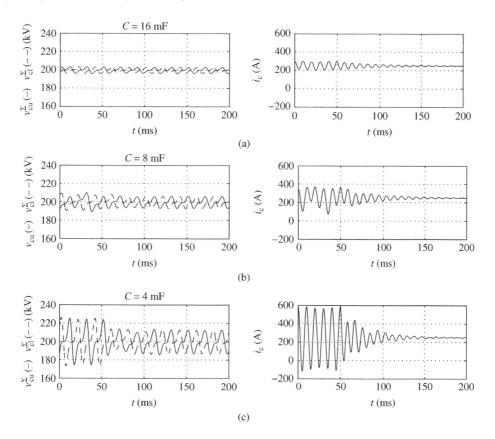

Figure 3.14 Sum capacitor voltages and circulating current for direct voltage control. Closed-loop circulating-current control is commenced at $t = 50$ ms.

steady-state operation has been reached. At t = 50 ms, closed-loop circulating-current control is commenced by stepping the proportional gain from $R_a = 0$ to $R_a = 10 \, \Omega$. In the R part, $\alpha_2 = 200$ rad/s is used. It can be observed in the right-hand subplots that for t < 50 ms there is a parasitic component of twice the fundamental frequency in the circulating current, even for the large C = 16 mF. For the smallest C = 4 mF, there is a significant reduction, not only of the mentioned parasitic component but also of the sum-capacitor-voltage ripples, when circulating-current control is commenced.

It can be concluded that, although direct voltage control adds parasitic components to the output and internal voltages, the effect of these components can be nullified by using PR output- and circulating-current controllers. On the other hand, an asymptotically stable system is obtained without the usage of explicit arm-energy control loops. The sum capacitor voltages converge to their desired mean value v_d without explicit control action.

3.5.3   Closed-Loop Voltage Control

In this scheme, the ideal insertion-index selection in Equation (3.25) is implemented as it stands. Unlike direct voltage control the output and internal voltages adhere to their references; no parasitic components at respectively ω_1 and $2\omega_1$ appear. On the other hand, the resulting internal dynamics will be marginally stable. This can be understood by substituting Equation (3.25) in Equation (3.18): $v_{cu,l}^{\Sigma}$ in the denominators of $n_{u,l}$ cancel in the expressions $n_{u,l} v_{cu,l}^{\Sigma}$. Thus, the sum capacitor voltages vanish altogether from the right-hand sides of the dynamic relations for the output and circulating currents. In addition, $v_{cu,l}^{\Sigma}$ are given as open-loop integrations; see Equation (3.29). Closed-loop arm-energy control is needed to achieve asymptotic stability. This is illustrated in Figure 3.15.

3.5.3.1   Arm-Energy Control

To stabilize the system, the sum capacitor voltages v_{cu}^{Σ} and v_{cl}^{Σ} need to be controlled to their desired common mean value (normally v_d) through the addition of control loops. The basic principles for this are now to be outlined. More details can be found in [14].

Instead of controlling the sum-capacitor voltages directly, it is convenient to do so via the total and imbalance arm energies. So, let us revisit the total sum and imbalance arm-energy dynamics given by Equation (3.30):

$$\frac{dW_{\Sigma}}{dt} = 2v_c^{\star} i_c - v_s^{\star} i_s \qquad \frac{dW_{\Delta}}{dt} = v_c^{\star} i_s - 2v_s^{\star} i_c. \qquad (3.124)$$

By controlling the mean values of W_{Σ} and W_{Δ} respectively to $W_{\Sigma 0}$ and 0—compare Equation (3.37)—the mean values of v_{cu}^{Σ} and v_{cl}^{Σ} are both controlled to v_d. How shall this control be performed? The circulating current facilitates the transfer of energy to and from both arms and between the upper and lower arms. A dc pulse in i_c will multiply with v_d on the right-hand side of the first relation in Equation (3.124) and integrate into a ramp change in W_{Σ}, so that the total energy changes. In the same fashion, a fundamental-frequency ac pulse in i_c will multiply with $v_s^{\star}$ on the right-hand side of the second relation in Equation (3.124) and integrate into a

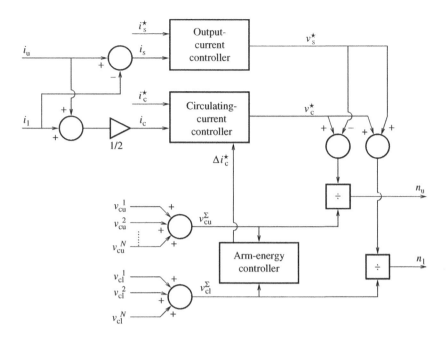

Figure 3.15 Closed-loop voltage control.

ramp change in W_Δ, so that the imbalance energy changes. This motivates adding an increment to the circulating-current reference, given as

$$\Delta i_c^\star = K_\Sigma(W_{\Sigma 0} - \text{LPF}\{W_\Sigma\}) - K_\Delta \text{LPF}\{W_\Delta\} \cos \omega_1 t. \qquad (3.125)$$

In Equation (3.125) there is a P controller for the total energy W_Σ and another P controller for the imbalance energy W_Δ. "LPF" indicates a low-pass filter; these are needed to suppress the arm-energy ripples and should have bandwidths below ω_1. The second term includes a "modulation": there is a factor that induces a fundamental-frequency oscillation in order to create an ac pulse in i_c for a nonzero W_Δ. Reference increment $\Delta i_c^\star$ is then used in a circulating-current controller similar to Equation (3.122), but now a P controller suffices:

$$v_c^\star = \frac{v_d}{2} - R i_c^\star - R_a(i_c^\star + \Delta i_c^\star - i_c). \qquad (3.126)$$

That is, $\alpha_2 = 0$ in Figure 3.11, where $\Delta i_c^\star$ should be added to the reference input.

Example 3.8 *The properties of closed-loop voltage control are evaluated by simulation of the Test Converter with $C = 4$ mF. Figure 3.16 shows transitions from direct voltage control to closed-loop voltage control at $t = 50$ ms. Closed-loop circulating-current control according to Equation (3.126) with $R_a = 10\ \Omega$ is used. As there is no R part, there is a strong parasitic component at twice the fundamental frequency in i_c for direct voltage control. This component vanishes once the transition to closed-loop voltage control is made, when also the amplitudes of the sum-capacitor-voltage ripples reduce. However, as the left-hand plot of Figure 3.16(a)*

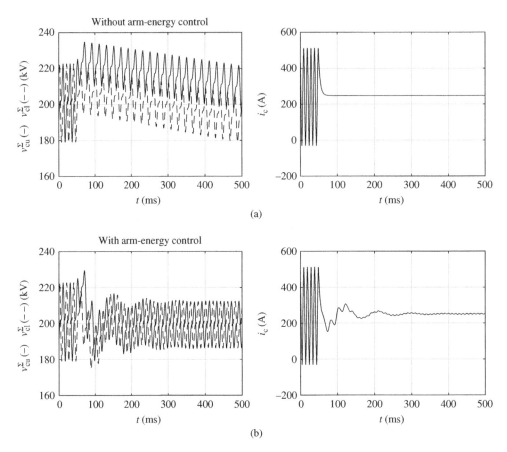

Figure 3.16 Sum capacitor voltages and circulating current for transition from direct voltage control to closed-loop voltage control at $t = 50$ ms.

shows, without arm-energy controllers an imbalance immediately appears; the mean values of v_{cu}^{Σ} and v_{cl}^{Σ} are no longer equal. In addition, both mean values drift slowly. As predicted, the system is marginally stable. Adding arm-energy controllers according to Equation (3.125) improves the situation tremendously. After a transient, both sum capacitor voltage reach a steady state with the correct mean values, as does i_c, the latter with just a trace of a parasitic component.

3.5.4   Open-Loop Voltage Control

This scheme can be considered as an enhancement of direct voltage control. The "trick" is to, instead of dividing by $v_{cu,l}^{\Sigma}$ in the insertion-index selections and thereby induce marginal stability, add to the desired mean value v_d an estimate of sum-capacitor-voltage ripple. This can be made by dividing with the sum-capacitor-voltage references $v_{cu,l}^{\Sigma\star}$, as shown in Equation

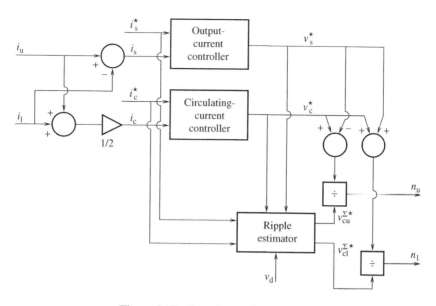

Figure 3.17 Open-loop voltage control.

(3.118), which gives the control scheme depicted in Figure 3.17. In the circulating-current controller—see Figure 3.11—the R part can be omitted.

3.5.4.1 Ripple Estimation Using Explicit Formulas

The ripple estimator block in Figure 3.17 is the core of the open-loop voltage control scheme. It can be realized according to at least two different principles. The first principle relies on the theory developed in Section 3.3.5, more precisely the explicit formulas for the sum-capacitor-voltage ripples derived there. These equations can be used almost as they stand for computing references $v_{cu,l}^{\Sigma\star}$. The steps involved are:

1. For each phase leg, Equations (3.35) and (3.36) are employed to compute online estimates ΔW_Σ and ΔW_Δ of the total and imbalance arm-energy ripples, as

$$\Delta W_\Sigma = -\frac{\hat{V}_s \hat{I}_s}{4\omega_1} \sin(2\omega_1 t - \varphi) \tag{3.127}$$

$$\Delta W_\Delta = \frac{v_d \hat{I}_s}{2\omega_1} \sin(\omega_1 t - \varphi) - \frac{2\hat{V}_s i_c^\star}{\omega_1} \sin \omega_1 t \tag{3.128}$$

where the reference $i_c^\star$ has been substituted in the place of i_c in Equation (3.128) in order to make the scheme open loop also with respect to the currents. (This substitution can be done, as the closed-loop circulating-current control is assumed to be accurate and with high bandwidth; see Section 3.5.1.) For a three-phase system the instantaneous peak values and the arguments of the sine functions that are used in these equations can be obtained from the reference signals in the output-current controller, by extracting their magnitudes and phase angles. With reference to the symbols that are used in Section 3.7

$$\hat{V}_{\mathrm{s}} = \frac{|\mathbf{v}_{\mathrm{s}}^{\star}|}{K} \qquad \hat{I}_{\mathrm{s}} = \frac{|\mathbf{i}_{\mathrm{s}}^{\star}|}{K} \qquad \varphi = \arg \mathbf{v}_{\mathrm{s}}^{\star} - \arg \mathbf{i}_{\mathrm{s}}^{\star}$$

$$(\omega_1 t)_a = \arg \mathbf{v}_{\mathrm{s}}^{\star \mathrm{s}} \qquad (\omega_1 t)_b = \arg \mathbf{v}_{\mathrm{s}}^{\star \mathrm{s}} - \frac{2\pi}{3} \qquad (\omega_1 t)_c = \arg \mathbf{v}_{\mathrm{s}}^{\star \mathrm{s}} - \frac{4\pi}{3}. \quad (3.129)$$

2. To the ripple estimates, the mean values given in Equation (3.37) are added, i.e.

$$W_{\Sigma 0} = \frac{C v_{\mathrm{d}}^2}{N} \qquad W_{\Delta 0} = 0. \qquad (3.130)$$

3. The arm-energy references are computed according to Equation (3.38), i.e.

$$W_{\mathrm{u}} = \frac{W_{\Sigma 0} + \Delta W_{\Sigma} + \Delta W_{\Delta}}{2} \qquad W_1 = \frac{W_{\Sigma 0} + \Delta W_{\Sigma} - \Delta W_{\Delta}}{2}. \qquad (3.131)$$

4. Finally, the sum-capacitor-voltage references $v_{\mathrm{cu,l}}^{\Sigma \star}$ are computed according to Equations (3.39) and (3.40), i.e.

$$v_{\mathrm{cu}}^{\Sigma \star} = \sqrt{\frac{2N}{C} W_{\mathrm{u}}} = \sqrt{\frac{N}{C}(W_{\Sigma 0} + \Delta W_{\Sigma} + \Delta W_{\Delta})} = \sqrt{v_{\mathrm{d}}^2 + \frac{N}{C}(\Delta W_{\Sigma} + \Delta W_{\Delta})} \qquad (3.132)$$

$$v_{\mathrm{cl}}^{\Sigma \star} = \sqrt{\frac{2N}{C} W_1} = \sqrt{\frac{N}{C}(W_{\Sigma 0} + \Delta W_{\Sigma} - \Delta W_{\Delta})} = \sqrt{v_{\mathrm{d}}^2 + \frac{N}{C}(\Delta W_{\Sigma} - \Delta W_{\Delta})}. \qquad (3.133)$$

Explicit formulas are often convenient for simulations, but they are not always practical for online implementation in a control system. Drawbacks are that imbalances among the three phases are not captured by Equation (3.129) and that several trigonometric and other nonlinear functions need to be computed online. An alternative is therefore presented next.

3.5.4.2 Ripple Estimation Using Band-Pass Filtering

In theory, the arm-energy references could be computed online as Equation (3.30) says, i.e. by direct integration of the right-hand sides of the two relations. In practice, though, this will invariably fail. An open-loop integration will gradually build up a bias, resulting in flawed operation. Using a low-pass filter instead of a pure integrator may seem as a feasible solution, but the bandwidth of the filter must be made very low to prevent unacceptably large phase errors from appearing. The dc-component suppression will then not be fast enough, resulting in deteriorating stability properties, owing to the tendency of bias buildup.

Band-pass filtering is a better approach [17]. A BPF for an order-h harmonic component is obtained as a special case of the resonant filter in Equation (3.65) with $K_h = \alpha_h$ and $\phi_h = 0$, giving

$$H_h(s) = \frac{\alpha_h s}{s^2 + \alpha_h s + (h\omega_1)^2}. \qquad (3.134)$$

As integration corresponds to division by s, a frequency component at $h\omega_1$ in W_Σ or in W_Δ can be extracted by filtering the signal corresponding to the right-hand sides of the relations in Equation (3.30) through

$$\frac{H_h(s)}{s} = \frac{\alpha_h}{s^2 + \alpha_h s + (h\omega_1)^2}. \tag{3.135}$$

Transfer function $H_h(s)/s$ can be obtained directly from Equation (3.65) by setting

$$K_h = \frac{\alpha_h}{h\omega_1} \qquad \phi_h = -\frac{\pi}{2} \tag{3.136}$$

which allows straightforward discrete-time realization, as shown in Section 3.4.2. The sum-capacitor-voltage references can now be obtained via the following steps.

1. The total and imbalance arm-energy ripples are estimated by filtering the right-hand sides of the relations in Equation (3.30). Since ΔW_Σ was found to have a frequency component at $2\omega_1$, whereas ΔW_Δ was found to have a frequency component at ω_1—compare Equations (3.35) and (3.36)—the following filters should be used:

$$H_1(s) = \frac{\alpha_f s}{s^2 + \alpha_f s + \omega_1^2} \qquad H_2(s) = \frac{\alpha_f s}{s^2 + \alpha_f s + (2\omega_1)^2} \tag{3.137}$$

where the same bandwidth α_f is used in both filters. This bandwidth should be selected significantly smaller than the fundamental angular frequency

$$\alpha_f \ll \omega_1 \tag{3.138}$$

e.g. $\alpha_f = 50$ rad/s. Substituting the references $i_s^\star$ and $i_c^\star$ for the measured currents in Equation (3.30), we obtain

$$\Delta W_\Sigma = \frac{H_2(s)}{s}(2v_c^\star i_c^\star - v_s^\star i_s^\star) \tag{3.139}$$

$$\Delta W_\Delta = \frac{H_1(s)}{s}(v_c^\star i_s^\star - 2v_s^\star i_c^\star). \tag{3.140}$$

This filtering rejects all frequency components—including any dc bias—except those of interest, which pass unaffected. Since the right-hand sides of Equations (3.139) and (3.140) are functions only of references and of v_d (which is constant in the steady state), time-delay compensation is not needed.

2. The mean reference $W_{\Sigma 0}$ as given by Equation (3.37) is added to ΔW_Σ. The arm-energy references are then computed using Equation (3.38).

3. Finally, the sum-capacitor-voltage references are obtained from the arm-energy references as

$$v_{cu,l}^{\Sigma\star} = \sqrt{\frac{2N}{C} W_{u,l}}. \tag{3.141}$$

A block diagram illustrating the method is shown in Figure 3.18.

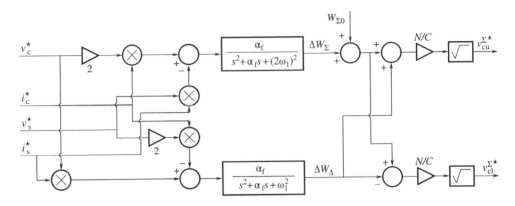

Figure 3.18   Ripple estimator in open-loop voltage control.

Example 3.9 *Evaluation is made for the Test Converter. At $t = 50$ ms, transition is made from direct voltage control to open-loop voltage control with ripple estimation using explicit formulas. Two different values of the circulating controller's P gain are tested. As can be seen in Figure 3.19, poor damping results if R_a is selected too small (in this example $R_a = 0$). There is also a remaining parasitic component in the circulating current. On the other hand, good performance for $R_a = 10\ \Omega$ is verified.*

3.5.5   Hybrid Voltage Control

In the fourth and final voltage control variant, we introduce a hybrid between open-loop and closed-loop control. Asymptotic stability is obtained without the usage of arm-energy controllers as well. Hybrid voltage control is therefore often preferable to closed-loop voltage control. The scheme is based on direct voltage control, i.e. the insertion-index selection in Equation (3.116), but sum-capacitor-voltage ripples are added. The latter are obtained using band-pass filtering of the measured sum capacitor voltages, as shown in Equation (3.119). Figure 3.20 shows the block diagram for the scheme. An R part is not required in the circulating-current controller.

3.5.5.1   Ripple Computation Using Band-Pass Filtering

The theory developed in Section 3.3.5—particularly Equations (3.41) and (3.42)—shows that the total sum capacitor voltage $v_c^\Sigma = v_{cu}^\Sigma + v_{cl}^\Sigma$ has a ripple component at $2\omega_1$, whereas the imbalance sum capacitor voltage $v_c^\Delta = v_{cu}^\Sigma - v_{cl}^\Sigma$ has a ripple component at ω_1. The ripples consequently can be computed online in a fashion similar to that for open-loop voltage control in Section 3.5.4.

1. The total and imbalance sum-capacitor-voltage ripples are computed by the following BPFs:

$$\Delta v_c^\Sigma = H_2(s)(v_{cu}^\Sigma + v_{cl}^\Sigma) \qquad \Delta v_c^\Delta = H_1(s)(v_{cu}^\Sigma - v_{cl}^\Sigma) \tag{3.142}$$

where $H_1(s)$ and $H_2(s)$ are given by Equation (3.137).

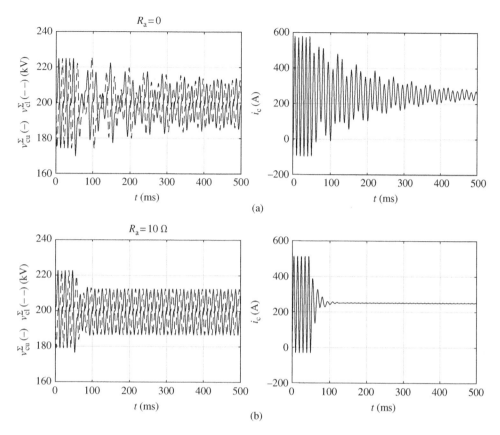

Figure 3.19 Sum capacitor voltages and circulating current for transition from direct voltage control to open-loop voltage control at $t = 50$ ms.

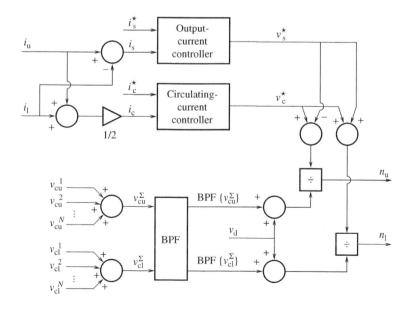

Figure 3.20 Hybrid voltage control.

2. The ripples $\text{BPF}\{v_{\text{cu,l}}^{\Sigma}\}$ in Equation (3.119) are then simply obtained as

$$\text{BPF}\{v_{\text{cu}}^{\Sigma}\} = \frac{\Delta v_{\text{c}}^{\Sigma} + \Delta v_{\text{c}}^{\Delta}}{2} \qquad \text{BPF}\{v_{\text{cl}}^{\Sigma}\} = \frac{\Delta v_{\text{c}}^{\Sigma} - \Delta v_{\text{c}}^{\Delta}}{2}. \tag{3.143}$$

A non-negligible total time delay, T_{d}, can be compensated by applying nonzero compensation angles in the BPFs as

$$H_h(s) = \frac{\alpha_{\text{f}}(s \cos \phi_h - h\omega_1 \sin \phi_h)}{s^2 + \alpha_{\text{f}}s + (h\omega_1)^2} \qquad \phi_h = h\omega_1 T_{\text{d}}. \tag{3.144}$$

As for open-loop voltage control, α_{f} should be selected according to Equation (3.138), e.g. $\alpha_{\text{f}} = 50$ rad/s. The so obtained BPF scheme is shown in Figure 3.21.

Example 3.10 *Evaluation is made for the Test Converter. At $t = 50$ ms, transition is made from direct voltage control to hybrid voltage control with $\alpha_{\text{f}} = 50$ rad/s. In the circulating-current controller, $R_{\text{a}} = 10\ \Omega$. Figure 3.22 shows satisfactory results. However, instability may result (not shown in the figure) if closed-loop circulating-current control is not used, i.e. $R_{\text{a}} = 0$, and/or the selection recommendation Equation (3.138) for α_{f} is not followed.*

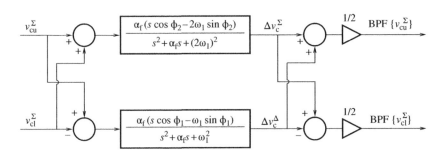

Figure 3.21   Band-pass filtering in hybrid voltage control.

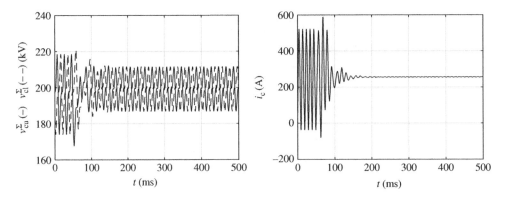

Figure 3.22   Sum capacitor voltages and circulating current for transition from direct voltage control to hybrid voltage control at $t = 50$ ms.

3.6   Three-Phase Systems

So far, modeling, analysis, and controller design has been made on a per-phase basis. Henceforth our focus is on three-phase systems and converters. This section gives a review of the theory for three-phase systems that serves as an introduction to Section 3.7, where vector output-current control is considered.

By a three-phase system it may be meant a three-phase ac network. Another meaning is a set of three oscillating quantities—the phase quantities—which all have the same angular frequency. The phases are here denoted a, b, and c. The quantities of phases b and c ideally lag $120°$ ($2\pi/3$ rad) and $240°$ ($4\pi/3$ rad) respectively behind that of phase a. A fundamental-frequency three-phase system can be expressed as

$$v_a = \hat{V}_a \cos(\omega_1 t + \varphi_a)$$
$$v_b = \hat{V}_b \cos(\omega_1 t - 2\pi/3 + \varphi_b) \qquad (3.145)$$
$$v_c = \hat{V}_c \cos(\omega_1 t - 4\pi/3 + \varphi_c).$$

If the quantities are voltages, then $v_{a,b,c}$ are called the *phase voltages*. $\hat{V}_{a,b,c}$ are the peak values and $\varphi_{a,b,c}$ are the phase angles.

3.6.1   Balanced Three-Phase Systems

In a balanced three-phase system the three peak values are equal and the phase shifts between the phase quantities are exactly $120°$. A balanced three-phase system has two important properties.

1. The sum of the instantaneous values is zero: $v_a + v_b + v_c = 0, \forall t$.
2. The sum of the squared instantaneous values is constant: $v_a^2 + v_b^2 + v_c^2 = \text{const.}, \forall t$.

The first property implies that a neutral current return path is not needed, only three phase connections. The second property implies that the three-phase power is constant, which, for example, allows three-phase motors to produce torque with a minimum of pulsations.

3.6.2   Imbalanced Three-Phase Systems

An imbalanced three-phase system as given by Equation (3.145) can be decomposed into three components, namely

- a positive-sequence component
- a negative-sequence component
- a zero-sequence component.

The positive-sequence component is a balanced three-phase system where the phase-b quantity lags $120°$ behind the phase-a quantity:

$$v_a^+ = \hat{V}_{+1} \cos(\omega_1 t + \varphi_{+1})$$
$$v_b^+ = \hat{V}_{+1} \cos(\omega_1 t - 2\pi/3 + \varphi_{+1}) \qquad (3.146)$$
$$v_c^+ = \hat{V}_{+1} \cos(\omega_1 t - 4\pi/3 + \varphi_{+1}).$$

The negative-sequence component is a balanced three-phase system where the phase-b quantity lags 240° behind the phase-a quantity:

$$v_a^- = \hat{V}_{-1}\cos(\omega_1 t - \varphi_{-1})$$
$$v_b^- = \hat{V}_{-1}\cos(\omega_1 t - 4\pi/3 - \varphi_{-1}) \tag{3.147}$$
$$v_c^- = \hat{V}_{-1}\cos(\omega_1 t - 2\pi/3 - \varphi_{-1}).$$

In the zero-sequence component the phase angles are equal for all phase quantities:

$$v_a^0 = \hat{V}_0\cos(\omega_1 t + \varphi_0)$$
$$v_b^0 = \hat{V}_0\cos(\omega_1 t + \varphi_0) \tag{3.148}$$
$$v_c^0 = \hat{V}_0\cos(\omega_1 t + \varphi_0).$$

A three-phase system shall ideally consist only of a fundamental-frequency positive-sequence component. In ac transmission systems, small negative-sequence voltage components arise because of an imbalance between the phase impedances. Larger negative-sequence components often appear during faults, which typically clear within hundreds of milliseconds [12]. See Chapter 4. Zero-sequence voltage components are prevented by star-point grounding of transformers, which is commonly made in high-voltage transmission systems. Zero-sequence current components can, on the other hand, be prevented by keeping the star point ungrounded, which is often made on the converter side of the transformer; see Figure 3.2. For these reasons, the control of zero-sequence components is not considered here.

In an imbalanced three-phase system that is devoid of a zero-sequence component, the sum of the instantaneous values is still zero: $v_a + v_b + v_c = 0, \forall t$. However, the sum of the squared instantaneous values is no longer constant.

3.6.3   Instantaneous Active Power

Constant sum of the squared instantaneous values as mentioned implies that the instantaneous power $P = v_a i_a + v_b i_b + v_c i_c$ is constant. To show this, let us consider a balanced load consisting of three identical impedances, Z, connected to a positive-sequence voltage, as given by Equation (3.146); see Figure 3.23(a). This gives a balanced set of three-phase currents:

$$i_a = \hat{I}_{+1}\cos(\omega_1 t - \varphi_{+1})$$
$$i_b = \hat{I}_{+1}\cos(\omega_1 t - 2\pi/3 - \varphi_{+1}) \tag{3.149}$$
$$i_c = \hat{I}_{+1}\cos(\omega_1 t - 4\pi/3 - \varphi_{+1}),$$

where $\hat{I}_{+1} = \hat{V}_{+1}/|Z(j\omega_1)|$ and $\varphi_{+1} = \arg Z(j\omega_1)$. Through straightforward but tedious trigonometric manipulations it is found that the total instantaneous active power, P, developed in the three impedances is given by

$$P = v_a i_a + v_b i_b + v_c i_c = \frac{3\hat{V}_{+1}\hat{I}_{+1}}{2}\cos\varphi = 3V_{+1}I_{+1}\cos\varphi, \tag{3.150}$$

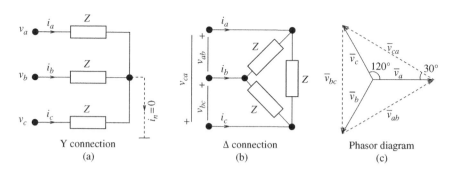

Figure 3.23   Balanced three-phase impedances.

where rms values are introduced as $V_{+1} = \hat{V}_{+1}/\sqrt{2}$ and $I_{+1} = \hat{I}_{+1}/\sqrt{2}$. Equation (3.150) is similar to that obtained for a single-phase system, except for the factor 3 and that in the single-phase case the power is not constant; the expression in that case only accounts for the mean value to which pulsations of twice the fundamental frequency add.

3.6.4   Wye (Y) and Delta (Δ) Connections

A three-phase load consisting of three impedances can be connected either in a Y (also called a *star*) or in a Δ, as illustrated in Figures 3.23(a) and (b), respectively. Whereas in the Y connection there is a phase voltage across each impedance, the Δ connection has a line-to-line voltage. From the geometry in Figure 3.23(c) it is found that the line-to-line voltages are given by

$$v_{ab} = v_a - v_b = \sqrt{3}\hat{V}_{+1}\cos(\omega_1 t + \pi/6)$$

$$v_{bc} = v_b - v_c = \sqrt{3}\hat{V}_{+1}\cos(\omega_1 t - \pi/2) \qquad (3.151)$$

$$v_{ca} = v_c - v_a = \sqrt{3}\hat{V}_{+1}\cos(\omega_1 t - 7\pi/6)$$

still with v_a as phase reference. The peak value of each line-to-line voltage is $\sqrt{3}$ times larger than the peak value of each phase voltage. Thus, the power developed in a three-phase load is increased three times when the connection is changed from a Y to a Δ: $P_\Delta = 3P_Y$.

It is unimportant from the standpoint of dynamics and control whether a load is connected in a Y or in a Δ. A balanced Δ-connected load can be treated as if it were connected in a Y, but with all impedances reduced to 1/3 of the actual values. This is called an *equivalent Y*.

3.6.5   Harmonics

As fundamental-frequency negative- and zero-sequence components, harmonics are undesirable. Major sources of harmonic pollution in an ac grid are diode rectifiers and thyristor converters. These act as balanced but nonlinear loads that draw balanced but nonsinusoidal currents. That is, the waveforms in the three phases are identical, only time shifted corresponding to 120° of the fundamental. The dominant harmonics thus form a balanced, multifrequency three-phase system, which can be expressed as the Fourier series expansion

$$
\begin{bmatrix} v_a \\ v_b \\ v_c \end{bmatrix} = \sum_{h=1,3,5,\ldots} \begin{bmatrix} \hat{V}_h \cos[h\omega_1 t + \varphi_h] \\ \hat{V}_h \cos[h(\omega_1 t - 2\pi/3) + \varphi_h] \\ \hat{V}_h \cos[h(\omega_1 t - 4\pi/3) + \varphi_h] \end{bmatrix}. \tag{3.152}
$$

The 3rd, 5th, and 7th harmonics (all with normalized amplitudes $\hat{V}_h = 1$ and with $\varphi_h = 0$) are shown in Figure 3.24. The 3rd harmonic has the same phase angle in all three phases, so it is a zero-sequence component. It is further seen that the 5th harmonic is a negative-sequence component and that the 7th harmonic is a positive-sequence component. The obvious extension of this is

$$
\begin{array}{lcccccccccc}
h = & 1 & 3 & 5 & 7 & 9 & 11 & 13 & 15 & \cdots \\
\text{seq.} & + & 0 & - & + & 0 & - & + & 0 & \cdots
\end{array}
$$

All harmonics that are multiples of 3 (so-called triplen harmonics) are zero-sequence components. Since zero-sequence components normally can be disregarded, the conclusion is that only non-triplen harmonics need to be considered, i.e. the orders

$$
-5 \quad +7 \quad -11 \quad +13 \quad -17 \quad +19 \quad -23 \quad +25 \quad \cdots \tag{3.153}
$$

If the fundamental positive- and negative-sequence components are included and the numbers are listed in order with sign, then the orders of the frequency components that normally appear in a three-phase system are

$$
\cdots \quad -17 \quad -11 \quad -5 \quad -1 \quad +1 \quad +7 \quad +13 \quad +19 \quad \cdots \tag{3.154}
$$

3.6.6 Space Vectors

Since zero-sequence components normally can be disregarded, it is possible to reduce a three-phase system to an equivalent two-phase system. The phase quantities thereof, denoted with the subscripts α and β, respectively, are 90° phase shifted. It is convenient to consider the equivalent two-phase system as projected on a complex plane—the $\alpha\beta$ plane—whose real and imaginary axes respectively represent the α and β quantities. With the complex two-phase representation $\mathbf{v}^s = v_\alpha + jv_\beta$, the three-phase/two-phase transformation—which is also known as the *Clarke transformation* [19]—is given by

$$
\mathbf{v}^s = v_\alpha + jv_\beta = \frac{2}{3}K(v_a + e^{j2\pi/3}v_b + e^{j4\pi/3}v_c) \tag{3.155}
$$

where K is a scaling constant to be selected. The complex quantity $\mathbf{v}^s$ is called a *space vector*. Figure 3.25 illustrates its construction. The contributions of the phase quantities are equipped with their respective directions (i.e. the directions defined by the unit vectors 1, $e^{j2\pi/3}$, and $e^{j4\pi/3}$), then added together, and, finally, scaled with $2K/3$. Space vector $\mathbf{v}^s$ is said to be expressed in stationary coordinates, or in the $\alpha\beta$ frame, which is indicated by the superscript ^{s}.

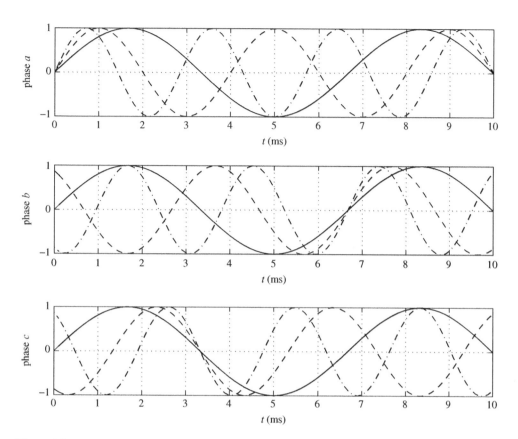

Figure 3.24 Three-phase harmonics: (solid) 3rd, (dashed) 5th, and (dash-dotted) 7th for a 50 Hz fundamental frequency.

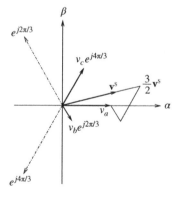

Figure 3.25 Construction of a space vector.

For the positive-sequence component of Equation (3.146), the corresponding space vector is given by

$$\mathbf{v}^s = K\hat{V}_{+1}e^{j(\omega_1 t + \varphi_{+1})}.$$

(3.156)

This vector rotates counterclockwise and is aligned with phase a. Transforming the negative-sequence component of Equation (3.147), we obtain

$$\mathbf{v}^s = K\hat{V}_{-1}e^{-j(\omega_1 t + \varphi_{-1})}$$

(3.157)

which rotates clockwise.

The seemingly odd name "space vector" originates from their first usage, namely the description of spatial flux distributions in ac machines (hence 'space'). "Vector" is used to distinguish it from the related complex phasor. Although space vectors were invented for the purpose of analysis (and subsequently control) of ac machines [20], they are equally useful for analysis and control of grid-connected converters. Space vectors and complex phasors are obviously related. The fundamental difference is that a space vector is not only an analysis tool, but that it (or rather, its components) actually can exist as signals in the control system of a three-phase converter. In addition, whereas a phasor only can describe steady-state conditions, space vectors are useful also for descriptions of transient phenomena.

3.6.6.1 Transformation Matrices

The three-phase/two-phase transformation given by Equation (3.155) can equivalently be expressed in matrix form as

$$\begin{bmatrix} v_\alpha \\ v_\beta \end{bmatrix} = K \underbrace{\begin{bmatrix} \frac{2}{3} & -\frac{1}{3} & -\frac{1}{3} \\ 0 & \frac{1}{\sqrt{3}} & -\frac{1}{\sqrt{3}} \end{bmatrix}}_{T_{32}} \begin{bmatrix} v_a \\ v_b \\ v_c \end{bmatrix}.$$

(3.158)

Transforming instead from a two-phase representation to a three-phase representation, the natural choice is

$$\begin{bmatrix} v_a \\ v_b \\ v_c \end{bmatrix} = \frac{1}{K} \underbrace{\begin{bmatrix} 1 & 0 \\ -\frac{1}{2} & \frac{\sqrt{3}}{2} \\ -\frac{1}{2} & -\frac{\sqrt{3}}{2} \end{bmatrix}}_{T_{23}} \begin{bmatrix} v_\alpha \\ v_\beta \end{bmatrix}$$

(3.159)

since then $T_{32}T_{23} = I$, where I is the 2×2 identity matrix.

3.6.6.2 Multiple Frequencies

Space vectors allow straightforward representation of multiple frequencies, i.e. the fundamental positive- and negative-sequence components plus balanced harmonics, as given by the orders listed in Equation (3.154). In increasing absolute order, we have

$$\mathbf{v}^s = K[\hat{V}_{+1}e^{j(\omega_1 t+\varphi_{+1})} + \hat{V}_{-1}e^{-j(\omega_1 t+\varphi_{-1})} + \hat{V}_{-5}e^{-j(5\omega_1 t+\varphi_{-5})} + \hat{V}_{+7}e^{j(7\omega_1 t+\varphi_{+7})} + \cdots].$$

(3.160)

3.6.6.3   Coordinate Transformations

Consider a positive-sequence space vector

$$\mathbf{v}^s = K\hat{V}_{+1}e^{j(\omega_1 t+\varphi_{+1})}.$$

(3.161)

The so-called *dq* or Park transformation [21]

$$\mathbf{v} = e^{-j\omega_1 t}\mathbf{v}^s = K\hat{V}_{+1}e^{j\varphi_{+1}}$$

(3.162)

removes the rotation of the vector, making it similar to a fixed complex phasor. The *dq* transformation can be regarded as observing the space vector from a coordinate system that rotates with the fundamental frequency, called *synchronous coordinates* or the *dq* frame. We denote a space vector in synchronous coordinates without a superscript, and its components with the subscripts d and q, as

$$\mathbf{v} = v_d + jv_q.$$

(3.163)

Since the fundamental positive-sequence component transforms to a complex dc quantity in the steady state, the *dq* frame is useful not only for analysis but also for implementation of control algorithms, as is shown in Section 3.7.

The inverse of Equation (3.162) is called the *αβ transformation* and is given as

$$\mathbf{v}^s = e^{j\omega_1 t}\mathbf{v}.$$

(3.164)

3.6.6.4   *dq* Transformation of Multiple Frequencies

The *dq* transformation shifts the fundamental component to dc, i.e. a frequency translation by $-\omega_1$. Applying this to multiple frequencies obviously yields the rule $h \to h-1$ for the order, which can be illustrated for Equation (3.154) as

$$\begin{array}{ccccccccc}
\alpha\beta & \cdots & -11 & -5 & -1 & +1 & +7 & +13 & \cdots \\
\downarrow & & \downarrow & \downarrow & \downarrow & \downarrow & \downarrow & \downarrow & \\
dq & \cdots & -12 & -6 & -2 & 0 & +6 & +12 & \cdots
\end{array}$$

(3.165)

or as applied to Equation (3.160):

$$\mathbf{v} = K[\hat{V}_{+1}e^{j\varphi_{+1}} + \hat{V}_{-1}e^{-j(2\omega_1 t+\varphi_{-1})} + \hat{V}_{-5}e^{-j(6\omega_1 t+\varphi_{-5})} + \hat{V}_{+7}e^{j(6\omega_1 t+\varphi_{+7})} + \cdots].$$

(3.166)

Interesting to note is that the 5th and 7th harmonics are both transformed to 6th harmonics in synchronous coordinates. Similarly, the 11th and 13th harmonics are both transformed to 12th harmonics. We find that, in the *dq* frame, balanced harmonics appear at the orders $h = \pm 6n$, $n = 1, 2, \ldots$ The fundamental negative-sequence component is transformed to $h = -2$.

3.6.6.5   *dq* Transformation of a Time Derivative

Let us consider the time derivative of an $\alpha\beta$-transformed space vector, e.g. a current $\mathbf{i}^s = e^{j\omega_1 t}\mathbf{i}$. According to the chain rule,

$$\frac{d\mathbf{i}^s}{dt} = \frac{d(e^{j\omega_1 t}\mathbf{i})}{dt} = e^{j\omega_1 t}\left(\frac{d\mathbf{i}}{dt} + j\omega_1\mathbf{i}\right). \tag{3.167}$$

It is found that a differentiation in the $\alpha\beta$ frame is transformed to a differentiation plus a multiplication by $j\omega_1$ in the dq frame. Thus, in Laplace (differential-operator) form dq transformation implies

$$s \rightarrow s + j\omega_1. \tag{3.168}$$

For example, the impedance of an inductor is transformed from sL in the $\alpha\beta$ frame to $(s + j\omega_1)L$ in the dq frame.

3.6.6.6   *dq* Transformation of a Time Delay

Consider an $\alpha\beta$-frame space vector $\mathbf{y}^s$, which equals a space vector $\mathbf{u}^s$ delayed the time T_d (e.g. the total time delay). That is, with explicit notation of time arguments

$$\mathbf{y}^s(t) = \mathbf{u}^s(t - T_d). \tag{3.169}$$

Introduction of dq-frame vectors in Equation (3.169) yields

$$e^{j\omega_1 t}\mathbf{y}(t) = e^{j\omega_1(t-T_d)}\mathbf{u}(t - T_d) \Rightarrow \mathbf{y}(t) = e^{-j\omega_1 T_d}\mathbf{u}(t - T_d). \tag{3.170}$$

We find that a time delay in the $\alpha\beta$ frame is transformed to a time delay together with a rotation factor $e^{-j\omega_1 T_d}$ in the dq frame. The latter causes an angular displacement $-\omega_1 T_d$ of the vector. This is normally a small angle, but yet an appropriate correction can be added to the control system, as is shown in Section 3.7.3. [It may be noted that the property given by Equation (3.170) also can be deduced by applying Equation (3.168) to the delay transfer function e^{-sT_d}.]

3.6.7   Instantaneous Power

Given the formula $P + jQ = \vec{v}\vec{i}^*$ for the single-phase, rms-value-scaled phasors $\vec{v}$ and $\vec{i}$, it may be conjectured that the three-phase complex power is given as

$$P + jQ \sim \mathbf{v}^s(\mathbf{i}^s)^* = e^{j\omega_1 t}\mathbf{v}(e^{j\omega_1 t}\mathbf{i})^* = \mathbf{v}\mathbf{i}^* \tag{3.171}$$

where $\sim$ indicates proportionality and where it should be noted that the expression is invariant of the choice of reference frame. Let us show that Equation (3.171) holds and determine the proportionality constant. From the definition of space vectors

$$\mathbf{v}^s(\mathbf{i}^s)^* = \left(\frac{2K}{3}\right)^2 (v_a + e^{j2\pi/3}v_b + e^{j4\pi/3}v_c)(i_a + e^{j2\pi/3}i_b + e^{j4\pi/3}i_c)^*. \tag{3.172}$$

Observing that $e^{j4\pi/3} = e^{-j2\pi/3}$, and assuming that there is no zero-sequence current component, i.e. $i_a + i_b + i_c = 0$, we obtain

$$\mathbf{v}^s(\mathbf{i}^s)^* = \left(\frac{2K}{3}\right)^2 \left[v_a i_a + v_b i_b + v_c i_c + \underbrace{\left(-\frac{1}{2} + j\frac{\sqrt{3}}{2}\right)}_{e^{j2\pi/3}}(v_a i_c + v_b i_a + v_c i_b) \right.$$

$$\left. + \underbrace{\left(-\frac{1}{2} - j\frac{\sqrt{3}}{2}\right)}_{e^{-j2\pi/3}}(v_a i_b + v_b i_c + v_c i_a) \right]$$

$$= \left(\frac{2K}{3}\right)^2 \left[v_a i_a + v_b i_b + v_c i_c - \frac{1}{2}\left(v_a \underbrace{(i_b + i_c)}_{-i_a} + v_b \underbrace{(i_a + i_c)}_{-i_b} + v_c \underbrace{(i_a + i_b)}_{-i_c} \right) \right.$$

$$\left. + j\frac{\sqrt{3}}{2}[v_a(i_c - i_b) + v_b(i_a - i_c) + v_c(i_b - i_a)] \right]$$

$$= \frac{2K^2}{3}\left[v_a i_a + v_b i_b + v_c i_c + j\frac{1}{\sqrt{3}}[v_a(i_c - i_b) + v_b(i_a - i_c) + v_c(i_b - i_a)] \right].$$

So, the instantaneous active power is given by

$$P = \frac{3}{2K^2}\text{Re}\{\mathbf{v}^s(\mathbf{i}^s)^*\} = \frac{3}{2K^2}\text{Re}\{\mathbf{vi}^*\} = v_a i_a + v_b i_b + v_c i_c \tag{3.173}$$

and it follows that the instantaneous reactive power is given as

$$Q = \frac{3}{2K^2}\text{Im}\{\mathbf{v}^s(\mathbf{i}^s)^*\} = \frac{3}{2K^2}\text{Im}\{\mathbf{vi}^*\}$$

$$= \frac{1}{\sqrt{3}}[v_a(i_c - i_b) + v_b(i_a - i_c) + v_c(i_b - i_a)]. \tag{3.174}$$

For two positive-sequence space vectors where the voltage is taken as phase reference

$$\mathbf{v}^s = K\sqrt{2}V_{+1}e^{j\omega_1 t} \qquad \mathbf{i}^s = K\sqrt{2}I_{+1}e^{j(\omega_1 t - \varphi)} \tag{3.175}$$

the instantaneous active and reactive powers are constant and given by

$$P = \frac{3}{2K^2}\text{Re}\{\mathbf{v}^s(\mathbf{i}^s)^*\} = 3V_{+1}I_{+1}\cos\varphi \qquad Q = \frac{3}{2K^2}\text{Im}\{\mathbf{v}^s(\mathbf{i}^s)^*\} = 3V_{+1}I_{+1}\sin\varphi \tag{3.176}$$

where the first relation agrees with Equation (3.150).

3.6.7.1 Instantaneous Power in a Voltage-Aligned dq Frame

An important special case of the dq-frame relation for the complex power

$$P + jQ = \frac{3}{2K^2}\mathbf{vi}^* \tag{3.177}$$

arises if the dq frame is aligned with the voltage vector, such that the latter is real valued

$$\mathbf{v} = v. \tag{3.178}$$

Then

$$P + jQ = \frac{3}{2K^2}v\mathbf{i}^* = \frac{3}{2K^2}v(i_d - ji_q) \tag{3.179}$$

which shows that the active power is proportional to i_d, whereas the reactive power is proportional to $-i_q$. The current components i_d and i_q respectively become active-power-producing and reactive-power-consuming. It may be noted that this is opposite to a field-oriented ac motor drive, where i_q and i_d respectively are the torque- and flux-producing current components [20]. This difference arises because for an ac motor drive the dq frame is oriented along a flux linkage, which is the integral of the corresponding voltage. A 90° phase shift is thereby introduced.

3.6.8 Selection of the Space-Vector Scaling Constant

The scaling constant K can be chosen arbitrarily. Depending on the application, one choice may be more convenient than another. There are three standard selections. Selecting $K = 1$ we obtain for a positive-sequence component

$$\mathbf{v}^s = \hat{V}_{+1}e^{j(\omega_1 t+\varphi)} \qquad \mathbf{v} = \hat{V}_{+1}e^{j\varphi} \tag{3.180}$$

so $K = 1$ gives peak-value scaling. $K = 1/\sqrt{2}$ yields

$$\mathbf{v}^s = \frac{\hat{V}_{+1}}{\sqrt{2}}e^{j(\omega_1 t+\varphi)} \qquad \mathbf{v} = \frac{\hat{V}_{+1}}{\sqrt{2}}e^{j\varphi}. \tag{3.181}$$

That is, rms-value scaling. The expression in Equation (3.173) for the instantaneous active power then reads $P = 3\text{Re}\{\mathbf{v}^s(\mathbf{i}^s)^*\} = 3\text{Re}\{\mathbf{vi}^*\}$. To remove the factor 3, $K = \sqrt{3/2}$ should instead be chosen, giving

$$P = \text{Re}\{\mathbf{v}^s(\mathbf{i}^s)^*\} = \text{Re}\{\mathbf{vi}^*\}. \tag{3.182}$$

$K = \sqrt{3/2}$ is thus the choice for power invariance. To reiterate:

$$\begin{aligned} &\text{Peak-value scaling:} \quad &K = 1. \\ &\text{RMS-value scaling:} \quad &K = 1/\sqrt{2}. \\ &\text{Power-invariant scaling:} \quad &K = \sqrt{3/2}. \end{aligned}$$

3.7 Vector Output-Current Control

For a three-phase converter there is much to be gained from using vector output-current control instead of per-phase output-current control. One reason is lower computational complexity, since the controller needs to handle only two parallel signal paths instead of three. This is a minor simplification, though. The major benefits are:

- In a dq frame aligned with the ac-bus voltage, the d and q output-current components respectively are proportional to the active-power generation and the reactive-power consumption, as given by Equation (3.179). That is, power control has a direct correspondence to control of the current components.
- Saturation (i.e. limitation to a maximum modulus) of a space vector is easier to make without waveform distortion than saturation of a phase quantity.
- Separate control of the fundamental positive- and negative-sequence components is facilitated; see Chapter 4.

Vector current control can be performed either in the $\alpha\beta$ frame or in the dq frame. It is sometimes argued that the former is preferable, because coordinate transformations are avoided. But that it not altogether true. For dq-frame control, two coordinate transformations are needed: for the controller input (i.e. current) and the output (i.e. voltage reference). However, for $\alpha\beta$-frame control the current reference must yet be transformed from the dq frame to the $\alpha\beta$ frame in order to maintain the aforementioned correspondence between the active and reactive powers and the current components. The computational savings are therefore marginal. Another benefit of dq-frame control appears in the case it is desired to control not only the fundamental component but also balanced harmonics (this is known as *active filtering*) [9]. As concluded in Section 3.4.1, R parts (in the form of SOGIs) that are tuned to each harmonic frequency are needed in order to nullify the static control error. In $\alpha\beta$-frame control this would require as many R parts as the number of controlled harmonics. But balanced harmonics appear at the orders $\pm6n$; see Equation (3.166). For dq-frame control this cuts in half the number of required R parts, since each R part, at $h = 6n$, controls a harmonic pair.

The per-phase output-current controller designed in Section 3.4.5 is here modified to a dq-frame vector current controller. One important difference is obviously that, whereas each signal path in the per-phase controller is scalar, each signal path in the vector controller is vectorial, i.e. it has two components: d and q. Vector current control is thus a multivariable control system with two inputs and two outputs. However, since complex vectors are used, the notation is similar to the scalar case, except that boldface variable names are employed to denote dq-frame vectors. The per-phase block diagram in Figure 3.8 is modified to the block diagram shown in Figure 3.26, whose details and differences to the per-phase case are now presented.

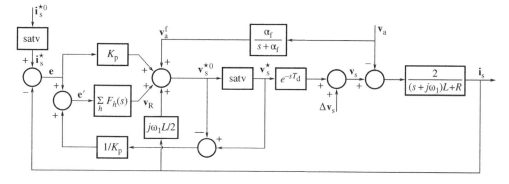

Figure 3.26   Vector output-current control loop in the dq frame.

3.7.1 PR (PI) Controller

It is normally desired that the output current should contain only a fundamental positive-sequence component. Thus, i_s^* is normally a constant space vector in the steady state. Its components are often determined by higher-level controllers, as is shown in Section 3.8. In order to track i_s^* and to suppress all other frequency components, given by the orders listed in Equation (3.165), the PR controller should have R parts for $h = 0$, $h = 2$, and $h = 6, 12, \ldots, N_H$, where N_H is the highest harmonic order (in the dq frame) to be suppressed. The R parts are placed in parallel, which in Figure 3.26 is denoted by $\sum_h F_h(s)$. We have

$$\sum_h F_h(s) = \sum_{h=0,2,6,12,\ldots}^{N_H} \frac{K_h(s\cos\phi_h - h\omega_1\sin\phi_h)}{s^2 + (h\omega_1)^2}. \tag{3.183}$$

Note that, since $F_h(\pm jh\omega_1) = \infty$, each R part for $h = 6n$, $n = 1, 2, \ldots$ controls a harmonic pair at $h = \pm 6m$. The gain selection of Equation (3.108) is still applicable

$$K_p = \frac{\alpha_c L}{2} \qquad K_h = 2\alpha_h K_p = \alpha_h\alpha_c L \tag{3.184}$$

with $\alpha_h \ll \alpha_c$, and $\phi_h = h\omega_1 T_d$ can be selected for a direct compensation of the total time delay [9].

If control neither of the fundamental negative-sequence component nor of the harmonics is required, then it is sufficient to include just the R part for $h = 0$, which reduces to an integrator

$$\sum_h F_h(s) = \frac{K_0}{s}. \tag{3.185}$$

The PR controller then reduces to a pure PI controller. For discrete-time implementation the I part can be realized using a simple Euler update

$$v_R(n + 1) = v_R(n) + T_s K_0 e'(n). \tag{3.186}$$

3.7.1.1 AC-Bus-Voltage Feedforward

Feedforward of the ac-bus voltage, introduced in Section 3.4.5 for per-phase output-current control, is useful also for vector current control. It is sufficient to feed forward only the fundamental positive-sequence component. A first-order low-pass filter is suitable [8]

$$v_a^f = H_0(s)v_a, \qquad H_0(s) = \frac{\alpha_f}{s + \alpha_f}. \tag{3.187}$$

Similarly to the I part—see Equation (3.186)—the low-pass filter can be Euler discretized as

$$v_a^f(n + 1) = v_a^f(n) + T_s\alpha_f[v_a(n) - v_a^f(n)]. \tag{3.188}$$

The selection of α_f is a double-edged sword. A large value—in the range of α_c—is good for the rejection of an ac-bus-voltage disturbance, as demonstrated in Example 3.5 for the per-phase case. On the other hand, a large α_f may result in detrimental feedforward of harmonics and possibly destabilization of grid resonances. See [8, 22, 23] for more information.

3.7.1.2 *dq* Decoupling

A time derivative transforms as $s \rightarrow s + j\omega_1$; compare Equation (3.168). Thus, the *dq*-frame correspondence to Equation (3.104) is

$$\frac{L}{2}\frac{d\mathbf{i}_s}{dt} = \mathbf{v}_s - \mathbf{v}_a - \frac{j\omega_1 L + R}{2}\mathbf{i}_s \Rightarrow \mathbf{i}_s = \frac{2}{(s + j\omega_1)L + R}(\mathbf{v}_s - \mathbf{v}_a). \tag{3.189}$$

The added term $\frac{j\omega_1 L}{2}\mathbf{i}$ introduces a cross coupling between the *d* and *q* axes; in component form, with $\mathbf{i}_s = i_{sd} + ji_{sq}$,

$$\frac{j\omega_1 L}{2}\mathbf{i}_s = \frac{j\omega_1 L}{2}(i_{sd} + ji_{sq}) = \frac{\omega_1 L}{2}(-i_{sq} + ji_{sd}). \tag{3.190}$$

The cross coupling can be decoupled by an inner positive feedback loop with gain $j\omega_1 L/2$, as shown in Figure 3.8. The decoupling feedback adds a term to the ideal voltage reference $\mathbf{v}^{\star 0}$, giving the control law

$$\mathbf{v}_s^{\star 0} = K_p\mathbf{e} + \mathbf{v}_R + \mathbf{v}_a^f + \frac{j\omega_1 L}{2}\mathbf{i}_s. \tag{3.191}$$

Let us assume that $\mathbf{v}_s^\star = \mathbf{v}_s^{\star 0}$, i.e. the vectorial saturation block "satv," is not active; see Section 3.7.2. Then, from Figure 3.26 it can found that

$$\mathbf{v}_s = e^{-sT_d}\left(K_p\mathbf{e} + \mathbf{v}_R + \mathbf{v}_a^f + \frac{j\omega_1 L}{2}\mathbf{i}_s\right) + \Delta\mathbf{v}_s. \tag{3.192}$$

Substitution of Equation (3.192) in Equation (3.189) yields

$$\frac{L}{2}\frac{d\mathbf{i}_s}{dt} = e^{-sT_d}\left(K_p\mathbf{e} + \mathbf{v}_R + \mathbf{v}_a^f + \frac{j\omega_1 L}{2}\mathbf{i}_s\right) + \Delta\mathbf{v}_s - \mathbf{v}_a - \frac{j\omega_1 L + R}{2}\mathbf{i}_s \tag{3.193}$$

which, with $\mathbf{v}_a^f$ given by Equation (3.187), can be simplified to

$$\frac{L}{2}\frac{d\mathbf{i}_s}{dt} = e^{-sT_d}(K_p\mathbf{e} + \mathbf{v}_R) + \Delta\mathbf{v}_s - \frac{s + (1 - e^{-sT_d})\alpha_f}{s + \alpha_f}\mathbf{v}_a - \frac{R}{2}\mathbf{i}_s - (1 - e^{-sT_d})\frac{j\omega_1 L}{2}\mathbf{i}_s. \tag{3.194}$$

The right-hand side of Equation (3.194) shows that the cross coupling is cancelled in the steady state, i.e. for $s = 0$, as then the imaginary (last) term vanishes. It can also be observed that the impact of $\mathbf{v}_a$ is cancelled in the steady state because of ac-bus-voltage feedforward. Decoupling and feedforward both help to improve the transient performance as well.

3.7.1.3 Anti-Windup

The suggested anti-windup scheme, of the back-calculation type, is identical to that introduced in Section 3.4 for the per-phase case. It subtracts the ideal voltage reference $\mathbf{v}^{\star 0}$ from the saturated voltage reference $\mathbf{v}^\star$, and feeds the difference, multiplied with $1/K_p$, back to the input of the R parts.

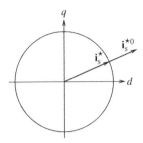

Figure 3.27   Vectorial saturation.

3.7.2   Reference-Vector Saturation

Since power transistors in general do not have an overcurrent capacity, it is important to limit $\mathbf{i}_s^\star$ to the maximum allowed modulus $K\hat{I}_{\max}$. Let us denote the ideal reference, whose components are often set by higher-level controllers (see Section 3.8), as $\mathbf{i}_s^{\star 0}$. The following vectorial saturation method preserves the angle of $\mathbf{i}_s^{\star 0}$ but limits its modulus to $K\hat{I}_{\max}$, as illustrated in Figure 3.27:

$$\mathbf{i}_s^\star = \mathrm{satv}(\mathbf{i}_s^{\star 0}) = \frac{\mathbf{i}_s^{\star 0}}{|\mathbf{i}_s^{\star 0}|}\min(|\mathbf{i}_s^{\star 0}|, K\hat{I}_{\max}). \tag{3.195}$$

A similar vectorial saturation scheme is used for the output-voltage reference: whenever $|\mathbf{v}_s^{\star 0}|$ exceeds its maximum value $K\hat{V}_{\max}$, the modulus of the saturated vector $\mathbf{v}_s^\star$ is limited to $K\hat{V}_{\max}$, whereas the angle of $|\mathbf{v}_s^{\star 0}|$ is preserved, i.e.

$$\mathbf{v}_s^\star = \mathrm{satv}(\mathbf{v}_s^{\star 0}) = \frac{\mathbf{v}_s^{\star 0}}{|\mathbf{v}_s^{\star 0}|}\min(|\mathbf{v}_s^{\star 0}|, K\hat{V}_{\max}). \tag{3.196}$$

If there is no fundamental negative-sequence component and no harmonic pollution, $\mathbf{v}_s^{\star 0}$ is a constant vector in the steady state. Application of Equation (3.196) then gives no waveform distortion. The low-order harmonics that can be observed in Figure 3.10 for per-phase output-current control are avoided.

3.7.3   Transformations

Figure 3.26 is simplified in the sense that transformations between the dq and $\alpha\beta$ frames as well as between the $\alpha\beta$ frame and the three-phase quantities are not shown explicitly. This simplification is obtained by instead transforming the circuit model to the dq frame, as shown in Equation (3.189). The transformations that in practice need to be made of the current controller's input and output signals are shown in Figure 3.28. Equation (3.158) can be applied to obtain the component-wise three-phase-to-two-phase transformation of the output current as

$$i_{s\alpha} = K\left(\frac{2}{3}i_{sa} - \frac{1}{3}i_{sb} - \frac{1}{3}i_{sc}\right) \tag{3.197}$$

$$i_{s\beta} = K\left(\frac{1}{\sqrt{3}}i_{sb} - \frac{1}{\sqrt{3}}i_{sc}\right). \tag{3.198}$$

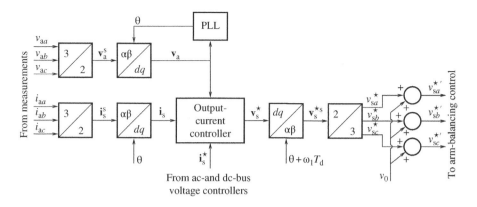

Figure 3.28 Transformations of the output-current controller's input and output signals.

The ac-bus-voltage components are transformed in a similar fashion. Transformations to the dq frame of the output-current and ac-bus-voltage vectors are made as

$$\mathbf{i}_s = e^{-j\theta}\mathbf{i}_s^s \qquad \mathbf{v}_a = e^{-j\theta}\mathbf{v}_a^s \tag{3.199}$$

where angle θ is selected such that the dq frame is in synchronism with the positive-sequence component of $\mathbf{v}_a^s$. This is performed by the PLL, as shown in Section 3.8.1. In component form, with $\mathbf{i}_s = i_{sd} + ji_{sq}$ and $\mathbf{i}_s^s = i_{s\alpha} + ji_{s\beta}$, the first relation in Equation (3.199) can be expressed as

$$i_{sd} + ji_{sq} = (\cos\theta - j\sin\theta)(i_{s\alpha} + ji_{s\beta}) \tag{3.200}$$

from which the real and imaginary parts can be identified as

$$i_{sd} = i_{s\alpha}\cos\theta + i_{s\beta}\sin\theta \tag{3.201}$$

$$i_{sq} = -i_{s\alpha}\sin\theta + i_{s\beta}\cos\theta. \tag{3.202}$$

The output-voltage reference is $\alpha\beta$ transformed as

$$\mathbf{v}_s^{\star s} = e^{j(\theta + \omega_1 T_d)}\mathbf{v}_s^\star \tag{3.203}$$

where the incremental vector rotation obtained by the added angle $\omega_1 T_d$ compensates for the displacement angle resulting from dq transformation of the total time delay; see Equation (3.170). In component form, with $\mathbf{v}_s^\star = v_{sd}^\star + jv_{sq}^\star$ and $\mathbf{v}_s^{\star s} = v_{s\alpha}^\star + jv_{s\beta}^\star$, Equation (3.203) can be expressed as

$$v_{s\alpha}^\star = v_{sd}^\star\cos(\theta + \omega_1 T_d) - v_{sq}^\star\sin(\theta + \omega_1 T_d) \tag{3.204}$$

$$v_{s\beta}^\star = v_{sd}^\star\sin(\theta + \omega_1 T_d) + v_{sq}^\star\cos(\theta + \omega_1 T_d). \tag{3.205}$$

The $\alpha\beta$-frame components are then transformed to three-phase components using Equation (3.159). In component form, this can be expressed as

$$v_a^\star = \frac{1}{K}v_{s\alpha}^\star + v_0 \tag{3.206}$$

$$v_b^\star = \frac{1}{K}\left(-\frac{1}{2}v_{s\alpha}^\star + \frac{\sqrt{3}}{2}v_{s\beta}^\star\right) + v_0 \tag{3.207}$$

$$v_c^\star = \frac{1}{K}\left(-\frac{1}{2}v_{s\alpha}^\star - \frac{\sqrt{3}}{2}v_{s\beta}^\star\right) + v_0 \tag{3.208}$$

where a zero-sequence component, v_0, can be added. The purpose and selection of this component is discussed next.

3.7.4 Zero-Sequence Injection

If the star point on the converter side of the transformer is ungrounded (alternatively that the dc-bus midpoint is ungrounded, though this is impossible in a symmetric monopole HVDC transmission), then a zero-sequence component, v_0, can be added to the phase output-voltage references, as shown in Figure 3.28, without driving a zero-sequence output-current component. This can be utilized to extend the voltage range to the obtainable maximum without entering saturation; compare Equation (3.196). The maximum nominal peak value of the output voltage is, for sinusoidal phase-voltage references, given by Equation (3.51). For the special case $Q = 0$ we get $\hat{V}_{max} \le v_d/2$. By adding a suitable v_0, this can be increased to $\hat{V}_{max} \le v_d/\sqrt{3}$, i.e. by approximately 15%. In other words, for a certain desired $\hat{V}_{max}$, the dc-bus voltage can be reduced to $\sqrt{3}/2 \approx 87\%$ of the value required for sinusoidal three-phase references. Such a downrating voltage wise implies a relatively significant cost saving for a high-voltage, high-power converter.

There are several methods for selecting v_0, of which we shall describe two.

3.7.4.1 Third-Harmonic Addition

The mentioned peak-value increase of the fundamental component of v_s, given a certain v_d, can be obtained by adding just a third harmonic to the phase-voltage references, as we shall now show. Consider the function

$$f(\theta, \gamma) = \hat{V}_s(\cos\theta - \gamma\cos 3\theta) \tag{3.209}$$

where $\theta = \omega_1 t$ and where the added third-harmonic component is phase shifted 180° relative to the fundamental in order to make the peak value of f smaller than $\hat{V}_{max}$ for $\gamma > 0$. The extremes of this function can be found by solving $f_\theta'(\theta, \gamma) = 0$ for θ. In addition to the trivial solutions $\theta = 0$ and $\theta = \pi$, we also find a solution at

$$\theta = \arctan\sqrt{\frac{9\gamma - 1}{3\gamma + 1}} \tag{3.210}$$

which corresponds to the global maximum of f. This solution is substituted in Equation (3.209) and $f_\gamma'(\theta, \gamma) = 0$ is solved for γ in order to minimize the peak value. This yields $\gamma = 1/6$. In turn, when this is substituted in Equation (3.210), $\theta = \pi/6$ is obtained. Equation (3.209) yields $f(\pi/6, 1/6) = \sqrt{3}\hat{V}_s/2$, so the optimal $\gamma = 1/6$, as predicted, reduces the peak

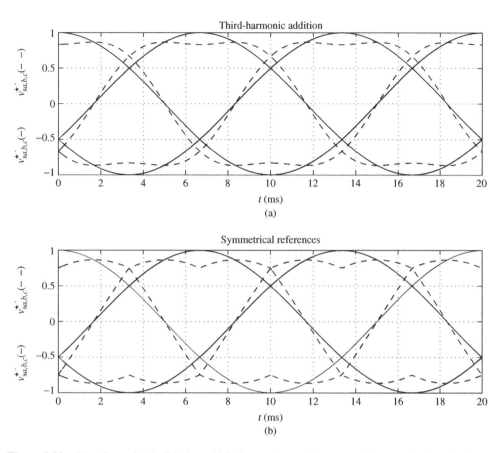

Figure 3.29 Waveforms for (solid) sinusoidal phase-voltage references with normalized peak value 1 and (dashed) with zero-sequence injection.

value of f from $\hat{V}_s$ to $\sqrt{3}\hat{V}_s/2$. Equating this with the nominal maximum output-voltage amplitude, $v_d/2$ yields $\hat{V}_s = v_d/\sqrt{3}$, i.e. the aforementioned increased maximum. Figure 3.29(a) illustrates the result. The peak value of the solid curve is 1, whereas the peak value of the dashed curve is $\sqrt{3}/2 \approx 0.87$. Yet, the peak value of the *fundamental component* of the dashed curve remains equal to 1. The added zero-sequence component can be obtained as

$$v_0 = -\frac{|\mathbf{v}_s^{\star s}|}{6}\cos(3\arg\mathbf{v}_s^{\star s}) = -\frac{|\mathbf{v}_s^{\star}|}{6}\cos[3(\theta + \omega_1 T_d + \arg\mathbf{v}_s^{\star})]. \tag{3.211}$$

3.7.4.2 Symmetrical References

In this alternative, v_0 is selected so that, at every time instant, the maximum and minimum values of the phase-voltage references are equal, but with opposite signs. The phase-voltage references become symmetrized, hence the name *symmetrical references*. This is obtained by selecting

$$v_0 = -\frac{\max(v_{sa}^\star, v_{sb}^\star, v_{sc}^\star) + \min(v_{sa}^\star, v_{sb}^\star, v_{sc}^\star)}{2}. \tag{3.212}$$

The peak value of the total waveform resulting from symmetrical references is the same as for third-harmonic addition, $\sqrt{3}\hat{V}_s/2$, given a peak value $\hat{V}_s$ for sinusoidal references. This can be deduced as follows. Consider the balanced set of references

$$v_{sa}^\star = \hat{V}_s \cos\theta \qquad v_{sb}^\star = \hat{V}_s \cos(\theta - 2\pi/3) \qquad v_{sc}^\star = \hat{V}_s \cos(\theta - 4\pi/3). \tag{3.213}$$

For $\theta = \pi/6$, symmetry is obtained without the addition of a zero-sequence component, since then $v_{sa}^\star = -v_{sc}^\star = \sqrt{3}\hat{V}_s/2$ and $v_{sb}^\star = 0$. It is easy to verify that the addition of v_0 given by Equation (3.212): $v_{sa,b,c}^\star{}' = v_{sa,b,c}^\star + v_0$, gives $|v_{sa,b,c}^\star{}'| \leq \sqrt{3}\hat{V}_s/2$ for $\theta \neq \pi/6$. The phase waveforms that result from symmetrical references are shown by the dashed curves in Figure 3.29(b). A comparison to the dashed curves in Figure 3.29(a) indicates that the two methods give fairly similar waveforms, but it can be deduced that symmetrical references add zero-sequence harmonics of higher orders, i.e. 9th, 15th, etc. harmonics in addition to the 3rd harmonic.

3.7.4.3 Effect on the Sum-Capacitor-Voltage Ripples

Zero-sequence injection has an effect on the waveforms of the sum-capacitor-voltage ripples. For third-harmonic addition smaller harmonic components at $3\omega_1$ and $4\omega_1$ are added to Equations (3.35) and (3.36), which can be generalized as [1]

$$\Delta W_\Sigma = -\frac{\hat{V}_s \hat{I}_s}{4\omega_1} \sin(2\omega_1 t - \varphi) + \frac{\gamma \hat{V}_s \hat{I}_s}{4\omega_1} \sin(2\omega_1 t + \varphi) + \frac{\gamma \hat{V}_s \hat{I}_s}{8\omega_1} \sin(4\omega_1 t - \varphi) \tag{3.214}$$

$$\Delta W_\Delta = \frac{v_d \hat{I}_s}{2\omega_1} \sin(\omega_1 t - \varphi) - \frac{2\hat{V}_s i_c}{\omega_1} \sin\omega_1 t + \frac{2\gamma \hat{V}_s i_c}{3\omega_1} \sin 3\omega_1 t \tag{3.215}$$

where $\gamma = 0$ for sinusoidal references and $\gamma = 1/6$ for third-harmonic addition. However, the effect is fairly insignificant, as shown in the following example. For symmetrical references yet more, but negligible, components add to the ripples.

Example 3.11 *Example 3.2 is repeated, i.e. the equations of the Test Converter, with $C = 4$ mF, $\hat{V}_s = 100$ kV, and $\hat{I}_s = 0.5$ kA, are employed to produce traces of the insertion indices and the sum-capacitor-voltage ripples. This time, though, third-harmonic addition is used. A comparison of Figures 3.3 and 3.30 shows that the peak values of the insertion indices are reduced by roughly 13% in the latter figure. Even for $\varphi = -90°$ is there now a margin to the maximum value 1.*

3.8 Higher-Level Control

In this section, four higher-level control loops are considered.

- The PLL, which synchronizes the dq frame of the converter control system with the ac-bus-voltage vector and produces the transformation angle θ.

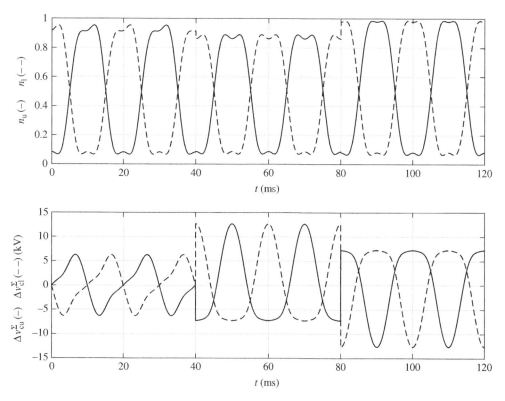

Figure 3.30 Insertion indices and sum-capacitor-voltage ripples for third-harmonic addition with $\varphi = 0$ (0 ms $\leq t <$ 40 ms), $\varphi = 90°$ (40 ms $\leq t <$ 80 ms), and $\varphi = -90°$ (80 ms $\leq t <$ 120 ms).

- Active- and reactive-power control, which can be made in an open-loop fashion.
- DC-bus-voltage control, which regulates the dc-bus voltage by adjusting the converter's active-power reference $P^\star$.
- Power-synchronization control, which is a useful alternative for converters connected to weak or passive networks.

Characteristic for all four loops is that their bandwidths are much lower than the bandwidth α_c of the output-current control loop, and that their outputs feed into this loop as θ and the components of $\mathbf{i}_s^{\star 0}$. The converter control system operates in a cascaded fashion.

3.8.1 Phase-Locked Loop

PLLs are commonly found in radio-frequency electronic systems, where they are used for synchronization and demodulation purposes [24]. Whereas such a PLL is often implemented in an analog or mixed analog/digital integrated circuit, the PLL in a converter control system is generally implemented in software on a DSP. Also here, the PLL performs synchronization. As illustrated in Figure 3.28, the PLL acts as a feedback loop around the dq transformation of $\mathbf{v}_a^s$. The PLL's output is the transformation angle θ (which, in addition, is used in the dq

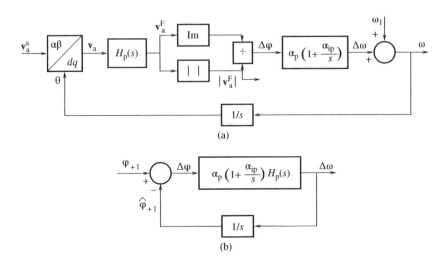

Figure 3.31   (a) Block diagram of the PLL. (b) Linearized PLL model.

transformation of $\mathbf{i}_s^s$ and in the $\alpha\beta$ transformation of $\mathbf{v}_s^\star$; see Figure 3.28). The details of the PLL are shown in Figure 3.31(a). The input signal is $\mathbf{v}_a$, which is low-pass filtered through the PLL filter $H_p(s)$, giving signal, $\mathbf{v}_a^F$. [This signal is different from $\mathbf{v}_a^f$, which is used for feedforward in the output-current controller. The filter for the latter, given by Equation (3.187), is different from $H_p(s)$.] A signal, $\Delta\varphi = \text{Im}\{\mathbf{v}_a^F\}/|\mathbf{v}_a^F|$ (its notation as an angle difference will be clarified momentarily), is then formed and fed to the PI controller $\alpha_p(1 + \alpha_{ip}/s)$, whose output is the angular frequency deviation $\Delta\omega$. To this, the nominal ω_1 is added, forming the instantaneous angular synchronous frequency ω. Finally, ω is integrated into θ. (This integration corresponds to the voltage-controlled oscillator for a PLL implemented in analog or mixed analog/digital electronics.) In the following, we shall analyze the PLL, determine its dynamic and static properties, and find suitable parameter selections.

3.8.1.1 Linearized PLL Model

Because of the *dq* transformation, the PLL is a nonlinear closed-loop system. As can be observed in Figure 3.31(a), θ is formed as the integral of the instantaneous angular synchronous frequency $\omega = \omega_1 + \Delta\omega$, giving

$$\theta = \int \omega \, dt = \omega_1 t + \underbrace{\int \Delta\omega \, dt}_{\hat{\varphi}_{+1}}. \tag{3.216}$$

Suppose that the ac-bus voltage contains multiple frequencies according to Equation (3.160)

$$\mathbf{v}_a^s = K[\hat{V}_{+1}e^{j(\omega_1 t + \varphi_{+1})} + \hat{V}_{-1}e^{-j(\omega_1 t + \varphi_{-1})} + \hat{V}_{-5}e^{-j(5\omega_1 t + \varphi_{-5})} + \hat{V}_{+7}e^{j(7\omega_1 t + \varphi_{+7})} + \cdots]. \tag{3.217}$$

Transformation to synchronous coordinates as $\mathbf{v}_a = e^{-j\theta}\mathbf{v}_a^S$, with θ given by Equation (3.216), yields

$$\mathbf{v}_a = K[\hat{V}_{+1}e^{j(\varphi_{+1}-\hat{\varphi}_{+1})} + \hat{V}_{-1}e^{-j(2\omega_1 t+\varphi_{-1}+\hat{\varphi}_{+1})} + \hat{V}_{-5}e^{-j(6\omega_1 t+\varphi_{-5}+\hat{\varphi}_{+1})}$$

$$+\hat{V}_{+7}e^{j(6\omega_1 t+\varphi_{+7}-\hat{\varphi}_{+1})} + \cdots]. \tag{3.218}$$

Filtering through $H_p(s)$ (ideally, at least) suppresses the components at $2\omega_1, 6\omega_1, \ldots$, giving

$$\mathbf{v}_a^F = K\hat{V}_{+1}e^{j(\varphi_{+1}-\hat{\varphi}_{+1})} \Rightarrow \Delta\varphi = \frac{\text{Im}\{\mathbf{v}_a^F\}}{|\mathbf{v}_a^F|} = \sin(\varphi_{+1} - \hat{\varphi}_{+1}). \tag{3.219}$$

The PLL can be assumed to accurately track the phase angle of the ac-bus-voltage vector. Hence, it can be assumed that $\varphi_{+1} - \hat{\varphi}_{+1}$ is small, at least during normal operation. This allows Equation (3.219) to be linearized as

$$\Delta\varphi = \varphi_{+1} - \hat{\varphi}_{+1} \tag{3.220}$$

which shows that in the linearized model, signal $\Delta\varphi$, is the phase tracking error, i.e. the difference between the phase angle of the fundamental component of the ac-bus voltage and its estimate $\hat{\varphi}_{+1}$ [the latter which is not an explicit variable in the PLL, but is present implicitly in θ; see Equation (3.216)]. This results in the linearized model shown in Figure 3.31(b).

3.8.1.2 Closed-Loop System and Parameter Selection

The open-loop transfer function of the system shown in Figure 3.31(b) is given by

$$G_k(s) = \frac{\alpha_p}{s}\left(1 + \frac{\alpha_{ip}}{s}\right)H_p(s) \tag{3.221}$$

which, in turn, gives the following closed-loop system from φ_{+1} to the phase tracking error $\Delta\varphi$:

$$G_c(s) = \frac{1}{1 + G_k(s)} = \frac{s^2}{s^2 + \alpha_p(s + \alpha_{ip})H_p(s)} \tag{3.222}$$

Let us at first neglect the dynamic impact of the PLL filter, assuming $H_p(s) \approx 1$. This yields

$$G_c(s) = \frac{s^2}{s^2 + \alpha_p s + \alpha_p \alpha_{ip}}. \tag{3.223}$$

For pure P control, $\alpha_{ip} = 0$, Equation (3.223) simplifies to $G_c(s) = s/(s + \alpha_p)$. Here, α_p is the bandwidth of the closed-loop system, which is of high-pass character. It is sufficient to select α_p below the fundamental angular frequency

$$\alpha_p < \omega_1 \tag{3.224}$$

e.g. $\alpha_p = 50$ rad/s, since too high a bandwidth may result in amplification of parasitic components which are not sufficiently suppressed by the PLL filter $H_p(s)$, or even destabilization of grid resonances [22].

For PI control, $\alpha_{ip} > 0$, the poles of $G_c(s)$ in Equation (3.223) are located at

$$s = \frac{\alpha_p}{2}\left(-1 \pm \sqrt{1 - \frac{4\alpha_{ip}}{\alpha_p}}\right).$$
(3.225)

If $\alpha_{ip} > \alpha_p/4$, complex-conjugated poles are obtained. To ensure good damping of the closed-loop dynamics, it should be avoided that the imaginary parts get bigger than the real part, absolute-value wise. Therefore, the following selection is recommended of the I-part bandwidth:

$$\alpha_{ip} < \frac{\alpha_p}{2}.$$
(3.226)

3.8.1.3   PLL Filter

PLL filter $H_p(s)$ needs to adequately suppress any multiple frequency components that may be present in v_a; see Equation (3.218). A low-pass filter is thus required, but its bandwidth must not be selected below (or even in the range of) α_p, since this would degrade the closed-loop PLL dynamics [the approximation $H_p(s) \approx 1$ could no longer be made in (3.223)]. An appropriate choice is a second-order Butterworth low-pass filter

$$H_p(s) = \frac{\alpha_b^2}{s^2 + \sqrt{2}\alpha_b s + \alpha_b^2}$$
(3.227)

with the bandwidth selection recommendation

$$\alpha_p < \alpha_b < 2\omega_1.$$
(3.228)

Transfer function $H_p(s)$ as given by Equation (3.227) can be obtained as a special case of the resonant filter in Equation (3.65), with

$$h = \frac{\alpha_b}{\omega_1} \qquad K_h = \alpha_b \qquad \alpha_h = \sqrt{2}\alpha_b \qquad \phi_h = -\frac{\pi}{2}.$$
(3.229)

In this case, h is generally not an integer.

3.8.1.4   Static Phase-Tracking Error

A constant phase angle, φ_{+1}, is tracked by the PLL without a static error even for pure P control. Owing to the integrator in the feedback path, we have $G_c(0) = 0$ even for $\alpha_{ip} = 0$. But this is not the case for a linearly increasing phase angle, which results if the angular frequency of the ac-bus-voltage deviates from the nominal ω_1 by the quantity $\Delta\omega_1$. Since the frequency in an ac grid varies slightly about its nominal value with the loading conditions, it cannot always be assumed that $\Delta\omega_1 = 0$. Let us consider a frequency step that occurs at $t = 0$, giving $\varphi_{+1} = \Delta\omega_1 t$. Applying the final value theorem allows the resulting static phase-tracking error to be calculated as

$$\lim_{t\to\infty} \Delta\varphi = \lim_{s\to 0} sG_c(s)\mathcal{L}\{\Delta\omega_1 t\} = \lim_{s\to 0} sG_c(s)\frac{\Delta\omega_1}{s^2} = \lim_{s\to 0} \frac{\Delta\omega_1 s}{s^2 + \alpha_p H_p(s)(s + \alpha_{ip})}$$

$$= \lim_{s\to 0} \frac{\Delta\omega_1 s}{\alpha_p(s + \alpha_{ip})} = \begin{cases} 0, & \alpha_{ip} > 0 \\ \frac{\Delta\omega_1}{\alpha_p}, & \alpha_{ip} = 0. \end{cases} \tag{3.230}$$

This result shows that PI control is needed to avoid a static phase-tracking error if $\Delta\omega_1 \neq 0$.

3.8.1.5 Discrete-Time Integrator Realization

Numerical integration of ω into θ can be made using a simple Euler update

$$\theta(n + 1) = \theta(n) + T_s\omega(n). \tag{3.231}$$

In addition, θ should be computed modulo 2π, i.e. be confined to the interval $[0, 2\pi]$, to prevent numerical overflow. Whenever θ exceeds 2π at any sampling instant, 2π should be subtracted

$$\text{if } \theta(n + 1) > 2\pi \text{ then } \theta(n + 1) \to \theta(n + 1) - 2\pi. \tag{3.232}$$

Example 3.12 *The tracking performance of a PLL with $\alpha_p = 50$ rad/s is studied. The ac-bus voltage contains just a positive-sequence component, so $\alpha_b = 2000$ rad/s is selected, in contradiction to Equation (3.228). P and PI control are tested, for a 5° phase-angle step at t = 0.1 s and a 5 rad/s frequency step at t = 0.5 s. Results are shown in Figure 3.32. In agreement with the theoretical results, only when PI control is used does the PLL manage to track the frequency step without a static phase error. It can also be observed that transients decay with the time constant $1/\alpha_p = 0.2$ s.*

3.8.2 Open-Loop Active- and Reactive-Power Control

Because of the PLL's synchronizing action, v_a is real valued (at least in the steady state): $v_a = |v_a|$. As a result, Equation (3.179) can be used to calculate the complex output power as

$$P + jQ = \frac{3}{2K^2} v_a i_s^* = \frac{3}{2K^2} |v_a|(i_{sd} - ji_{sq}) \tag{3.233}$$

Thus, the components of $i_s^{\star 0}$ can be computed from $P^\star$ and $Q^\star$ by solving respectively for i_{sd} and i_{sq} in Equation (3.233). One modification should be made here. To suppress multiple frequency components, $|v_a|$ needs to be replaced by the filtered $|v_a^F|$, which is tapped as an output of the PLL, as shown in Figure 3.31. We get

$$i_{sd}^{\star 0} = \frac{2K^2}{3|v_a^F|}P^\star \qquad i_{sq}^{\star 0} = -\frac{2K^2}{3|v_a^F|}Q^\star. \tag{3.234}$$

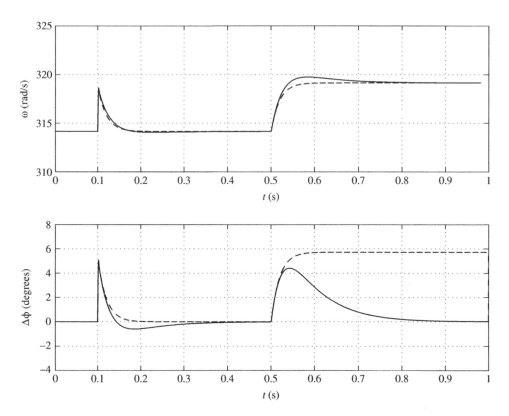

Figure 3.32 Responses to phase-angle and frequency steps for a PLL with (solid) PI control (α_{ip} = 10 rad/s) and (dashed) P control (α_{ip} = 0).

Active- and reactive-power control can thus be made in an open-loop fashion with good accuracy (although not with guaranteed zero static error) and with bandwidth α_b of the PLL filter. The open-loop control can be augmented with outer closed loops. Particularly, P^* can be determined by the dc-bus-voltage controller, as shown next.

3.8.3   DC-Bus-Voltage Control

Regulating the dc-bus voltage is straightforward. First, let us revisit Equation (3.61), which is the model for the dc bus including the contribution from the submodule capacitors

$$\underbrace{\left(C_d + \frac{2MC}{N}\right)}_{C'_d}\frac{dv_d}{dt} = i_d - \frac{P}{v_d}. \tag{3.235}$$

Multiplying both sides by v_d yields

$$C'_d v_d \frac{dv_d}{dt} = v_d i_d - P \Rightarrow \frac{C'_d}{2}\frac{dv_d^2}{dt} = v_d i_d - P. \tag{3.236}$$

We now introduce the effective dc-bus energy $W_d = C'_d v_d^2/2$ and the dc-bus input power $P_d = v_d i_d$, obtaining the compact relation

$$\frac{dW_d}{dt} = P_d - P. \tag{3.237}$$

Because the output-current control loop is significantly faster than the higher-level control loops, it can often (though not always, particularly not for low SCRs) be assumed that the active output power will respond near instantaneously—as observed from the dc bus—to a reference change. The PI controller

$$P^\star = \alpha_d \left(1 + \frac{\alpha_{id}}{s} \right) (W_d - W_d^\star) \tag{3.238}$$

yields, when combined with Equation (3.237) and the assumption $P = P^\star$, the following closed-loop system:

$$W_d = \frac{\alpha_d(s + \alpha_{id})}{s^2 + \alpha_d s + \alpha_d \alpha_{id}} W_d^\star + \frac{s}{s^2 + \alpha_d s + \alpha_d \alpha_{id}} P_d. \tag{3.239}$$

As long as $\alpha_{id} > 0$, the transfer function from P_d to W_d has zero static gain, showing that W_d will track $W_d^\star$ with zero static error regardless of P_d.

The denominators of the transfer functions in Equation (3.239) are formally identical to that in Equation (3.223) for the PLL. Consequently, the selection recommendation for α_{id} can be similar to Equation (3.226)

$$\alpha_{id} < \frac{\alpha_d}{2}. \tag{3.240}$$

Moreover, α_d can be selected similarly to Equation (3.228)

$$\alpha_d < \omega_1 \tag{3.241}$$

e.g. $\alpha_d = 50$ rad/s. Reference $i_{sd}^{\star 0}$ is obtained from $P^\star$ via the first relation in Equation (3.234).

A block diagram of the proposed dc-bus-voltage controller is shown in Figure 3.33. There are two additions to what already has been described:

- A filter $H_p(s)$ in the path for v_d is added. Its purpose is similar to the PLL filter, i.e. to suppress ripple and other high-frequency disturbances. As indicated, it can be identical to the PLL filter—see Equation (3.227)—(except that its input and output signals are scalar quantities).
- The ideal reference $i_s^{\star 0}$ is formed by adding the q-direction reference [computed e.g., for example, using the second relation in Equation (3.234)]. Vectorial saturation according to Equation (3.195) (Figure 3.27) is then applied, giving $i_s^\star = \text{satv}(i_s^{\star 0})$.
- An anti-windup feedback of the error $i_{sd}^\star - i_{sd}^{\star 0}$, $i_{sd}^\star = \text{Re}\{i_s^\star\}$ is included.

Example 3.13 *For the Test Converter, circulating-current control and direct voltage control according to Section 3.5, vector output-current control according to Section 3.7, a PLL according to Section 3.8.1, and dc-bus-voltage control according to Section 3.8.3 are implemented. The converter is operated as a rectifier with the output power 100 MW. The ac-bus voltage has the peak value 95 kV. It is a pure positive-sequence component. Two cases are evaluated,*

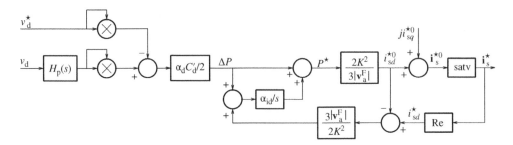

Figure 3.33   DC-bus-voltage controller, where $\Delta P = \alpha_{\mathrm{d}}(W_{\mathrm{d}} - W_{\mathrm{d}}^{\star})$.

which differ in the bandwidth selection of the identical PLL and dc-bus-voltage-controller low-pass filters $H_{\mathrm{p}}(s)$: high and low bandwidth, see Table 3.2. Three transients are imposed:

- *At $t = 0.1$ s, a 100 ms balanced voltage sag down to the peak value 80 kV occurs.*
- *At $t = 0.3$ s, the dc-side input power step increases to $P_{\mathrm{d}} = 120$ MW.*
- *At $t = 0.6$ s, the dc-bus-voltage reference step increases to $v_{\mathrm{d}}^{\star} = 204$ kV.*

For Case 1, the following conclusions can be drawn from Figure 3.34.

- *Current component i_{sd} and the active output power P track their references with just minor transient deviations.*
- *The voltage sag is ridden through with minor disturbances in the active power and in the dc-bus voltage. Since the converter operates below its current rating, the output current can be increased during the duration of the sag, thus maintaining the level of power transfer. The increased current is reflected in a slight increase of the sum-capacitor-voltage ripples (the sum capacitor voltages are shown for both arms of phase a).*
- *The step in P_{d} causes a transient in v_{d} as well as in $v_{\mathrm{cu,la}}$. Notice that the mean value of $v_{\mathrm{cu,la}}$ equals v_{d} in the steady state, whereas on the transient there is a deviation.*
- *The step in $v_{\mathrm{d}}^{\star}$ results in a transient reduction of P, allowing the capacitors to charge from the dc-side input power.*

Case 2—see Figure 3.35—shows similar performance, but with a slight lagging effect. There is now a noticeable transient in the output power during the voltage sag.

3.8.4   Power-Synchronization Control

The standard MMC control method described so far, which is based on output-current control with synchronization to the ac-bus voltage using a PLL, works well as long as SCR—given by Equation (3.105)—is large enough. However, roughly for SCR < 2, stability becomes hard to achieve, at least for active-power transfer at the rated level. A low SCR—also called a *weak grid*—typically results when the PCC is located electrically far from the nearest major clusters of synchronous generators.

Table 3.2 Bandwidths in rad/s.

	Case 1	Case 2
α_c	2000	2000
α_f	1000	1000
α_0	50	50
α_p	50	50
α_d	50	50
α_{id}	25	25
α_b	2000	200

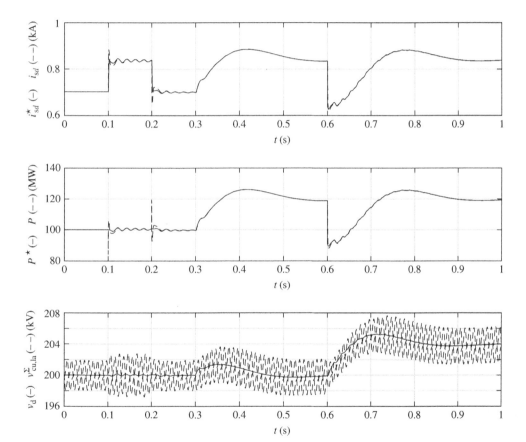

Figure 3.34 Case 1: Response to transients with high-bandwidth filters.

The standard control method fails too when the converter is connected to a passive network. This is the situation when a VSC-HVDC transmission is used for restoring an ac network after a blackout, called *black start*. This is also the situation for an offshore wind-power-collecting ac network that is connected to an onshore grid via a VSC-HVDC transmission.

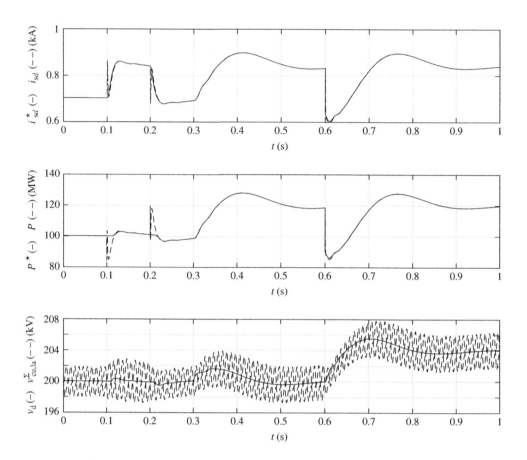

Figure 3.35   Case 2: Response to transients with low-bandwidth filters.

In the described situations, an alternative to the standard control method must be relied upon. Such an alternative is the power-synchronization control method [25]. The idea is to emulate the dynamic characteristics of a synchronous machine, as first described in [26]. A similar approach is known as a *synchronverter* [27]. While holding the magnitude of the ac-bus voltage constant, active power is controlled by adjusting the phase angle of the ac-bus voltage. To illustrate in more detail the principle, we let $\mathbf{v}_a = v_a e^{j\delta_a}$ and model the grid impedance as a reactance X; see Figure 3.36(a). This reactance is inversely proportional to the SCR. The stiff grid voltage, $\mathbf{v}_g$, behind this reactance is given by $\mathbf{v}_g = v_g e^{j\delta_g}$. In the steady state, the output current is obtained as

$$\mathbf{i}_s = \frac{\mathbf{v}_a - \mathbf{v}_g}{jX} = \frac{v_a e^{j\delta_a} - v_g e^{j\delta_g}}{jX} \tag{3.242}$$

which yields the following complex output power:

$$P + jQ = \frac{3}{2K^2}\mathbf{v}_a\mathbf{i}_s^* = \frac{3}{2K^2}v_a e^{j\delta_a}\frac{v_a e^{-j\delta_a} - v_g e^{-j\delta_g}}{-jX} = j\frac{3}{2K^2}\frac{v_a[v_a - v_g(\cos\tilde{\delta} + j\sin\tilde{\delta})]}{X} \tag{3.243}$$

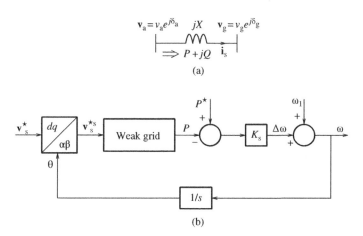

Figure 3.36 (a) Weak grid modeled as a reactance X. (b) Closed-loop system formed by power-synchronization control.

where $\tilde{\delta} = \delta_a - \delta_g$. From Equation (3.243), the active and reactive powers are identified as

$$P = \frac{3}{2K^2}\frac{v_a v_g \sin\tilde{\delta}}{X} \qquad Q = \frac{3}{2K^2}\frac{v_a(v_a - v_g \cos\tilde{\delta})}{X}. \qquad (3.244)$$

It is found that P can be adjusted primarily by varying $\tilde{\delta}$, which can be made by varying δ_a. Q on the other hand can be adjusted primarily by varying v_a.

The ac-bus voltage is not directly available for manipulation, but can be adjusted indirectly via $\mathbf{v}_s^\star$. Consider the block diagram for vector output-current control shown in Figure 3.26. Instead of letting the ideal voltage reference $\mathbf{v}_s^{\star 0}$ be the current-controller output, we could let

$$\mathbf{v}_s^{\star 0} = \mathbf{v}_a^\star + \frac{j\omega_1 L + R}{2}\mathbf{i}_s \qquad (3.245)$$

where $\mathbf{v}_a^\star = v_a^\star e^{j\delta_a}$. Assuming that $\mathbf{v}_s = \mathbf{v}_s^{\star 0}$, substitution of Equation (3.245) in Equation (3.189) yields

$$\frac{L}{2}\frac{d\mathbf{i}_s}{dt} = \mathbf{v}_a^\star + \frac{j\omega_1 L + R}{2}\mathbf{i}_s - \mathbf{v}_a - \frac{j\omega_1 L + R}{2}\mathbf{i}_s$$

$$= \mathbf{v}_a^\star - \mathbf{v}_a \qquad (3.246)$$

In the steady state, i.e. for $d\mathbf{i}_s/dt = 0$, we get $\mathbf{v}_a = \mathbf{v}_a^\star$, as desired. But Equation (3.246) shows that Equation (3.245) cannot be used as it stands. Because $\mathbf{i}_s$ does not appear explicitly on the right-hand side of Equation (3.246)—only indirectly, as $\mathbf{v}_a$ via the grid impedance is a function of $\mathbf{i}_s$—there is a risk that a poorly damped or even unstable system will result. This can be remedied by modifying Equation (3.245) to

$$\mathbf{v}_s^{\star 0} = \mathbf{v}_a^\star + \frac{j\omega_1 L + R}{2}\mathbf{i}_s^f + \frac{R_s}{2}(\mathbf{i}_s^f - \mathbf{i}_s) \qquad \mathbf{i}_s^f = \frac{\alpha_1}{s + \alpha_1}\mathbf{i}_s \qquad (3.247)$$

where a low-pass filtered variant $\mathbf{i}_s^f$ is used in place of $\mathbf{i}_s$ in the second term. A third term is added, with gain $R_s/2$. When substituted in Equation (3.189), this modified control law results in an asymptotically stable second-order system (if the dynamic dependency between $\mathbf{v}_a$ and $\mathbf{i}_s$ is disregarded)

$$\frac{L}{2}\frac{d\mathbf{i}_s}{dt} = \mathbf{v}_a^\star + \frac{j\omega_1 L + R + R_s}{2}(\mathbf{i}_s^f - \mathbf{i}_s) - \mathbf{v}_a \tag{3.248}$$

$$\frac{d\mathbf{i}_s^f}{dt} = \alpha_1(\mathbf{i}_s - \mathbf{i}_s^f). \tag{3.249}$$

Similarly to the P gain of the circulating-current controller in Section 3.5.1, R_s in Equation (3.247) acts as an "active resistance." selecting

$$R_s \gg R \qquad \alpha_1 < \frac{R_s}{L} \tag{3.250}$$

ensures that the system given by Equations (3.248) and (3.249) becomes well damped. In the steady state we still get the desired $\mathbf{v}_a = \mathbf{v}_a^\star = v_a^\star e^{j\delta_a}$.

3.8.4.1 Black-Start (Power-Collecting) Operation

Suppose that we let $\delta_a = 0$, giving the constant reference vector $\mathbf{v}_a^\star = v_a^\star$. This results in a feedforward-type control that maintains the magnitude of the ac-bus voltage at $|\mathbf{v}_a| = v_a^\star$ in the steady state. It follows from Equation (3.244) that the active output power is determined primarily by the angle δ_g of the grid voltage. The converter respectively provides or accepts the power that respectively is consumed or produced in the grid.

This mode of operation is used in the aforementioned black-start scenario. Suppose that an MMC-HVDC connects a small ac grid to a larger one. Should a blackout occur in the small grid, the MMC-HVDC transmission can be used to re-energize the small grid from the larger grid. The converter in the small grid ramps up the voltage using power-synchronization control in the black-start mode of operation. Once the voltage is restored, generators in the small grid can be restarted. Once a sufficient number of generators are online, then loads can be reconnected, and the MMC-HVDC terminal in the small grid can switch to the standard control method. The method can be applied also for the re-energization of the affected area in a partial blackout within one synchronous grid.

Blackouts occur seldom, though (preferably never). A much more common situation is when an MMC-HVDC transmission connects an offshore wind farm (or possibly an onshore, but remotely located, wind farm) to the main grid. The offshore converter station then needs to accept whatever power that is supplied from the wind generators (up to the rating of the converter). It is also the responsibility of the offshore converter to provide a stiff voltage for the wind turbines to synchronize against and deliver power. This mode of operation is identical to the black-start mode, except that the direction of power transfer is the opposite. It may therefore be called *power-collecting operation*.

3.8.4.2 Active-Power-Control Operation

For connection to weak grids a constant ac-bus voltage $|\mathbf{v}_a| = v_a^\star$ can be maintained, but closed-loop active-power control should be added. This means that phase angle δ_a has to be adjusted, which can be done by pure I control

$$\delta_a = \frac{K_s}{s}(P^\star - P).$$ (3.251)

In practice, Equation (3.251) is implemented in place of the PLL. While letting $\delta_a = 0$ in Equation (3.247), an angular frequency deviation is formed as

$$\Delta\omega = K_s(P^\star - P)$$ (3.252)

to which ω_1 is added and integrated to form the dq-frame angle θ; see Figure 3.37(a). This replaces the computation in the PLL block diagram shown in Figure 3.31.

The described principle explains the name power-synchronization control: instead of using a PLL for synchronizing against the ac-bus voltage by computation of the dq-frame angle θ, this angle is adjusted so that the active output power tracks its reference based on the first relation in Equation (3.244).

Equation (3.251) may be used also in black-start operation to introduce small frequency variations in a droop fashion relative a nominal power $P^\star$. In this case, K_s is normally significantly smaller than in active-power-control operation.

3.8.4.3   Closed-Loop System and Gain Selection

The closed-loop system formed by power-synchronization control is depicted in Figure 3.36(b). The block "weak grid" includes the steady-state relation between angle and active power given by the first relation in Equation (3.244), as well as the dynamics of the grid. If the latter are neglected, the closed-loop system that results from power-synchronization

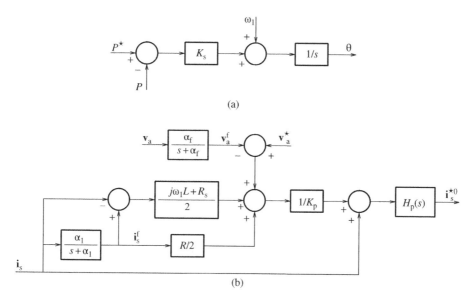

(a)

(b)

Figure 3.37   Block diagrams for power-synchronization control: (a) active-power control and (b) voltage-magnitude control realized via the output-current controller.

control can be calculated by linearizing the first relation in Equation (3.244) as $\sin \tilde{\delta} \approx \tilde{\delta}$. Substituting this relation in Equation (3.251) yields

$$P = \frac{\alpha_s}{s + \alpha_s} P^\star \qquad \alpha_s = \frac{3K_s v_a v_g}{2K^2 X}. \qquad (3.253)$$

To give the closed-loop system bandwidth α_s (which should be chosen as $\alpha_s \ll \alpha_c$), the following gain selection should be applied:

$$K_s = \frac{2\alpha_s K^2 X}{3 v_a v_g}. \qquad (3.254)$$

That is, the gain shall be selected proportional to the grid reactance X, i.e. inversely proportional to the SCR. In other words, the gain needs to be selected higher the weaker the grid is.

3.8.4.4   Selection of the AC-Bus-Voltage Reference

In many situations it is not desired to maintain the ac-bus voltage at a constant value. Rather, the PCC voltage, i.e. the voltage on the grid side of the transformer in Figure 3.2(a), should be kept constant. This can be accomplished by selecting

$$\mathbf{v}_a^\star = v_{PCC}^\star + jX_T \mathbf{i}_s^f \qquad (3.255)$$

where $v_{PCC}^\star$ is the PCC-voltage reference and X_T is the transformer reactance at the fundamental frequency. This adds a compensation for the voltage drop across the transformer, giving an open-loop-type voltage control. If needed, a term proportional to the integral of the PCC-voltage control error, $\int (v_{PCC}^\star - v_{PCC}) \, dt$, can be added to Equation (3.255) in order to remove the residual error that undoubtedly would result from the open-loop control.

The control according to Equation (3.255) may need to be modified in some situations. For example, if several VSC terminals are connected at the PCC, they would all strive to maintain the PCC voltage at its reference. This could lead to an exchange of reactive power among the converters, giving unnecessarily high converter output currents and, as a result, increased losses. The problem can be remedied by instead controlling the PCC voltage in a droop (i.e. proportional) fashion, for example as

$$\mathbf{v}_a^\star = v_{PCC}^\star + K_v(v_{PCC}^\star - v_{PCC}) \qquad (3.256)$$

where K_v is the droop gain. This is similar to how automatic voltage regulation usually is performed in synchronous generators.

3.8.4.5   Realization via the Output-Current Controller

Equation (3.247) can be realized in a control system as it stands, but this would have some drawbacks. There would be no direct way of preventing overcurrent during abnormal conditions, such as voltage sags or swells. In addition, multiple frequency components cannot be suppressed from the output current. These drawbacks can be circumvented by implementing the power-synchronization control law via the output-current controller; see Section 3.7. The control law realized in Figure 3.26 is

$$v_s^{\star 0} = K_p(i_s^{\star} - i_s) + v_R + v_a^f + \frac{j\omega_1 L}{2}i_s. \tag{3.257}$$

By equating this with Equation (3.247), $i_s^{\star}$ can be solved

$$i_s^{\star} = i_s + \frac{1}{K_p}\left[v_a^{\star} + \frac{j\omega_1 L + R_s}{2}(i_s^f - i_s) + \frac{R}{2}i_s^f - v_a^f - v_R\right]. \tag{3.258}$$

This equation forms the basis for how the output-current reference can be set by the power-synchronizing controller, but three steps remain.

1. Instead of $i_s^{\star}$, $i_s^{\star 0}$ shall be selected according to Equation (3.258) and then be passed through the vectorial saturation of Equation (3.195) to give $i_s^{\star}$. Overcurrent is thereby prevented.
2. Multiple frequencies shall be suppressed from $i_s^{\star 0}$ by applying a filter similar or identical to the PLL filter; see Equation (3.227). This results in

$$i_s^{\star 0} = H_p(s)\left\{i_s + \frac{1}{K_p}\left[v_a^{\star} + \frac{j\omega_1 L + R_s}{2}(i_s^f - i_s) + \frac{R}{2}i_s^f - v_a^f\right]\right\} \tag{3.259}$$

where v_R has been omitted as compared to Equation (3.258), since this component anyhow would be suppressed by $H_p(s)$. The block diagram shown in Figure 3.37(b) is obtained.

3.9   Control Architectures

The implementation of an MMC control system for HVDC applications can be very challenging, owing to a large number of submodules and the hierarchical control structure shown in Figure 3.1. Such a complex control structure can be practically implemented using two different architectures:

- centralized
- decentralized.

The centralized control architecture implies the usage of only one main controller, which carries out all the necessary processing and control operations. Therefore, this method gives very high computational burden and large number of input/output units. A very large number of fiber-optic cables is necessary as well, in order to connect each submodule to the central controller. This is a challenge in terms of cost and complexity. A conceptual scheme for centralized control is shown in Figure 3.38(a). For simplicity, only the gate signals for the upper and lower arms (G_u^i, G_l^i, $i = 1, 2, \ldots, N$) and the submodule capacitor-voltage measurement signals (v_{cu}^i, v_{cl}^i, $i = 1, 2, \ldots, N$) are shown as being exchanged between the main controller and each submodule.

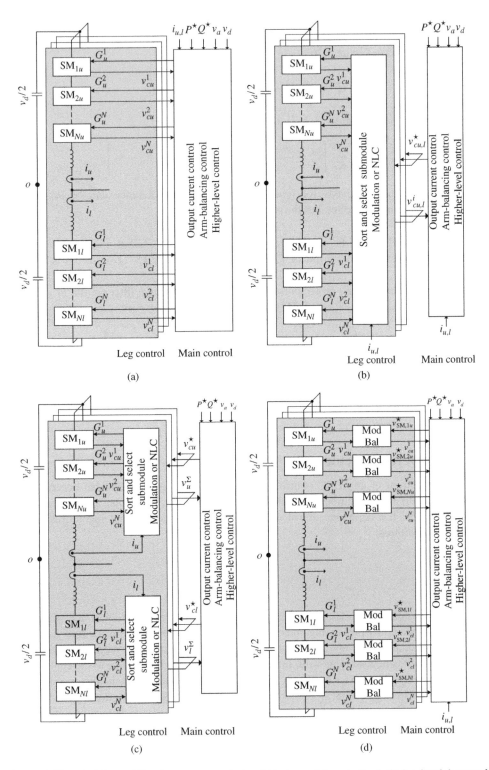

Figure 3.38 Modular multilevel converter control architectures: (a) centralized, (b) leg-level decentralized, (c) arm-level decentralized, and (d) submodule-level decentralized.

The decentralized architecture implies the use of multiple controllers with different structures. It has emerged from the physical block construction of an MMC in HVDC application. Each block is an arm or a section of an arm. Depending on the task sharing among controllers, the following decentralized architectures are possible:

- leg-level
- arm-level
- submodule-level.

The leg-level structure presented in Figure 3.38(b) involves a leg controller for each phase. The arm-balancing control is performed in the main controller. By high-speed communication, the sum capacitor voltages (v_{cu}^{Σ}, v_{cl}^{Σ}) are calculated in the leg controller and are sent to the main controller. In turn, the main controller sends the upper-arm and lower-arm sum-capacitor-voltage references ($v_{cu}^{\Sigma\star}$, $v_{cl}^{\Sigma\star}$) to the local leg controllers, which are responsible for PWM and individual balancing of the submodules across the whole leg.

The arm-level structure depicted in Figure 3.38(c) is one level higher in decentralization where the same functionality is implemented at the arm level. For large systems, where the arm is physically built in several sections, arm-section decentralization is possible.

Finally the submodule-level structure shown in Figure 3.38(d) is completely decentralized. Each submodule controller receives the voltage reference, but individually carries out PWM and balancing.

3.9.1 Communication Network

In all variants of decentralized control architecture, a high-speed real-time communication network is required. The processing burden is distributed among the different controllers, as the central controller does not perform all the control tasks alone. As shown in Figure 3.39, the communication network can be configured in two ways:

- star.
- ring.

For the star configuration, the highest communication speed is achieved at the cost of a large number of fiber-optic cables. Additionally, the leg communication hub needs to be provided with $2N$ communication ports per leg.

For the star configuration, the communication speed is reduced along with the length requirements for fiber-optic cables. Additionally, the leg communication hub needs to be provided with only one communication port.

In [28], the real-time industrial Ethernet protocol is reported for use with a decentralized control implementation of an MMC system with low number of submodules ($N = 4$). The updated period is in the range of $100\,\mu s$.

EtherCAT is an open-source protocol [29] based on Ethernet which allows full duplex communication and uses the classical master/slave configuration. It has been widely used in multi-axis, high-performance motion control. Only the master of the network is allowed to send an EtherCAT frame. Each EtherCAT slave reads and writes data "on the fly." When the master sends a telegram, it goes to the first slave of the network, which processes the data and then sends the telegram further to the next slave. This process goes on until the last slave of the

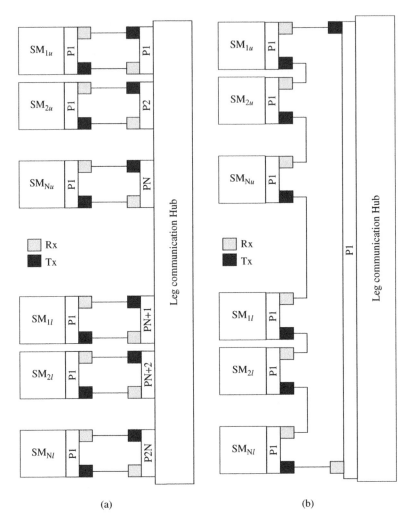

Figure 3.39   Communication network architectures: (a) star and (b) ring.

segment is reached. The last slave will send the message back to the master. The telegram will be delayed by the wire propagation delay and the processing delay introduced by the slaves.

The distributed clocks mechanism is an EtherCAT protocol feature used for high-precision clock synchronization and generation of synchronous output signals. Each EtherCAT slave has an internal clock. A difference between the slave clocks may exist because of the following two reasons. First, when the slaves are turned on, the internal register holding the current time is set to zero. However, this does not occur at the same time in all the slaves and an initial offset between the clocks is present. Second, a small difference between the frequencies of the internal oscillators of the slaves always exists. One of the clocks, usually the clock of the first slave, is used as a reference. The distributed clocks algorithm is in charge of synchronizing the EtherCAT master and slaves clocks with the reference clock. This is done by calculating and compensating the propagation delay between each slave, the initial time offset, and the

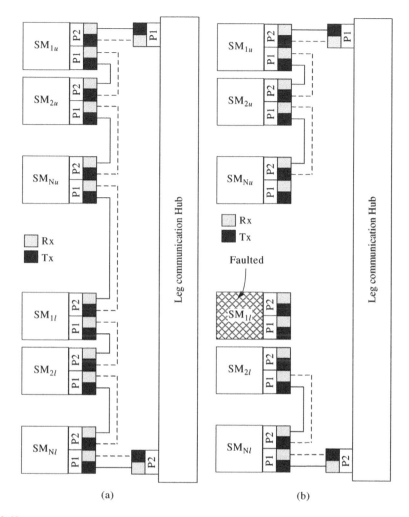

Figure 3.40 Fault-redundant ring communication network with EtherCat. (a) Normal operation with redundant fiber-optic cables. (b) Reconfigured communication in case of a submodule or fiber-optic-cable failure.

local clock drifts. Generation of synchronized output signals is also possible, owing to the distributed clocks mechanism. As the internal clocks are synchronized, all the slaves in an EtherCAT network will be able to generate a synchronized output with a jitter down to 20 ns [30]. This is a key feature allowing proper synchronization of the MMC submodules.

3.9.2 Fault-Tolerant Communication Networks

EtherCAT is a fault-tolerant communication protocol. In case of a local controller failure (an "$N - 1$ situation"), the submodules automatically reconfigure in order to reclose the communication loop and continue the operation. This is depicted in Figure 3.40(b). The main controller seamlessly reconfigures the distributed clocks to ensure synchronization in the $N - 1$ situation. Also the voltage reference for the individual submodules is adapted to $N - 1$.

Another important EtherCAT feature is that it allows communication redundancy in the network using a ring configuration, as shown in Figure 3.40. Each EtherCAT slave device has two communication ports: P1 and P2. The master sends the telegram at the same time through both ports; hence, if the communication cable between two submodules is broken, the MMC can continue normal operation as the master is still able to communicate with all slaves. This provides communication failure redundancy to the system.

3.10 Summary

This chapter considers modeling of the MMC using the principle of averaging. The obtained model serves as the basis for the design of the output-current controller as well as the controllers needed for arm balancing. The arm balancing includes control of the sum capacitor voltages and the circulating current. Four different alternatives for arm balancing are considered. The arm-balancing control is made on a per-phase basis. An introduction to the theory for three-phase systems and space vectors is given. Based on this theory, the fundamentals of vector output-current control are reviewed. Finally, the outer control loops that feed in a cascaded fashion into the output-current control loop are considered, which is followed by a review of control architectures.

References

[1] L. Harnefors, A. Antonopoulos, S. Norrga, L. Ängquist, and H.-P. Nee, "Dynamic analysis of modular multilevel converters," *IEEE Transactions on Industrial Electronics*, vol. 60, no. 7, pp. 2526–2537, Jul. 2013.

[2] J. He, Y. W. Li, F. Blaabjerg, and X. Wang, "Active harmonic filtering using current-controlled, grid-connected DG units with closed-loop power control," *IEEE Transactions on Power Electronics*, vol. 29, no. 2, pp. 642–653, Feb. 2014.

[3] A. J. Isaksson and S. F. Graebe, "Derivative filter is an integral part of PID design," *IEE Proceedings—Control Theory and Applications*, vol. 149, no. 1, pp. 41–45, Jan. 2002.

[4] A. J. Isaksson and S. F. Graebe, "Analytical PID parameter expressions for higher order systems," *Automatica*, vol. 35, no. 6, pp. 1121–1130, Jun. 1999.

[5] T. M. Rowan and R. J. Kerkman, "A new synchronous current regulator and an analysis of current-regulated PWM inverters," *IEEE Transactions on Industry Applications*, vol. IA-22, no. 4, pp. 678–690, Jul. 1986.

[6] L. Harnefors, A. G. Yepes, A. Vidal, and J. Doval-Gandoy, "Passivity-based stabilization of resonant current controllers with consideration of time delay," *IEEE Transactions on Power Electronics*, vol. 29, no. 12, pp. 6260–6263, Dec. 2014.

[7] R. Teodorescu, F. Blaabjerg, M. Liserre, and P. C. Loh, "Proportional-resonant controllers and filters for grid-connected voltage-source converters," *IEE Proceedings: Electric Power Applications*, vol. 153, no. 5, pp. 750–762, Sep. 2006.

[8] L. Harnefors, L. Zhang, and M. Bongiorno, "Frequency-domain passivity-based current controller design," *IET Power Electronics*, vol. 1, no. 4, pp. 455–465, Dec. 2008.

[9] S. Buso and P. Mattavelli, *Digital Control in Power Electronics*. Morgan & Claypool, 2006.

[10] L. Harnefors, "Implementation of resonant controllers and filters in fixed-point arithmetic," *IEEE Transactions on Industrial Electronics*, vol. 56, no. 4, pp. 1273–1281, Apr. 2009.

[11] L. Harnefors and H.-P. Nee, "Model-based current control of ac machines using the internal model control method," *IEEE Transactions on Industry Applications*, vol. 34, no. 1, pp. 133–141, Jan. 1998.

[12] M. H. Bollen, *Understanding Power Quality Problems: Voltage Sags and Interruptions*. Wiley–IEEE Press, 1999.

[13] L. Harnefors, K. Pietiläinen, and L. Gertmar, "Torque-maximizing field-weakening control: Design, analysis, and parameter selection," *IEEE Transactions on Industrial Electronics*, vol. 48, no. 1, pp. 161–168, Feb. 2001.

[14] A. Antonopoulos, L. Ängquist, and H.-P. Nee, "On dynamics and voltage control of the modular multilevel converter," *Proceedings of the 13th European Conference on Power Electronics and Applications*, pp. 1–10, Sep. 2009.

[15] L. Ängquist, A. Antonopoulos, D. Siemaszko, K. Ilves, M. Vasiladiotis, and H.-P. Nee, "Open-loop control of modular multilevel converters using estimation of stored energy," *IEEE Transactions on Industry Applications*, vol. 47, no. 6, pp. 2516–2524, Nov./Dec. 2011.

[16] A. Antonopoulos, L. Ängquist, L. Harnefors, K. Ilves, and H.-P. Nee, "Global asymptotic stability of modular multilevel converters," *IEEE Transactions on Industrial Electronics*, vol. 61, no. 2, pp. 603–612, Feb. 2014.

[17] L. Harnefors, A. Antonopoulos, K. Ilves, and H.-P. Nee, "Global asymptotic stability of current-controlled modular multilevel converters," *IEEE Transactions on Power Electronics*, vol. 30, no. 1, pp. 249–258, Jan. 2015.

[18] F. Briz del Blanco, M. W. Degner, and R. D. Lorenz, "Dynamic analysis of current regulators for ac motors using complex vectors," *IEEE Transactions on Industry Applications*, vol. 35, no. 6, pp. 1424–1432, Nov. 1999.

[19] W. C. Duesterhoeft, M. W. Schulz, and E. Clarke, "Determination of instantaneous currents and voltages by means of alpha, beta, and zero components," *Transactions of the American Institute of Electrical Engineers*, vol. 70, no. 2, pp. 1248–1255, Jul. 1951.

[20] B. Bose, *Power Electronics and Variable Frequency Drives: Technology and Applications*. IEEE Press, 1997.

[21] R. H. Park, "Two-reaction theory of synchronous machines generalized method of analysis: Part I," *Transactions of the American Institute of Electrical Engineers*, vol. 48, no. 3, pp. 716–727, Jul. 1929.

[22] L. Harnefors, M. Bongiorno, and S. Lundberg, "Input-admittance calculation and shaping for controlled voltage-source converters," *IEEE Transactions on Industrial Electronics*, vol. 54, no. 6, pp. 3323–3334, Dec. 2007.

[23] L. Harnefors, "Analysis of subsynchronous torsional interaction with power electronic converters," *IEEE Transactions on Power Systems*, vol. 22, no. 1, pp. 305–313, Feb. 2007.

[24] G.-C. Hsieh and J. C. Hung, "Phase-locked loop techniques: A survey," *IEEE Transactions on Industrial Electronics*, vol. 43, no. 6, pp. 609–615, Dec. 1996.

[25] L. Zhang, L. Harnefors, and H.-P. Nee, "Power-synchronization control of grid-connected voltage-source converters," *IEEE Transactions on Power Systems*, vol. 25, no. 2, pp. 809–820, May 2010.

[26] H.-P. Beck and R. Hesse, "Virtual synchronous machine," *Proceedings of the 9th International Conference on Electrical Power Quality and Utilisation*, pp. 1–6, Oct. 2007.

[27] Q.-C. Zhong and G. Weiss, "Synchronverters: Inverters that mimic synchronous generators," *IEEE Transactions on Industrial Electronics*, vol. 58, no. 4, pp. 1259–1267, Apr. 2011.

[28] P. Burlacu, L. Mathe, and R. Teodorescu, "Synchronization of the distributed pwm carrier waves for modular multilevel converters," *Proceedings of the International Conference on Optimization of Electrical and Electronic Equipment*, pp. 553–559, May 2014.

[29] "Anonymous EtherCAT Group," [Online]. Available: http://www.ethercat.org.

[30] C. Toh and L. Norum, "Implementation of high speed control network with fail-safe control and communication cable redundancy in modular multilevel converter," *Proceedings of the 15th European Conference on Power Electronics and Applications*, pp. 1–10, Sep. 2013.

4

Control under Unbalanced Grid Conditions

4.1 Introduction

In this chapter, the most recent grid codes for wind integration are evaluated and the need of both positive and negative sequence reactive current is highlighted as a challenge. An enhanced control structure for MMC is derived and the performance is evaluated.

4.2 Grid Requirements

The grid codes (GC) in countries with high wind-power penetration rates (e.g. Germany, Denmark, the UK, Spain, and Ireland) require high-voltage direct current (HVDC) stations connected to the point of common coupling (PCC) to remain connected for a limited time (typically a minimum of 150 ms) during balanced and unbalanced grid faults, and to inject during the fault reactive current for supporting the voltage recovery (see Figure 9.1 in Chapter 9). This is the so-called low-voltage ride-through (LVRT) requirement and includes two conditions:

- Prevent tripping of the converter, which can be due to output overcurrent or dc overvoltage.
- Inject capacitive positive sequence reactive current proportional to the positive sequence voltage drop (Figure 4.1, left):

$$i_{q+} = k_+(0.9 - V_+) \tag{4.1}$$

where V_+ is the positive-sequence voltage in p.u. As can be observed, a $\pm10\%$ non-injection band has been considered and the droop factor k_+ is considered to be (typically) 2.5 p.u., but could be within the interval (0..10) if requested by the local system operators. In the following this requirement will be referred to as positive-sequence injection low-voltage ride-through (PSI-LVRT).

In the case of unbalanced faults, the positive-sequence reactive-current injection can potentially increase the negative-sequence voltage of the grid and thereby increase the

Design, Control, and Application of Modular Multilevel Converters for HVDC Transmission Systems, First Edition.
Kamran Sharifabadi, Lennart Harnefors, Hans-Peter Nee, Staffan Norrga, and Remus Teodorescu.
© 2016 John Wiley & Sons, Ltd. Published 2016 by John Wiley & Sons, Ltd.
Companion Website URL: www.wiley.com/go/Sharifabadi/ModularConverters

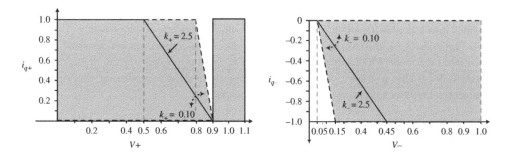

Figure 4.1 LVRT reactive current injection in Germany: (left) PSI-LVRT, (right) NSI-LVRT.

unbalance. Recently, the injection of negative-sequence current has been imposed in Germany in VDE-AR-N 4120:2015-01: Technical requirements for the connection and operation of customer installations to the high voltage network (TAB) [1]. The condition imposed here is:

- Inject inductive negative-sequence reactive current proportional to the negative sequence-voltage (Figure 4.1, right):

$$i_{q^-} = -k_-(V_- - 0.05) \tag{4.2}$$

where V_- is the negative-sequence voltage in p.u.

Here, a 5% non-injection band has been considered and the droop factor k_- is considered to be (typically) 2.5 p.u., but could be within the interval (0..10) if requested by the local system operator. This behavior is closely related to that of synchronous generators, which generate negative-sequence current during unbalanced grid faults in a more uncontrolled way. In the following, this requirement will be referred to as negative-sequence injection low-voltage ride-through (NSI-LVRT).

Thus, during LVRT both positive- and negative-sequence reactive current have to be controlled independently. This requirement is challenging for the conventional control of the modular multilevel converter (MMC) described in Chapter 3. In this chapter, two injection strategies are investigated: PSI-LVRT and mixed-sequence injection low-voltage ride-through (MSI-LVRT) (i.e. PSI and NSI simultaneously), first with the conventional control and later with an improved control structure.

4.3 Shortcomings of Conventional Vector Control

The conventional vector output-current control described in Section 3.7.2, Figure 3.28, is using the output current reference based on the instantaneous power theory (IPT) described in Equation (3.234) (for $K = 1$ and assuming $v_{ad} = |\mathbf{v}_a^F|$, i.e. the low-pass filtered ac-bus voltage aligned with d axis):

$$\begin{cases} i_{sd}^\star = \dfrac{2}{3}\dfrac{P^\star}{v_{ad}} \\[2mm] i_{sq}^\star = -\dfrac{2}{3}\dfrac{Q^\star}{v_{ad}} \end{cases} \tag{4.3}$$

During unbalanced grid conditions, the presence of the negative sequence in the grid voltage translates into a double-frequency ripple in the d-direction voltage component v_{ad}. According to Equation (4.3), this ripple transfers to the d-current-component reference $i^\star_{sd}$. As the proportional-resonant (PR) controller ($h = 0$ and $h = 2$) for the output current that is described in Section 3.7 is able to track both dc and the double-frequency components, this ripple will be present in the injected current, as shown in Example 4.1.

Example 4.1 *The Test Converter considered in Example 3.8.2 with control bandwidths shown in Table 3.2 case 2, operating in rectifier mode with the output power equal to 100 MW, is subjected to an unbalanced grid fault at time $T = 0.5$ s when the positive-sequence component's peak value decreases from 95 kV to 80 kV peak value and the negative-sequence component increases from zero to the 20 kV peak value. The output current controller has an added resonant integrator ($h = 2$) with $\alpha_2 = 50$ rad/s. The circulating current controller has a resonant integrator ($h = 2$) with $\alpha_{c2} = 200$ rad/s and $R_a = 20\ \Omega$. The results are shown in Figure 4.2 and the following observations can be made for the behavior of the converter during the unbalanced fault:*

- *As expected, the current reference $i^\star_{sd}$ exhibits double frequency ripple which is to some extent limited due to the action of the low-pass filter (LPF) in the phase-locked loop (PLL) ($\alpha_b = 200$ rad/s).*
- *The PR output current controller is able to follow the reference, because of the resonant integrator.*
- *The double-frequency oscillation in the i_{sd} verifies the presence of negative sequence current injection which at this point is not desirable.*
- *Both the dc-bus voltage and circulating current are controlled to the reference after a small transient (only circulating current in phase b is shown where the voltage is reduced most).*
- *The sum-capacitor voltages experience an increase ripple.*

It can be concluded that this control structure fails to deliver an acceptable current injection, owing to the non-intentional injection of negative sequence current. This is due to the PLL limitations in accurately tracking the angle and amplitude of the positive sequence of the grid voltage in the presence of a negative sequence.

4.3.1 PLL with Notch Filter

A solution to this problem is to add a notch filter in the PLL input in cascade with the existing LPF that can reject the double-frequency component at $2\omega_1$:

$$H_n(s) = \frac{s^2 + (2\omega_1)^2}{s^2 + \alpha_n s + (2\omega_1)^2}. \tag{4.4}$$

The notch-filter transfer function can be transformed as

$$\frac{s^2 + (2\omega_1)^2}{s^2 + \alpha_n s + (2\omega_1)^2} = 1 - \frac{\alpha_n s}{s^2 + \alpha_n s + (2\omega_1)^2} = 1 - H_2(s) \tag{4.5}$$

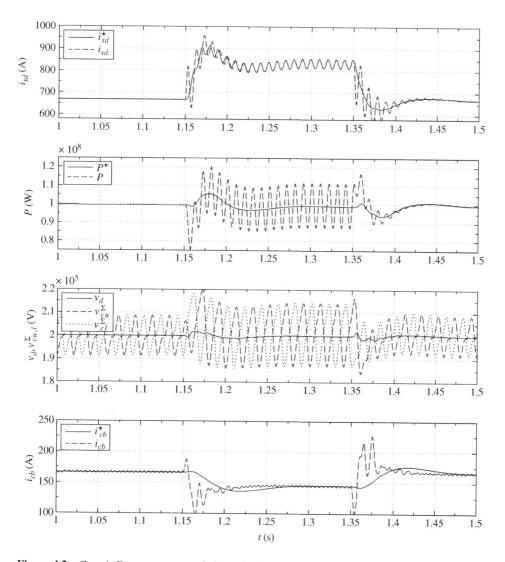

Figure 4.2 Case 1: Response to an unbalanced voltage fault with conventional dq-frame control.

where $H_2(s)$ is the resonant filter given in Equation (3.65) with

$$h = 2, \quad K_h = \alpha_n, \quad \alpha_h = \alpha_n, \quad \phi_h = 0. \tag{4.6}$$

The resulting pre-filter for the PLL is

$$H_p(s) \, H_n(s) \tag{4.7}$$

where $H_p(s)$ is the second-order Butterworth LPF described by Equation (3.227).

The test case in Example 4.1 is repeated with the addition of the notch filter ($\alpha_n = 100$ rad/s) in the PLL input filter. The results are shown in Figure 4.3.

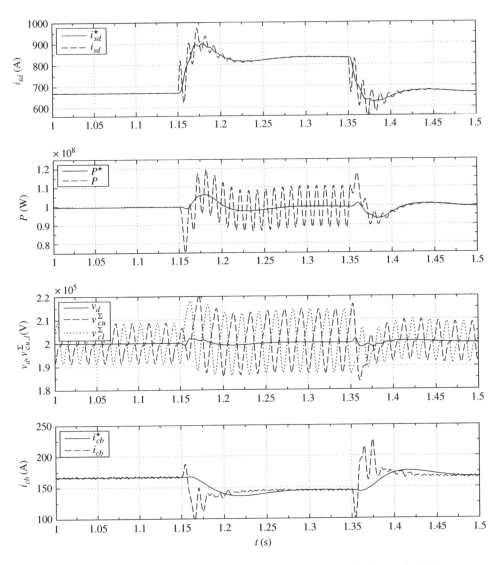

Figure 4.3 Response to an unbalanced voltage fault with notch filter in the PLL.

The following improvements can be observed:

- The double-frequency ripple from both the reference and the injected current is removed after a transient period, proving that only positive-sequence injection (PSI) is performed.

It can be concluded that the notch filter in the PLL structure is an effective method for the correct tracking of the positive-sequence angle and thus it can ensure proper positive-sequence current injection. However, this solution is not able to comply with the NSI-LVRT, which requires controlled injection of negative-sequence current simultaneously. In order to allow proper operation during unbalanced grid conditions, the vector output current control described in Figure 3.28 can be enhanced by the following elements:

- positive/negative-sequence extraction (PNSE);
- non-filtered PLL locked on the positive sequence voltage;
- component-based output current controller able to control both positive and negative sequences to their desired references in either stationary or synchronous reference frame;
- reference generator for reactive current positive and negative sequence and negative active current depending on the desired injection strategy.

4.4   Positive/Negative-Sequence Extraction

PNSE is necessary to achieve accurate tracking of the positive and negative sequences during unbalanced conditions. The symmetrical components theory first published in [2] is used. Two different techniques are briefly introduced and compared [3]:

- decoupled double synchronous reference frame (DDSRF) performed in dq synchronous coordinates
- double second-order generalized integrator (DSOGI) performed in $\alpha\beta$ stationary coordinates.

4.4.1   DDSRF-PNSE

This technique is an enhancement of the well-established synchronous reference frame (SRF)-PLL. It is based on two synchronous reference frames, rotating with the positive and negative synchronous angular frequencies, respectively. The positive and negative sequences are extracted as dc quantities after rotational transformations and decoupling.

Assuming a generic three-phase voltage vector $\mathbf{v}$, expressed here as a regular vector rather than a complex vector, with no zero-sequence component (i.e. a three-wire connection), the positive- and negative-sequence components in a dq frame rotating with the synchronous frequency ω_1 oriented along the positive-sequence voltage vector V^+ can be expressed as

$$\mathbf{v}^s = \begin{bmatrix} v_\alpha \\ v_\beta \end{bmatrix} = \mathbf{v}_{\alpha\beta}^+ + \mathbf{v}_{\alpha\beta}^- = \hat{V}_{+1} \begin{bmatrix} \cos(\omega_1 t + \varphi_{+1}) \\ \sin(\omega_1 t + \varphi_{+1}) \end{bmatrix} + \hat{V}_{-1} \begin{bmatrix} \cos(-\omega_1 t + \varphi_{-1}) \\ \sin(-\omega_1 t + \varphi_{-1}) \end{bmatrix}$$

$$\mathbf{v}_{dq+1} = \begin{bmatrix} v_{d+1} \\ v_{q+1} \end{bmatrix} = [T_{dq+1}] \cdot \mathbf{v}_{\alpha\beta} = \hat{V}_{+1} \begin{bmatrix} 1 \\ 0 \end{bmatrix} + \hat{V}_{-1} \begin{bmatrix} \cos(-2\omega_1 t) \\ \sin(-2\omega_1 t) \end{bmatrix} \qquad (4.8)$$

$$\mathbf{v}_{dq-1} = \begin{bmatrix} v_{d-1} \\ v_{q-1} \end{bmatrix} = [T_{dq-1}] \cdot \mathbf{v}_{\alpha\beta} = \hat{V}_{+1} \begin{bmatrix} \cos(2\omega_1 t) \\ \sin(2\omega_1 t) \end{bmatrix} + \hat{V}_{-1} \begin{bmatrix} 1 \\ 0 \end{bmatrix}$$

where the transformation matrix for a rotation with a generic angle θ is

$$[T_{dq+1}] = [T_{dq-1}]^T = \begin{bmatrix} \cos(\theta) & \sin(\theta) \\ -\sin(\theta) & \cos(\theta) \end{bmatrix}. \qquad (4.9)$$

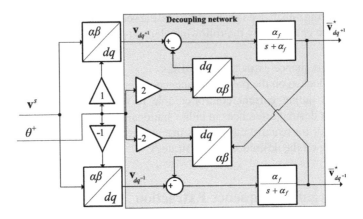

Figure 4.4   DDSRF-based PNSE with decoupling network.

As can be observed in Equation (4.8), there is a coupling between the positive and negative sequences in the rotating frame, owing to the fact that they rotate in opposite directions. The negative sequence appears in the dq^+ as an ac component oscillating with $-2\omega_1$, whereas the positive sequence oscillates in the dq^- frame with $2\omega_1$. Here it has to be noticed that the assumption that the angle of the negative sequence is equal in magnitude to the angle of the positive sequence is considered. In order to eliminate the oscillating component and efficiently use PI current controllers, a decoupling network as introduced in [3] can be used, as shown in Figure 4.4, along with an LPF.

The transfer function of the estimated decoupled positive sequence component is

$$\overrightarrow{\mathbf{v}}_{dq^{+1}}^{\star} = \begin{bmatrix} \overrightarrow{v}_{d^{+1}}^{\star} \\ \overrightarrow{v}_{q^{+1}}^{\star} \end{bmatrix} = [F]\{\mathbf{v}_{dq^{+1}} - [T_{dq^{+2}}]\overrightarrow{\mathbf{v}}_{dq^{-1}}^{\star}\}$$

$$\overrightarrow{\mathbf{v}}_{dq^{-1}}^{\star} = \begin{bmatrix} \overrightarrow{v}_{d^{-1}}^{\star} \\ \overrightarrow{v}_{q^{-1}}^{\star} \end{bmatrix} = [F]\{\mathbf{v}_{dq^{-1}} - [T_{dq^{-2}}]\overrightarrow{\mathbf{v}}_{dq^{+1}}^{\star}\} \tag{4.10}$$

where

$$[T_{dq^{+2}}] = [T_{dq^{-2}}]^T = \begin{bmatrix} \cos(2\omega_1 t) & \sin(2\omega_1 t) \\ -\sin(2\omega_1 t) & \cos(2\omega_1 t) \end{bmatrix} \tag{4.11}$$

$$[F] = \begin{bmatrix} \frac{\alpha_f}{s+\alpha_f} & 0 \\ 0 & \frac{\alpha_f}{s+\alpha_f} \end{bmatrix} \tag{4.12}$$

and α_f is the desired bandwidth in [rad/s] for the LPF ($\alpha_f < \omega_1$).

For an acceptable tradeoff between time response and oscillation damping during step response, the bandwidth of the LPF can be selected as

$$\alpha_f = \omega_1/\sqrt{2} \tag{4.13}$$

The transfer function of the detected positive sequence component with respect to the grid voltage in the stationary reference frame can be expressed as

$$v_s^+ = \frac{\alpha_f}{s^2 + 2\alpha_f s + \omega_1^2} \begin{bmatrix} s & -\omega_1 \\ \omega_1 & s \end{bmatrix} v_s. \tag{4.14}$$

Equation (4.14) is important as it defines the dynamics of the PNSE. The positive-sequence voltage once estimated is used to feed a conventional PLL for tracking the positive-sequence voltage angle. As the LPF in the DDSRF-PNSE is in the path of the positive-sequence voltage, the presence of an input filter in the PLL becomes unnecessary.

The usage of this double synchronous reference frame allows decoupling the effect of the negative-sequence voltage component on the dq signals detected by the synchronous reference frame rotating with positive angular speed, and vice versa, which makes possible an accurate grid synchronization under unbalanced grid faults.

4.4.2 DSOGI-PNSE

An alternative structure is the double second-order generalized integrator positive/negative-sequence extraction (DSOGI-PNSE), first introduced in [4] and depicted in Figure 4.5. It can extract the positive and negative sequences in a stationary reference frame using the 90-degrees-lagging phase-shifting operator $q = e^{-j\frac{\pi}{2}}$ for creating quadrature signals of the voltage components according to

$$v^{s+} = [T_{\alpha\beta+1}]v^s \quad ; \quad [T_{\alpha\beta+1}] = \frac{1}{2}\begin{bmatrix} 1 & -q \\ q & 1 \end{bmatrix}$$

$$v^{s-} = [T_{\alpha\beta-1}]v^s \quad ; \quad [T_{\alpha\beta-1}] = \frac{1}{2}\begin{bmatrix} 1 & q \\ -q & 1 \end{bmatrix} \tag{4.15}$$

Two second-order generalized integrator quadrature signal generator (SOGI-QSG) with a structure depicted in Figure 4.6 (left) are capable of producing the direct and in-quadrature signals for the α and β components of the input vector, i.e. $v'_\alpha, v'_\beta, qv'_\alpha, qv'_\beta$, with same dynamics determined by the SOGI gain k. These signals are provided as inputs for the PNSE, as shown in Figure 4.6 (right) for calculation of the positive and negative sequence in stationary reference frame according to Equation (4.15).

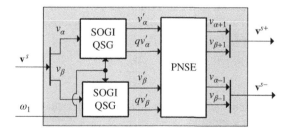

Figure 4.5   DSOGI-PNSE.

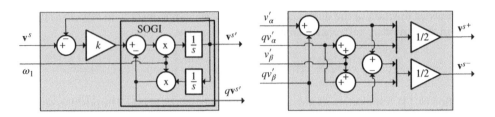

Figure 4.6   SOGI-QSG (left), PNSE (right).

It is worth noticing that the symmetric components calculated by SOGI-PNSE are not coupled as in the case of DDSRF and thus there is *no need* for a decoupling network.

The following equations describe the dynamics of the SOGI-QSG:

$$\frac{v'(s)}{v(s)} = \frac{k\omega_1 s}{s^2 + k\omega_1 s + \omega_1^2}$$

$$\frac{qv'(s)}{v(s)} = \frac{k\omega_1^2}{s^2 + k\omega_1 s + \omega_1^2} \tag{4.16}$$

As can be observed in Equation (4.16), both the direct and quadrature components have the same amplitude, provided that the grid frequency ω_1 is correctly estimated by the PLL. The gain k of the SOGI-QSG determines the dynamics in terms of settling time. For an acceptable tradeoff between the time response and step-response oscillation damping, the recommended value is [4]:

$$k = \sqrt{2} \tag{4.17}$$

The filtering characteristic of the SOGI-QSG attenuates the effect of the distorting high-order harmonics from the input to the output and make the requirement for the input filter in the PLL to become unnecessary.

The transfer function of the detected positive sequence component with respect to the grid voltage in stationary reference frame can be expressed as

$$\mathbf{v}^{s+} = \frac{1}{2} \frac{k\omega_1}{s^2 + k\omega_1 s + \omega_1^2} \begin{bmatrix} s & -\omega_1 \\ \omega_1 & s \end{bmatrix} \mathbf{v}^s \tag{4.18}$$

A quick look at the expression of $\mathbf{v}_s^+$ in Equations (4.14) and (4.18) reveals that, even if DDSRF-PNSE and DSOGI-PNSE seem very different in structure, they are actually equivalent in dynamic response provided the condition

$$k = 2\frac{\alpha_f}{\omega_1} \tag{4.19}$$

It is worth mentioning that the recommended tuning in Equations (4.13) and (4.17) by default gives the equivalence condition.

Example 4.2 *DSOGI-PNSE and DDSRF-PNSE are both simulated with parameters:* $K = \sqrt{2}$, $\alpha_f = \omega_1 \frac{1}{\sqrt{2}}$, *respectively, for the grid fault considered in Example 4.1. In both*

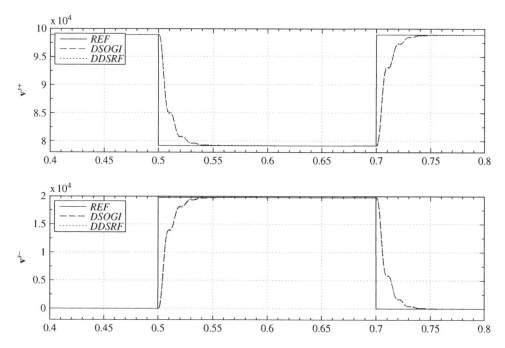

Figure 4.7   Step-down response of $\mathbf{v}^{s+}$ (top) and down-down response of $\mathbf{v}^{s-}$ (bottom) for DSOGI and DDSRF based PNSE with parameters: $k = \sqrt{2}, \alpha_f = \omega_1 \frac{1}{\sqrt{2}}$.

cases a classic PLL without input filtering is attached to the detected positive sequence ($\alpha_p = 50\ rad/s$, $\alpha_{ip} = 10\ rad/s$). *The response is shown in Figure 4.7 and the following observations can be made:*

- *The response of the two synchronization systems are dynamically identical.*
- *The dynamics for positive and negative sequence extraction are also identical.*

In summary, the main differences between dynamic equivalent DDSRF-PNSE and DSOGI-PNSE can be summarized as following:

- DSOGI exhibits increased robustness to grid disturbances as it uses the PLL-estimated frequency ω_1 as input, whereas the PLL-estimated phase angle θ^+ is used for DDSRF, which exhibits a higher dynamics during faults.
- DSOGI needs no rotational transformations opposed to DDSRF, which requires four.
- DSOGI needs no decoupling network.

4.5   Injection Reference Strategy

A possible strategy is the flexible positive/negative sequence control (FPNSC) introduced in [4] that defines a generic current reference injection into a grid with arbitrary generic voltage $\mathbf{v}$ in terms of positive and negative sequence for given P and Q reference as

$$i^{\star} = P \cdot \underbrace{\left(\frac{k_1}{|v^+|^2} \cdot v^+ + \frac{(1-k_1)}{|v^-|^2} \cdot v^- \right)}_{i_p^{\star}} + Q \cdot \underbrace{\left(\frac{k_2}{|v^+|^2} \cdot v_{\perp}^+ + \frac{(1-k_2)}{|v^-|^2} \cdot v_{\perp}^- \right)}_{i_q^{\star}} \qquad (4.20)$$

where k_1 and k_2 can be interpreted as droop factors regulating the ratio between the positive and negative sequence content of active and reactive current, v^+ and v^- are the positive sequence and negative sequence respectively and $v_{\perp}$ is an orthogonal version (90-degrees leaded) of the original grid voltage vector v defined in order to express reactive power in same way as active power using dot-product and not cross-product ($q = v_{\perp} \cdot i$):

$$\underset{abc}{v_{\perp}} = \frac{1}{\sqrt{3}} \begin{bmatrix} 0 & 1 & -1 \\ -1 & 0 & 1 \\ 1 & -1 & 0 \end{bmatrix} v_{abc} \quad ; \quad \underset{\alpha\beta}{v_{\perp}} = \begin{bmatrix} 0 & -1 \\ 1 & 0 \end{bmatrix} v_{\alpha\beta} \quad ; \quad \underset{dq}{v_{\perp}} = \begin{bmatrix} 0 & -1 \\ 1 & 0 \end{bmatrix} v_{dq}. \quad (4.21)$$

By means of regulating k_1 in within the range from 0 to 1, it is possible to change the proportion in which the positive- and negative-sequence components of the active currents injected into the grid are participating in exchanging a given amount of active power P to the grid. For instance, by making $k_1 = 1$, balanced positive injections is achieved, while by making $k_1 = 0$, balanced negative-sequence currents will be injected into the grid to deliver the active power P. In some special cases, k_1 might be in the [0–1] range. In such cases, one of the sequence components of the injected currents would be draining active power from the grid, whereas the other sequence component would be delivering as much active power as necessary to balance the system and make the total active power delivered to the grid equal to P.

Positive-sequence reactive-power injection is generally used to help the PCC-voltage recovery after faults in inductive lines. This grid support feature can be achieved by the setting: $k_2 = 1$. On the other hand, the negative-sequence voltage at the PCC can be reduced by injecting negative-sequence reactive currents. This can be achieved by the setting: $k_2 = 0$ Thus, k_2 can decide the type of application as follows:

- $k_2 = 1$: Injection of positive sequence reactive power = Grid voltage support.
- $k_2 = 0$: Injection of negative sequence reactive power = Grid voltage unbalance compensation.
- $0 < k_2 < 1$: Injection of simultaneous of positive and negative sequence reactive power.

To derive the reference of the current in the dq rotating frame, the IPT is used. Assuming a generic converter operating in rectifying mode under unbalanced grid voltage conditions, the instantaneous active and reactive powers can be expressed as

$$p = P + \underbrace{P_{c2} \cos(2\omega_1 t) + P_{s2} \sin(2\omega_1 t)}_{\tilde{p}}$$

$$q = Q + \underbrace{Q_{c2} \cos(2\omega_1 t) + Q_{s2} \sin(2\omega_1 t)}_{\tilde{q}}. \qquad (4.22)$$

As it can be observed, the instantaneous power consist in dc terms P and Q and ac terms oscillating at double fundamental frequency P_{c2}, P_{s2}, Q_{c2} and Q_{s2}. As there only four degrees of freedom that can be controlled $[i_{d+1}, i_{q+1}, i_{d-1}, i_{q-1}]$, it means that only four of the six power terms can be effectively controlled. As in the case of MMC, the capacitor unbalance is associated with active power unbalance within the arms, the following assumption is made:

- The oscillating terms Q_{c2} and Q_{s2} are neglected from the objective of the control
  The remaining four power terms can be expressed as

$$
\begin{bmatrix} P \\ Q \\ P_{c2} \\ P_{s2} \end{bmatrix} = \frac{3}{2} \cdot \underbrace{\begin{bmatrix} v_{d+1} & v_{q+1} & v_{d-1} & v_{q-1} \\ v_{q+1} & -v_{d+1} & v_{q-1} & -v_{d-1} \\ v_{d-1} & v_{q-1} & v_{d+1} & v_{q+1} \\ v_{q-1} & -v_{d-1} & -v_{q+1} & v_{d+1} \end{bmatrix}}_{A} \cdot \begin{bmatrix} i_{d+1} \\ i_{q+1} \\ i_{d-1} \\ i_{q-1} \end{bmatrix}.
\tag{4.23}
$$

Thus, it is possible to solve for the current reference:

$$
\begin{bmatrix} i_{d+1}^{\star} \\ i_{q+1}^{\star} \\ i_{d-1}^{\star} \\ i_{q-1}^{\star} \end{bmatrix} = A^{-1} \cdot \frac{2}{3} \cdot \begin{bmatrix} P \\ Q \\ P_{c2} \\ P_{s2} \end{bmatrix}.
\tag{4.24}
$$

Using FPNSC the instantaneous power delivered to the grid can be expressed as $p = P + \tilde{p}$ where:

$$
\tilde{p} = \left(\frac{P \cdot k_1}{|v^+|^2} + \frac{P \cdot (1 - k_1)}{|v^-|^2} \right) \mathbf{v}^+ \cdot \mathbf{v}^- + \left(\frac{Q \cdot k_2}{|v^+|^2} - \frac{Q \cdot (1 - k_2)}{|v^-|^2} \right) \mathbf{v}_\perp^+ \cdot \mathbf{v}^-.
\tag{4.25}
$$

As can be seen, there are two oscillating terms: one related to P and the other one to Q. Solving analytically the equation $\tilde{p} = 0$ for k_1 and k_2 is not possible as there are two unknowns.

The FPNSC is a generic current reference generator having the advantage that can be easily particularized for the different injection strategies as it follows.

4.5.1   PSI with PSI-LVRT Compliance

The goal in this strategy is to inject balanced sinusoidal currents with only positive-sequence component. Positive-sequence active-current injection for maintaining the load P at the dc side and positive-sequence reactive-current injection according to PSI-LVRT are performed simultaneously. As it deals only with positive sequence, this method can be easily implemented with conventional dq synchronous controllers providing that the PLL is able to estimate accurately the phase-angle of the positive-sequence component of the grid voltage during unbalanced conditions (see Section 4.3.1).

The general positive balanced current reference can be obtained from FPNSC by setting $k_1 = k_2 = 1$ in (4.20) resulting in

$$
\mathbf{i}^{\star} = P \underbrace{\frac{\mathbf{v}^+}{|\mathbf{v}^+|^2}}_{\mathbf{i}_p^{\star}} + Q \underbrace{\frac{\mathbf{v}_\perp^+}{|\mathbf{v}^+|^2}}_{\mathbf{i}_q^{\star}}
\tag{4.26}
$$

In a rotating dq frame we can assume that the PLL block orientate the dq frame along with the d component of the voltage results in $v_{d+1} = \hat{V}_{+1}, v_{q+1} = 0$ and the references simplifies to

$$
\begin{cases}
i^\star_{d+1} = \dfrac{2}{3}\dfrac{P}{v_{d+1}} \\[4mm]
i^\star_{q+1} = -\dfrac{2}{3}\dfrac{Q}{v_{d+1}}
\end{cases}
\tag{4.27}
$$

Note that the same result can be obtain by using the condition $i^\star_{d-1} = i^\star_{q-1} = 0$ in (4.23).

The reactive current reference can be adapted to the PSI-LVRT requirement and the reference becomes:

$$
\begin{cases}
i^\star_{d+1} = \dfrac{2}{3}\dfrac{P}{v_{d+1}} \\[4mm]
i^\star_{q+1} = k_+(0.9 - V_+)
\end{cases}
\tag{4.28}
$$

Balanced injection will not cancel the oscillating terms P_{c2}, P_{s2} which in turn, will generate double frequency ripple in both P and Q.

4.5.2 MSI-LVRT Mixed Positive- and Negative-Sequence Injection with both PSI-LVRT and NSI-LVRT Compliance

The goal in this strategy is to inject positive active current to maintain the load, positive reactive current according with the PSI-LVRT and negative sequence reactive currents according with the NSI-LVRT. Thus the complete reference for the MSI injection strategy case becomes:

$$
\begin{cases}
i^\star_{d+1} = \dfrac{2}{3}\dfrac{P}{v_{d+1}} \\[3mm]
i^\star_{q+1} = k_+(0.9 - V_+) \\[3mm]
i^\star_{d-1} = 0 \\[3mm]
i^-_q = -k_-(V_- - 0.05)
\end{cases}
\tag{4.29}
$$

This solution ensures compliance with both PSI-LVRT and NSI-LVRT. There are oscillations in P that could challenge the internal balance of the MMC. Oscillations in Q are present, but they normally do not have a negative impact on the grid.

4.6 Component-Based Vector Output-Current Control

In the following, two component-based control strategies for the output-current vector control are described, based on the DDSRF and DSOGI PNSE methods, respectively. A comparison follows.

4.6.1 DDSRF-PNSE-Based Control

The conventional dq control strategy shown in Figure 3.28 is enhanced by the use of the PNSE and positive/negative sequence reference generator (PNSRG), as shown in Figure 4.8.

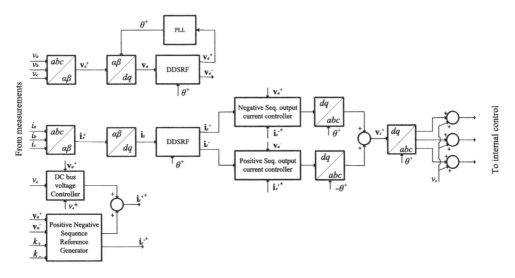

Figure 4.8   DDSRF-PNSE-based vector output-current control.

4.6.2   DSOGI-PNSE-Based Control

This strategy designed for stationary $(\alpha\beta)$ reference-frame control is depicted in Figure 4.9. The main advantage is that the positive- and negative-sequence components can be added algebraically in the stationary frame into a single current reference without the need for rotational transformations. Additionally, a single PR controller $(h = 1)$ will be able to regulate both the positive- and negative-sequence components embedded in the unique reference. There is of course the need for rotational transformations in order to translate the current reference given by the dc-bus voltage controller and the references calculated by the PNSRG into the stationary frame.

Overall, it can be said that the two control strategies are dynamically equivalent. In order to see differences, one has to look into implementation details such as

- discretization (SOGI require more attention than a simple integrator);
- accuracy of grid frequency and angle estimation of the PLL;
- computational burden (stationary control requires less transformations and only one current controller).

In the following example, the DSOGI-PNSE method with stationary reference current control is used in order to evaluate two injection strategies.

Example 4.3 *The Test Converter in rectifier mode operating in stationary conditions with 100 MW power on dc-bus is subjected to same grid fault conditions as in Example 4.1. For synchronization, a DSOGI-PNSE with the parameters: $k = \sqrt{2}$ and conventional PLL ($\alpha_p = 50$ rad/s, $\alpha_{ip} = 10$ rad/s) is considered. The PR output current controller in stationary reference frame is using $\alpha_c = 2000$ rad/s and $\alpha_1 = 50$ rad/s with an LPF for the feedforward ($\alpha_f = 1000$ rad/s). The dc-bus control and circulating current control are like in Example 4.1. During the fault, the following injection strategies are considered:*

- *PSI-LVRT with $k_+ = 2.5$*
- *MSI-LVRT with $k_+ = 2.5$ and $k_- = 2.5$*

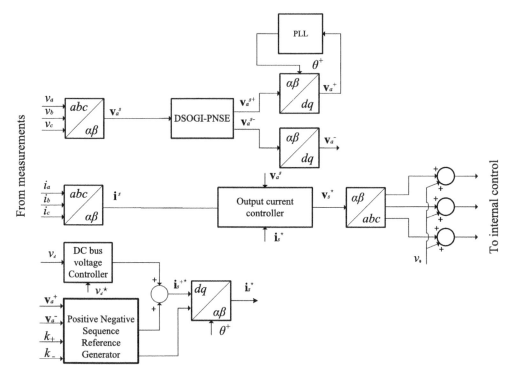

Figure 4.9 DSOGI-PNSE-based vector output-current control.

Observing the results in Figure 4.10 and 4.11, the followings remarks can be made:

- In both cases, the ripple on the dc-bus and circulating current are eliminated (after a short transient period), confirming the finding in Section 3.3.7, Equation (3.53).
- The resonant control of circulating current is proven effective during unbalanced grid conditions being able to control the circulating current to a dc value after a transient.
- MSI-LVRT is exhibiting larger ripple in the power and in the sum-capacitor voltage than PSI-LVRT, which is to be expected, owing to the unbalanced injected currents.
- Owing to the negative reactive current injection, the MSI-LVRT strategy is able to reduce the grid voltage negative sequence from 0.2 to 0.15, which is a desirable effect.

4.7 Summary

In summary, the following conclusions can be drawn related to the control of MMC during unbalanced grid conditions:

- The recent GCs for large-scale generation connected to the transmission systems require independent control of both positive- and negative-sequence reactive-current injection for improved grid support.

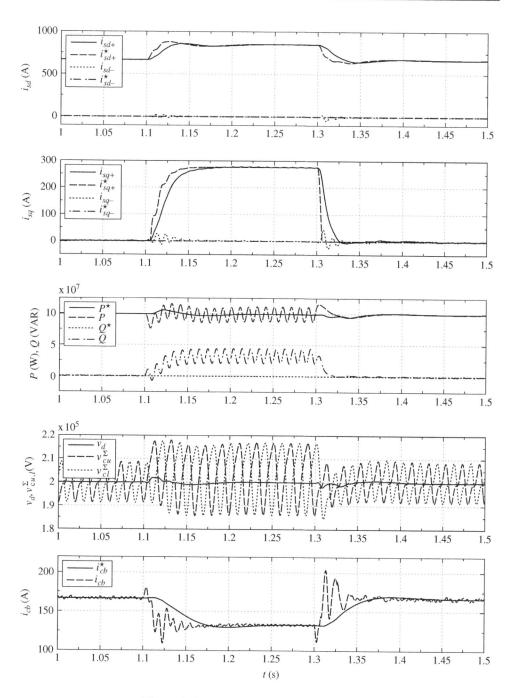

Figure 4.10   PSI with PSI-LVRT compliance.

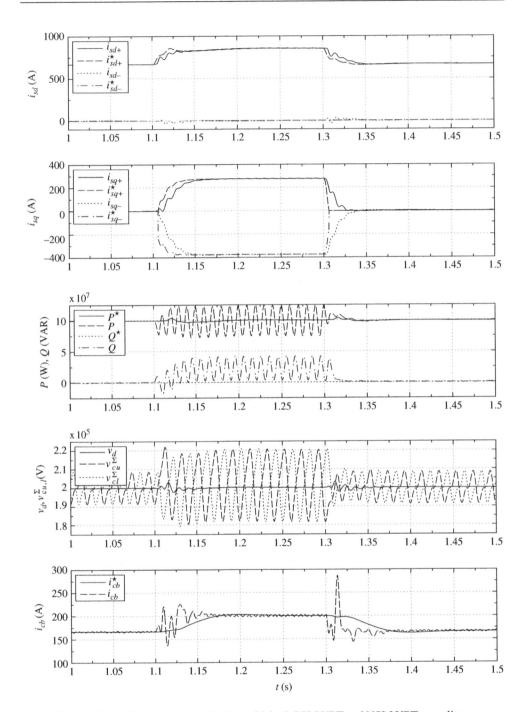

Figure 4.11 Mixed-sequence injection with both PSI-LVRT and NSI-LVRT compliance.

- The positive-sequence reactive-current injection requirement PSI-LVRT can be fulfilled with conventional vector control enhanced with a notch filter in the PLL structure able to properly extract the positive-sequence angle.
- The negative-sequence reactive-current injection requirement NSI-LVRT leads to the need of improving the conventional control by adding PNSE and component-based current control structures.
- DDSRF-PNSE and DSOGI-PNSE can both be successfully used and it has been demonstrated that they can be designed to be dynamically equivalent.
- It has been also demonstrated that stationary and synchronous control strategies can be designed to be dynamically equivalent, but stationary control is less complex (fewer transformations and controllers, no need of decoupling) and more robust during unbalanced faults, as it uses the grid frequency instead of the grid angle.
- Also, current reference limitation is less complex in stationary frame as the sequence components can be added algebraically in contrast to the synchronous frame where they are defined in different reference frames.
- For circulating current suppression, the resonant controller ($h = 2$) and $R_a = 20\,\Omega$ proves to be efficient during unbalanced grid faults as it is capable of controlling the circulating current to a dc value.
- It has been demonstrated that MMC-HVDC with PNSE and resonant control can comply with both PSI-LVRT and NSI-LVRT simultaneously without violating the limits for injected current, dc-bus voltage, or sum-capacitor voltages. Of course, this study is using the average arm model, assuming that all capacitor voltages in the arm are identical. For a complete analysis, it is necessary to use a detailed switching model, as described in Chapter 6.
- In recent studies [5] it has been shown that by injecting unbalanced current and by adding a double fundamental component to the circulating current, the oscillations in P can be significantly reduced at the cost of an increased capacitor voltage ripple in the faulted phases, which in practice will mean bigger capacitors. This feature has to be evaluated from the perspective of the application in order to decide whether increasing the size of the capacitors for balancing the ac power is a cost-effective solution.

References

[1] T. Wijnhoven, G. Deconinck, T. Neumann, and I. Erlich, "Control aspects of the dynamic negative sequence current injection of type 4 wind turbines," *PES General Meeting: Conference & Exposition*, pp. 1–5, 27–31 Jul. 2014.

[2] C. L. Fortescue, "Method of symmetrical coordinates applied to the solution of polyphase networks," *Transactions of the American Institute of Electrical Engineers*, vol. 37, pp. 1027–1140, Jul. 1918.

[3] P. Rodriguez, J. Pou, J. Bergas, J. I. Candela, R. P. Burgos, and D. Boroyevich, "Decoupled double synchronous reference frame PLL for power converters control," *IEEE Transactions on Power Electronics*, vol. 22, no. 2, pp. 584–592, Mar. 2007.

[4] R. Teodorescu, M. Liserre, and P. Rodriguez, "Grid converters for photovoltaic and wind power systems," Wiley-IEEE, doi: 10.1002/9780470667057, 2011.

[5] M. Vasiladiotis, N. Cherix, and A. Rufer, "Impact of grid asymmetries on the operation and capacitive energy storage design of modular multilevel converters," *IEEE Transactions on Industrial Electronics*, vol. 62, no. 11, pp. 6697–6707, Nov. 2015.

5

Modulation and Submodule Energy Balancing

5.1 Introduction

Previous chapters mostly treat the strings of submodules, which make up a modular multilevel converter (MMC), at an aggregated level, where the switching of individual submodules is not considered. In this chapter, the inner workings of the strings is studied in more detail. Solutions to two fundamental issues related to the timing of the submodule switchings are discussed. The first is the modulation of the output voltages, i.e. what number of cells to insert at each time instance to provide the voltages required for the operation of the converter. The second is the balancing of the submodule capacitor energies within each string. It can be achieved by appropriately selecting which submodules to insert or bypass.

A switching power converter cannot provide a continuously varying output. Instead, the output voltage can only assume a number of discrete levels. The purpose of any pulse-width modulation (PWM) method is, therefore, to choose the instances for switching between the levels to ensure that the short-time average of the ac-side voltage coincides with that of a reference signal over a switching cycle. For a sinusoidal reference this is equivalent to ensuring that the magnitude and phase of the fundamental component of the switched waveform should coincide with those of the reference. In addition to the desired low-frequency reference voltage the switching process generally also produces higher-order harmonics, which are undesirable. These have a number of negative consequences for the equipment connected to the converter, and their influence therefore needs to be handled in a proper way. In most cases increasing the switching frequency makes this easier.

As discussed in previous chapters, a transition to a multilevel converter topology allows for much improved harmonic performance at a given switching frequency. The harmonics can be made to appear at higher frequencies, and their amplitude can be reduced. In modular multilevel converters, where hundreds of submodules can be used, resulting in a corresponding number of voltage levels, this improvement is even more pronounced. The harmonic requirements may therefore not be decisive in determining the required switching frequency.

Design, Control, and Application of Modular Multilevel Converters for HVDC Transmission Systems, First Edition.
Kamran Sharifabadi, Lennart Harnefors, Hans-Peter Nee, Staffan Norrga, and Remus Teodorescu.
© 2016 John Wiley & Sons, Ltd. Published 2016 by John Wiley & Sons, Ltd.
Companion Website URL: www.wiley.com/go/Sharifabadi/ModularConverters

Still, there is a need for suitable methods for selecting the number of submodules inserted at each instant.

However, the operation of MMCs also poses some additional challenges since the dc capacitance is distributed among the submodules. As discussed in previous chapters, the energy stored in the submodule capacitors of an MMC is fluctuating significantly during operation. This fluctuation can be split into two parts. One is common to an entire submodule string since it is related to the overall energy exchange with the string. It corresponds to a fluctuation in the sum of all capacitor voltages in the string, i.e. the total direct voltage available in the string. It consists of low-order harmonic components, mainly fundamental and second. The control system needs to compensate for this variation, as described in Chapter 3.

There is also a fluctuation that is individual to each submodule, which is related to the switching of the submodules. An additional purpose of the control scheme for an MMC is therefore to achieve voltage balancing between the submodule capacitors over time. This is generally made by appropriately selecting which submodules to insert or bypass. Furthermore, the losses should preferably be uniformly distributed among all the semiconductor valves in the converter.

Both the modulation and the submodule balancing are concerned with the issue of when to switch the individual submodules, and are therefore intimately linked. This chapter is devoted to describing, and to some extent analyzing, methods for modulation and submodule energy balancing for modular multilevel converters. First, a couple of fundamental concepts related to modulation are introduced. Thereafter, carrier-based modulation, where the switching instances are made at the intersections of the reference and a high-frequency carrier, is described and analyzed. This treatment starts with two-level modulation, since it forms the basis also for various multilevel modulation schemes. It is later shown how an extension to several levels can be made, and how this affects the harmonic properties of the modulated voltages. Also, another modulation scheme, suitable for MMCs—nearest-level control—is reviewed. Finally, different submodule energy balancing methods are reviewed and compared.

5.2 Fundamentals of Pulse-Width Modulation

5.2.1 Basic Concepts

The purpose of any pulse-width modulation (PWM) method is to synthesize a certain voltage signal, labeled the *reference voltage* in the output. This is made by switching an available direct voltage, maintained by capacitors, between two or more discrete levels. In a multilevel converter there are several dc capacitors, permitting several levels in the output voltage.

A vital concept relevant to most forms of PWM in the steady-state is the *modulation index* (sometimes referred to as *amplitude modulation ratio*), which relates the magnitude of the desired low-frequency alternating voltage component at the converter output to half of the pole-to-pole dc-side voltage:

$$m_a = \frac{2\hat{v}_{s,1}}{V_d}. \tag{5.1}$$

It will generally assume values of between zero and one, but for three-phase converters slightly higher values are possible by adding a zero-sequence component to the references, as is discussed below.

In addition to the desired replication of the reference signal the switching process also always gives rise to undesired harmonics in the output. The desired outcome of modulation is generally to move these harmonics to higher frequencies where their negative effects are less pronounced, and where they can be more easily filtered out. In a multilevel converter also the magnitude of the harmonics can be reduced. The average frequency at which a single phase leg is operating is referred to as the *switching frequency*. This parameter is vital to the operation of the system since it determines the switching losses of the power semiconductors, which commonly make up a significant portion of the overall power losses. The *frequency ratio* relates the switching frequency to the frequency of the desired alternating voltage component

$$m_f = \frac{f_{sw}}{f_1}. \tag{5.2}$$

It will obviously always exceed one. For a multilevel converter the *pulse frequency*, i.e. the frequency at which the step-wise variations of the output voltages occur, is also of interest since this number will be decoupled from the switching frequency. This is the case since generally only a fraction of the switching elements are involved in each transition of the output voltage.

5.2.2   Performance of Modulation Methods

A key property for any voltage source converter is the harmonic distortion of the output voltage and current waveforms, i.e. the extent to which the waveforms deviate from a pure sinusoid. Depending on the application, the requirements in this context may take different forms.

For some applications the additional losses caused by harmonics in equipment connected to the converter may be most important. This is often the case in drive systems where the converter is connected to an electrical machine (a motor or a generator). These losses are caused by harmonic currents driven by the voltage harmonics.

For grid-connected systems, for instance for connecting renewable energy sources to the ac grid or in HVDC stations, certain standards (such as IEC 61000-3-2/4/6 and IEEE 519) regulate the maximum permissible harmonic content of the ac-side currents and voltages. In these cases fulfillment of the norms will generally imply a balance between the design of the ac-side filter, the properties of the modulation algorithm, and the choice of switching frequency.

In some applications, it may be vital to avoid specific frequencies, whereas harmonics at other frequencies are less critical. This applies, for instance, to line side converters in electric railway traction systems, owing due to strict requirements of non-interference with signalling systems. To be able to compare the merits of different modulation schemes with respect to the harmonic content there is a need for performance indicators that summarize the overall content into a single figure. One such figure is the total harmonic distortion (THD), defined as the combined rms value of the harmonics divided by the rms value of the fundamental

$$\text{THD} = \frac{1}{V_1} \sqrt{\sum_{h=2}^{\infty} V_h^2}. \tag{5.3}$$

In this expression V_h is the rms value of harmonic component of order h. In most applications the harmonic content of the phase *currents* is a more important property of a converter than that of the voltage harmonics. It is mainly the current harmonics that cause additional losses in ac-side equipment connected to the converter. Furthermore, most norms and standards stipulate limits for current rather than voltage harmonics. For this reason the figure of merit chosen should preferably provide information about the impact on the current harmonics caused by the switched voltage waveforms. This causes some difficulties in the general case as the phase currents also depend on the nature of the circuitry on the ac side of the converter. However, as this circuitry generally shows a more or less inductive behaviour at high frequencies, the current harmonics will be scaled with the reciprocal of the harmonic order, compared to the corresponding voltage harmonics

$$I_h = \frac{V_h}{h\omega_1 L}.$$ (5.4)

The inductance L represents, for instance, a line filter, or the total leakage inductance of an induction machine, and V_h is the rms value of the hth harmonic in the voltage applied to this inductance. This dependency has given rise to the use of *weighted total harmonic distortion (WTHD)* for the characterization of PWM voltage signals

$$\text{WTHD} = \frac{1}{V_1}\sqrt{\sum_{h=2}^{\infty}\left(\frac{V_h}{h}\right)^2}.$$ (5.5)

The use of the rms value of the fundamental for normalization has certain drawbacks. First, as the fundamental is proportional to the modulation index WTHD will tend to overstate the distortion at low modulation indices. Furthermore, when the modulation index approaches zero the fundamental voltage will also do so, whereby the expression will tend to infinity and thus become meaningless. To avoid these drawbacks a constant normalization factor can instead be used. A natural choice of normalization is the rms value of the fundamental obtained at $m_a = 1$.

For a converter with several phase legs the ac-side harmonic currents will be driven by the switched voltages appearing between the phases. Zero-sequence voltage harmonics, which are common to the phases, will generally not impact the circuitry on the ac side. In these cases the harmonics of the line-to-line voltages, rather than the phase voltages, are the most relevant for assessing different modulation schemes.

5.2.3   Reference Third-Harmonic Injection in Three-Phase Systems

A basic modulation scheme for a three-phase voltage source converter (VSC) can be achieved by applying any modulation method, with sinusoidal and symmetrically phase-shifted reference waveforms, to the three phase legs. Under such circumstances the maximum possible modulation index in the linear range equals one, which corresponds to a peak output voltage equal to half the dc-link voltage. If the reference amplitude is increased further the direct voltage will be insufficient for linear modulation during the entire fundamental cycle, and overmodulation will occur.

Under symmetric conditions a third harmonic component in the reference voltages will be a zero-sequence component. The addition of such a third harmonic component to all the references makes it possible to use a higher reference fundamental without overmodulating the converter. The third harmonic reduces the peak value of the reference so that it does not exceed the dc-link voltage; see Figure 5.1. In many three-phase applications the midpoint of the ac-side source or load the converter is attached to is not connected. This is the case in motor drives

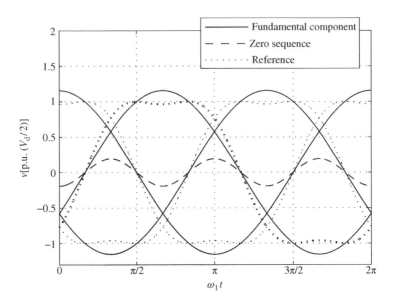

Figure 5.1   Impact of reference third-harmonic injection on three-phase modulation.

and in many grid-connected applications that use a transformer to interface the converter to the grid. The load or source is then not affected by the zero-sequence component which does not appear in the line-to-line voltages.

In [1], for example, it is explained that a sinusoidal zero-sequence with an amplitude of $1/6$ of the fundamental is optimal with respect to increasing the maximum modulation index. With this choice m_a can be increased to $2/\sqrt{3}$, which approximately amounts to a 15% increase. The zero-sequence does not need to be sinusoidal as long as its period is one-third of the fundamental period. The addition of a zero sequence, as described, is highly useful as it implies that higher output voltage, and thereby more power, can be achieved with a given converter. It is therefore widely used, both in motor drives and in grid-connected applications.

5.3   Carrier-Based Modulation Methods

An important class of modulation methods is the *carrier-based methods*, where the switching instances are determined by the intersections of a voltage reference and a carrier of higher frequency. These methods also lend themselves well to frequency-domain analysis, which makes it possible to clearly understand the benefits of multilevel operation.

5.3.1   Two-Level Carrier-Based Modulation

To begin the treatment of carrier-based methods the modulation of one phase leg of a two-level converter is first introduced. It serves as a basis for the discussion of polyphase converters and multilevel converters in the following sections. The precondition for the waveform synthesis is

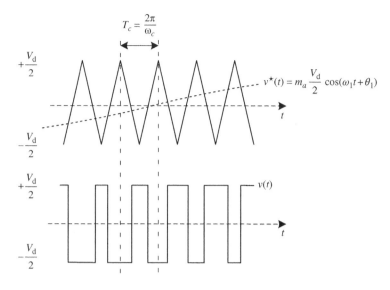

Figure 5.2 Carrier-based PWM.

in this case that the ac-side voltage is switched between two fixed dc levels: $+V_d/2$ and $-V_d/2$. See Figure 1.2 for a description of how this is achieved in practice. The modulation method is illustrated in Figure 5.2. As seen in the figure there is a reference wave (in this case a sinusoid) and a carrier (in this case a triangle wave). The magnitude of the carrier is $V_d/2$, implying that the carrier spans the entire range between the dc poles in a cycle. The modulated waveform is switched to the upper dc terminal $(+V_d/2)$ when the reference exceeds the carrier, and to the lower terminal otherwise. Thus, the commutations occur at the intersections of the two waveforms and the switching frequency equals the carrier frequency. From the figure it can intuitively be understood that the average values over a carrier cycle of the reference and the modulated function should be approximately equal. For instance, if the reference stays close to $+V_d/2$ the output will be switched to the upper dc terminal during most of the cycle. It can furthermore be understood that when the frequency ratio is an integer the modulated voltage $v(t)$ will be periodic, with the reference frequency as fundamental. This is the case since an integer number of carrier cycles will then fit into a fundamental cycle. As is discussed in the next section, however, using a non-integer frequency ratio is not necessarily a drawback.

5.3.2 Analysis by Fourier Series Expansion

Analyzing stationary PWM waveforms in the frequency domain is of interest to understand their harmonic content, which is vital for any design efforts aimed at achieving compliance with harmonic norms and regulations. For a periodic signal, $f(t)$, the *Fourier series expansion* can be used for this purpose. The complex Fourier series takes the following form:

$$f(t) = \sum_{h=-\infty}^{\infty} C_h e^{jh\omega_1 t} \tag{5.6}$$

where $\omega_1 = 2\pi/T$ is the angular velocity of the fundamental oscillation and C_h is the complex Fourier coefficient of order h. These coefficients can be computed by evaluating the Fourier integral:

$$C_h = \frac{1}{T} \int_0^T f(t) e^{-jh\omega_1 t} dt. \tag{5.7}$$

For a real-valued signal $f(t)$ the following relationship can easily be derived for the Fourier coefficients:

$$C_{-h} = \overline{C_h}. \tag{5.8}$$

Using this relation and basic trigonometric identities an alternative form of the Fourier series expansion can be derived:

$$f(t) = C_0 + 2 \sum_{h=1}^{\infty} |C_h| \cos(h\omega_1 t + \arg[C_h]) \tag{5.9}$$

It is clear that C_0 is the dc component of the signal. From this expression it is also obvious that the complex Fourier coefficients with $h \geq 1$—apart from a factor of two—can be interpreted as phasors, directly providing information of the magnitude and relative phase displacement of the harmonic components at each order.

However, as mentioned in the previous section, a carrier-based PWM signal is not necessarily periodical if the ratio of the carrier frequency to the reference fundamental frequency is a non-integer. Also, even for integer frequency ratios, computing the Fourier coefficients analytically is challenging, and may not provide any deeper insights into the frequency-domain behavior of the modulation scheme. Another methodology, which treats the periodicity of the carrier and reference signals separately, is more useful in these cases [2]. According to this methodology, the modulated signal is considered a function of the running phase angles $\omega_1 t$ and $\omega_c t$ of the reference and the carrier respectively. Thus, it will be a two-dimensional function $f(x, y)$ where

$$x = \omega_c t + \theta_c \tag{5.10}$$

$$y = \omega_1 t + \theta_1. \tag{5.11}$$

Importantly, the modulated signal will now be periodic in both x and y with a period of 2π rad. As such, it can be expanded into a *double Fourier series expansion* analogously to the case of how a one-dimensional periodical function can be expressed by a conventional Fourier series. The sum of complex exponentials in Equation (5.6) will then be replaced by a double sum as

$$f(x, y) = \sum_{m=-\infty}^{\infty} \sum_{n=-\infty}^{\infty} C_{mn} e^{j(mx+ny)}, \tag{5.12}$$

and the Fourier coefficients are given by a double integral as

$$C_{mn} = \frac{1}{4\pi^2} \int_{-\pi}^{\pi} \int_{-\pi}^{\pi} f(x, y) e^{-j(mx+ny)} dx dy. \tag{5.13}$$

Again, for a real-valued function $f(x, y)$, changing the sign of the indices of a Fourier coefficient will render the complex conjugate of the coefficient

$$C_{-m-n} = \overline{C_{mn}}. \tag{5.14}$$

Using this relation, Equation (5.12) can be rewritten as (analogously to Equation (5.9) for the one-dimensional case),

$$
\begin{aligned}
f(t) = \; & C_{00} + \\
& 2 \sum_{n=1}^{\infty} |C_{0n}| \cos(n\omega_1 t + \arg[C_{0n}]) + \\
& 2 \sum_{m=1}^{\infty} |C_{m0}| \cos(m\omega_c t + \arg[C_{m0}]) + \\
& 2 \sum_{m=1}^{\infty} \sum_{n=\pm 1}^{\pm\infty} |C_{mn}| \cos(m\omega_c t + n\omega_1 t + \arg[C_{mn}])
\end{aligned} \tag{5.15}
$$

This expression allows us to understand the frequency content of the switched function as an infinite number of cosine components at frequencies that are linear combinations of the reference and carrier frequencies. The different terms of the expression, corresponding to different classes of harmonics, will now be discussed in detail. Again, the Fourier coefficients, when multiplied by two, can be interpreted as phasors for the harmonics. First, the coefficient C_{00} corresponds to the average of the signal, i.e. the dc component. The next part corresponds to oscillations at multiples of the reference waveform frequency; these are generally referred to as *basebands*. Thereafter, the summation over positive values of m, with $n = 0$, contains the *carrier harmonics*, which are oscillations at multiples of the carrier frequency. Typically, these are the dominant components of the spectrum. The final summation in the expression is made over all positive values of m and nonzero values of $|n|$ and gives rise to the *sideband harmonics*. These appear in clusters centered around the carrier frequency multiples. Figure 5.3 shows a typical amplitude spectrum of a carrier-based PWM signal, indicating the different classes of harmonics resulting from the Fourier analysis, and the values of the two indices m and n corresponding to a certain harmonic. In polyphase systems and multilevel systems

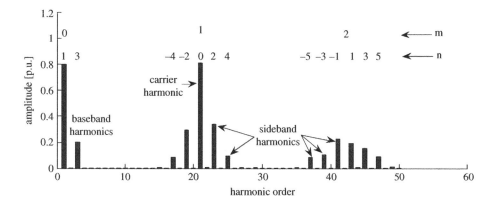

Figure 5.3   Typical amplitude spectrum of carrier-based PWM voltage.

the sum or difference of PWM voltage waveforms is frequently the relevant quantity in terms of harmonic performance. Then, the phase displacement of individual harmonics becomes of interest. Harmonic components which are in phase will not be present when subtracted, and components that are anti-phase (phase-shifted π rad) will cancel out when added. In this context the following identity is immensely useful:

$$C_{mn}(\theta_c, \theta_1) = e^{j(m\theta_c + n\theta_1)} C_{mn}(0, 0). \tag{5.16}$$

It relates the complex Fourier coefficients for a case where the reference and carrier waveforms have been phase shifted by θ_c and θ_1, respectively, to the case without such phase shifts. Evidently, only the phase and not the magnitude is affected. Furthermore, base-band harmonics are only affected by the reference phase-shift, whereas the carrier harmonics are only impacted by the carrier phase-shift.

The shape and nature of the two-dimensional switching function $f(x, y)$ obviously needs to be determined in order to compute the Fourier coefficients. When deriving this function one should simply consider what point in the (x, y)-plane a certain point in time corresponds to, and which dc terminal the ac-side phase voltage will be switched to at this point. Figure 5.4 shows that function, corresponding to the case of a cosine reference and a triangular carrier. Note that the figure depicts a function surface, where the shaded region is raised by $V_d/2$ and the unshaded region is lowered by the same amount with respect to the plane of the paper. The function is shown over the region $x, y \in [-\pi, \pi]$, which corresponds to one unit cell, but since it is periodic it repeats itself in both $x-$ and $y-$ directions to cover the entire plane. Analytically computing the Fourier coefficients by Equation (5.13) can be challenging and may often lead to integrals where no primitive functions exist. However, by employing a mathematical identity

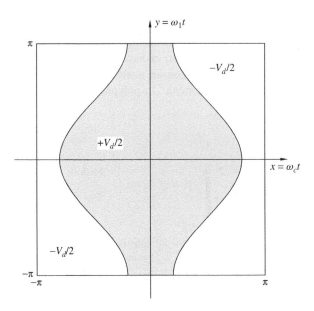

Figure 5.4 Two-dimensional switching function corresponding to cosine-triangle carrier-based modulation. Unit submodule.

called the *Jacobi–Anger expansion* the Fourier integral can in some cases be written in closed form. For cosine-triangle modulation, as depicted in Figure 5.2 and Figure 5.4, the Fourier coefficients can thus be derived as

Baseband harmonics $(m = 0, n > 0)$

$$C_{0n} = m_a \frac{1}{2} \frac{V_d}{2} \quad \text{for } n = 1, \quad \text{and zero otherwise.} \tag{5.17}$$

Carrier and sideband harmonics $(m > 0, |n| \geq 0)$

$$C_{mn} = \frac{V_d}{m\pi} J_n \left(m \frac{\pi}{2} m_a \right) \sin \left((m + n) \frac{\pi}{2} \right). \tag{5.18}$$

A first important observation to be made is that the only baseband present is a component identical to the reference, i.e the reference is fully reproduced in the switching function, as desired. Note that the factor $\frac{1}{2}$ in the expression for C_{0n} is canceled out by the factor two in Equation (5.15) so that the resulting amplitude at the reference frequency is indeed $m_a V_d/2$. In the expressions for the carrier and sideband harmonics $J_n(\xi)$ is the *Bessel function* of order n. Its appearance is a result of using the Jacobi–Anger expansion for evaluating the Fourier integrals. At small values of the argument ξ this function drops off rapidly toward zero with increasing $|n|$, implying that the sidebands around the first carrier-frequency multiples will be narrow. With increasing values of m, however, the sidebands will widen so that they eventually start overlapping each other. Furthermore, by examining the sine function in Equation (5.18), it is clear that only odd-order carrier harmonics (i.e. where m is odd) will appear in the modulated voltage. Likewise, only even-order sidebands (i.e. where $|n|$ is even) will appear around odd carrier harmonics and only odd-order sidebands will appear around even carrier harmonics.

Importantly, the pulse pattern is now completely defined by the unit cell together with the frequencies and phase displacements of the carrier and the reference, respectively. Altering the frequencies will only affect the location of the harmonic components in the spectrum, not their amplitudes, which are determined by the unit cell.

Furthermore, the double Fourier series methodology also works for non-integer frequency ratios. When the carrier frequency is not an integer multiple of the reference frequency, the resulting PWM waveform will generally not be periodic. This is the case because there will not be an integer number of carrier cycles during a cycle of the reference waveform. Therefore, each reference cycle will begin at a different phase angle with regards to the carrier, i.e. a different value of $\omega_c t + \theta_c$. Thus, there will be no periodicity at the reference frequency. Still, thanks to the separation of the two oscillations made possible by the double Fourier series expansion, Equation (5.15) will give a complete description of the pulse pattern.

In the derivations above, the voltage was measured from the midpoint of the dc link, i.e. it was switched between $-V_d/2$ and $+V_d/2$. In an MMC, half-bridges are used that instead switch the output voltage between zero and the submodule capacitor voltage $v_{cu,l}^i$. Notably, the only impact of this in the frequency domain is the appearance of a dc component of $v_{cu,l}^i/2$, i.e. $C_{00} = v_{cu,l}^i/2$. This becomes evident by observing that shifting the voltage reference level only amounts to a corresponding shift in the two levels of the unit cell. For the half-bridge the inner raised area of the unit cell in Figure 5.4 would be at level $v_{cu,l}^i$ whereas the surrounding area is shifted to the zero level. In the evaluation of the double Fourier components this only affects the dc component C_{00}. For all other components the change of dc level will integrate to zero, owing to the multiplication with a complex exponential.

5.3.3 Polyphase Systems

In a polyphase two-level voltage source converter—see Figure 1.1(b)–(c)—it is most common to use a single carrier for all phase legs. This way, the carrier harmonics will cancel out in the line-to-line voltages. It becomes obvious by studying Equation (5.16), which shows that the carrier harmonics of the phase legs will be identical with such a choice. The reference waveforms, on the other hand, are determined for each phase leg depending on which voltages should be provided between them. For a three-phase inverter in stationary operation, such as in Figure 1.1 (c), the simplest choice is to use sinusoidal symmetric reference voltages as follows:

$$v_a^\star = m_a \frac{V_d}{2} \cos(\omega_1 t)$$

$$v_b^\star = m_a \frac{V_d}{2} \cos(\omega_1 t - 2\pi/3) \tag{5.19}$$

$$v_c^\star = m_a \frac{V_d}{2} \cos(\omega_1 t + 2\pi/3).$$

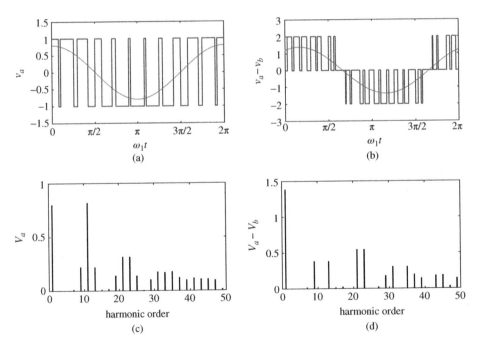

Figure 5.5 Carrier-based modulation of a three-phase converter. The graphs show (a) phase voltage, (b) line-to-line voltage, (c) phase voltage amplitude spectrum, and (d) line-to-line voltage amplitude spectrum. All values are normalized by the half of the dc link voltage $V_d/2$.

Figure 5.5 shows one phase voltage and one line-to-line voltage of a three-phase VSC modulated in this fashion. Both time-domain waveforms and amplitude spectra are displayed. As expected, the carrier harmonics do not appear in the line-to-line spectrum. Looking carefully, it can also be noted that the triplen sidebands, i.e. those harmonics that appear at an integer multiple of ±3 times the reference frequency from a carrier frequency multiple, are absent. This can also be understood from Equation (5.16). According to (5.19), the reference phase shifts (θ_1) of the three phases are 0, $-2\pi/3$, and $+2\pi/3$. For a triplen sideband these figures will be multiplied by $\pm 3p$, with p being an integer, and multiples of $\pm 2\pi$ will result. Therefore, the triplen sideband components will be common to the three phases and not appear in the line-to-line voltages.

A grid-connected converter, for instance in an HVDC station, is in most cases connected to the grid via a three-phase transformer where the converter-side windings lack any path to the dc link. They can be either delta-connected or whye-connected without a connected midpoint. This implies that a zero-sequence component in the phase voltages cannot drive any currents to the grid side. As discussed in Section 5.2.3, the inclusion of such a component in the phase-voltage references can thus be allowed without affecting any circumstances on the grid side. Furthermore, it can allow an extension of the upper limit of the modulation index from 1.0 to $2/\sqrt{3} \approx 1.155$, i.e. the ac voltage of the converter at a given direct voltage can be increased by more than 15% by this measure. Given that the maximum ac-side current is generally determined by the rating of the semiconductor switches, this amounts to a corresponding increase in the power-handling capability. The methodology of adding a zero sequence component is equally applicable to multilevel three-phase converters. Also in this case there is a limited total direct voltage range within which the reference has to stay at all times.

5.4   Multilevel Carrier-Based Modulation

There are several different ways of adapting the carrier-based modulation methods described in the previous section to allow them to be used with multilevel converters. Generally, the methods are based on splitting the multilevel waveform into a sum of a number of two-level PWM waveforms, each with its own carrier, while using a single reference voltage as follows:

$$v(t) = \sum_{i=1}^{N} v^i(t). \tag{5.20}$$

Depending on how the carriers and reference are arranged, different methods have been conceived. Some of these are described and analyzed below.

5.4.1   Phase-Shifted Carriers

In this case the multilevel waveform is made up of the sum of N two-level PWM waveforms according to Equation (5.20), where the carriers are symmetrically phase shifted; see Figure 5.6. This scheme is labeled *phase-shifted carrier* (PSC) modulation.

The carrier phase shifts of the different two-level waveforms can thus be formulated as

$$\theta_c^i = 2\pi \frac{i-1}{N} \quad \text{where } i = 1, \dots, N. \tag{5.21}$$

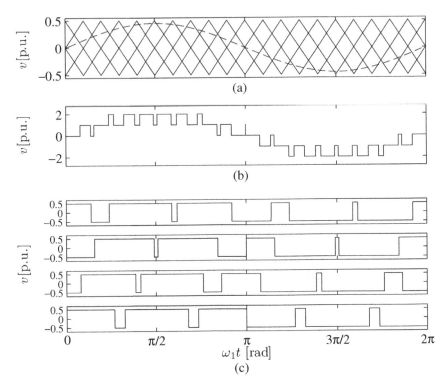

Figure 5.6 Phase-shifted carrier multilevel modulation. The graphs show, from top to bottom, (a) carriers and reference, (b) multilevel modulated waveform, and (c) constituent two-level modulated waveforms.

The frequency content of each of the two-level waveforms can, according to Equation (5.12), be described by

$$v^i(t) = \sum_{m=-\infty}^{\infty} \sum_{n=-\infty}^{\infty} C_{mn}^i e^{j(m\omega_c + n\omega_1)t} \tag{5.22}$$

The Fourier coefficients C_{mn} of each such waveform can, by using Equation (5.16), be expressed in terms of the coefficients of the first waveform (which lacks phase shift):

$$C_{mn}^i = e^{jm2\pi\frac{i-1}{N}} C_{mn}^1. \tag{5.23}$$

The multilevel waveform can now be described in the frequency domain as

$$v(t) = \sum_{i=1}^{N} \sum_{m=-\infty}^{\infty} \sum_{n=-\infty}^{\infty} C_{mn}^i e^{j(m\omega_c + n\omega_1)t}. \tag{5.24}$$

By changing the summation order and substituting Equation (5.23) into the resulting expression the following is obtained:

$$v(t) = \sum_{m=-\infty}^{\infty} \sum_{n=-\infty}^{\infty} \left\{ \sum_{i=1}^{N} e^{jm2\pi \frac{i-1}{N}} C_{mn}^1 \right\} e^{j(m\omega_c + n\omega_1)t}.$$
(5.25)

The expression within curly brackets can be identified as the double Fourier coefficients of the multilevel waveform (see Equation (5.12)):

$$C_{mn} = \sum_{i=1}^{N} e^{jm2\pi \frac{i-1}{N}} C_{mn}^1.$$
(5.26)

This is a geometric series. For positive values of m that are not multiples of N, it will sum to zero, i.e.

$$C_{mn} = \frac{1 - e^{jm2\pi}}{1 - e^{jm2\pi/N}} C_{mn}^1 = 0 \text{ for } 0 > m \neq pN$$
(5.27)

where p is a positive integer.

For integer multiples of N, on the other hand, the complex exponential will be equal to one for all i and we obtain

$$C_{(pN),n} = N C_{(pN),n}^1.$$
(5.28)

It can thus be concluded that *the only harmonics that will appear in the multilevel waveform are those where m is an integer multiple of the number of carriers (the number of two-level waveforms)*. The first group of harmonics will thus appear around N times the carrier frequency. Note that N carriers correspond to $N + 1$ levels in the waveform.

5.4.1.1 Application to MMCs

The carrier-based modulation described in the previous subsection can be employed for the submodule strings of an MMC. As mentioned previously, the strings of the upper- and lower-phase arms should provide voltages as follows:

$$v_u = V_d/2 - v_s$$
$$v_l = V_d/2 + v_s$$
(5.29)

where V_d is the direct voltage to appear between the dc poles and v_s is the inner emf (electromotive force) seen from the ac side.

As concluded earlier in this section, the fact that half-bridges are used, where the output voltage is switched between zero and $v_{cu,1}^i$, implies that the dc component $v_{cu,1}^i/2$ will appear at the terminal of each submodule. Thus, if the submodule capacitor voltages are all set to

$$v_{cu,1}^i = \frac{V_d}{N}$$
(5.30)

the dc component of the arm voltages will assume the desired value according to Equation (5.29). This leaves the ac components of v_s, which are to be implemented by the modulation.

Taking as example the case where, for simplicity, v_s only contains a single sinusoidal component with magnitude $m_a V_d/2$ the reference waveforms for each submodule of the lower- and upper-phase arms can be expressed as

$$v_1^{i\star} = \frac{m_a V_d}{2N} \cos(\omega_1 t)$$

$$v_u^{i\star} = \frac{m_a V_d}{2N} \cos(\omega_1 t + \pi).$$

(5.31)

Thus, the magnitudes of the references are the same, whereas their phase shifts are opposite according to:

$$\theta_{1,u} = \pi$$

$$\theta_{1,l} = 0.$$

(5.32)

The next issue is to determine the carriers for the submodules of a phase leg. To allow for harmonic canceling *within* the phase arms the carriers are symmetrically phase shifted. The carrier phase shifts of the submodules in the lower arm are thus set to

$$\theta_{c,l}^i = 2\pi \frac{i-1}{N} \quad \text{where} \quad i = 1, \dots, N.$$

(5.33)

The phase displacement of the carriers of the upper arm are additionally offset by an amount, $\delta\theta_c$

$$\theta_{c,u}^i = \theta_{c,l}^i + \delta\theta_c = 2\pi \frac{i-1}{N} + \delta\theta_c \quad \text{where} \quad i = 1, \dots, N.$$

(5.34)

The voltages that matter to the operation of the MMC are the ac-side inner emf v_s and the inner voltage v_c. These two quantities determine the harmonic distortion on the ac and dc sides of the converter, which is evident from the equivalent schematics in Figure 1.18. The symmetric phase shifting of the carriers means that neither v_s nor v_c will contain any harmonics apart from sidebands around multiples of the number of submodules, N, $2N$ etc., multiplied by the carrier frequency. It is of interest whether the phase shift between the carriers of the upper and lower submodule strings $\delta\theta_c$ can allow for further harmonic cancellation. A couple of alternatives with regard to the choice of this parameter will be discussed.

The upper and lower submodule string voltages can, by the derivations of the previous sections, be written as

$$v_u^i(t) = \sum_{m=-\infty}^{\infty} \sum_{n=-\infty}^{\infty} C_{u,mn}^i e^{j(m\omega_c + n\omega_1)t}$$

$$v_1^i(t) = \sum_{m=-\infty}^{\infty} \sum_{n=-\infty}^{\infty} C_{1,mn}^i e^{j(m\omega_c + n\omega_1)t}$$

(5.35)

where $C_{1,mn}^i$ and $C_{u,mn}^i$ are the double Fourier coefficients of the voltage of submodule i in the lower and upper submodule strings, respectively. According to the previous reasoning, the direct components ($m = n = 0$) will amount to

$$C_{u,00}^i = C_{1,00}^i = \frac{v_{cu,l}^i}{2}$$

(5.36)

whereby the direct component of v_c will assume the desired value of

$$V_{c,00} = Nv^i_{cu,1} = V_d/2. \tag{5.37}$$

And the first baseband, which corresponds to the desired harmonic at the frequency of the reference signal,

$$C^i_{1,01} = -C^i_{u,01} = \frac{1}{2}m_a\frac{V_d}{2N}. \tag{5.38}$$

The coefficient $1/2$ originates from the fact that the absolute value of any complex Fourier coefficient other than the direct component is half of the magnitude of the corresponding harmonic component. By employing Equation (1.19) the inner emf at the same harmonic order therefore equals

$$V_{s,01} = \frac{2NC^i_{1,01} - 2NC^i_{u,01}}{2} = m_a\frac{V_d}{2}. \tag{5.39}$$

As for the carrier harmonics and sidebands, by Equation (5.28) the corresponding coefficients of the lower arm for $m = pN$ can first be written as

$$C_{1,(pN)n} = NC^1_{1,(pN)n}. \tag{5.40}$$

The coefficients for all other values of m will be zero, as discussed previously. As before, p assumes positive integer values since the harmonics not canceled out between the submodules will appear around integral multiples of the number of carriers, i.e. the number of submodules per arm. For the upper arm, the additional carrier and reference phase shifts modify the coefficients as follows:

$$C_{u,(pN)n} = e^{j(pN\delta\theta_c + n\pi)}NC^1_{1,(pN)n}. \tag{5.41}$$

Using Equations (5.40) and (5.41), the carrier and sideband harmonics of v_s and v_c can now be written as

$$V_{s,(pN)n} = N[1 - e^{j(pN\delta\theta_c + n\pi)}]C^1_{1,(pN)n} \tag{5.42}$$

$$V_{c,(pN)n} = N[1 + e^{j(pN\delta\theta_c - n\pi)}]C^1_{1,(pN)n} \tag{5.43}$$

A first option for carrier phase shift is $\delta\theta_c = 0$, i.e. the sets of carriers for the upper and lower arms are identical. That is, the pulse patterns will only differ in terms of the reference waveform. In this case, the expression within square brackets in Equation (5.42) will be zero for even values of n and 2 for odd values. Whereas for v_c the corresponding expression in Equation (5.43) will be zero for odd values and 2 for even values of N. Now it has to be kept in mind that with a triangular carrier there will only be even-order sidebands (including carrier harmonics at $n = 0$) around odd multiples of the switching frequency (odd m) and odd-order sidebands around even multiples of the switching frequency (even m). Therefore, for a converter with an odd number of submodules per arm N there will only be harmonics remaining in v_s around multiples pN of the switching frequency where p is even. Every second group of harmonics is thus canceled out. Conversely, for v_c, harmonics will appear for odd values of p. For an even number of submodules, on the other hand, there will be no cancellation of harmonics in v_s at all, whereas all sidebands and carrier harmonics will be eliminated in v_c.

A further alternative with regard to the choice of carrier phase shift between the upper and lower arms is $\delta\theta_c = \pi/N$ rad, i.e. half of the phase shift between two consecutive submodules in an arm. Thereby, there will be a total of $2N$ different carriers in a phase leg. With this choice Equations (5.42) and (5.43) change into

$$V_{s,(pN)n} = [1 - e^{j(p+n)\pi}]NC^1_{1,(pN)n} \tag{5.44}$$

and

$$V_{c,(pN)n} = [1 + e^{j(p+n)\pi}]NC^1_{1,(pN)n}. \tag{5.45}$$

By similar reasoning as above it can be concluded that when N is odd all harmonics from $v_{u,l}$ will remain in v_s, whereas all harmonics will be eliminated in v_c. For an even number of submodules per arm, on the other hand, the harmonics will remain in v_s for even values of p, and in v_c when p is odd.

Another option is to let $\delta\theta_c = \pi$ rad. This means that for any submodule in the upper arm there will be a submodule in the lower arm where the carrier is in antiphase. Given that the references of the upper and lower arms are also phase shifted π rad with respect to each other, according to Equation (5.32), the switching functions of these two submodules will be complementary. That is, when a submodule in the upper arm is inserted a submodule in the lower arm is bypassed and vice versa. Since this applies to all the submodules, pairwise, also the two submodule string voltages will be complementary. Thus, the inner voltage v_c will remain constant at $V_d/2$ whereas the ac-side inner emf v_s will have the same harmonic content as the string voltages $v_{u,l}$.

In Table 5.1 the different options for harmonic cancellation between the phase arms are summarized. It shows at which multiples pN of the switching frequency harmonics will remain in v_s and v_c. From the table it is evident that if the aim is to reduce harmonic distortion on the ac bus the carrier phase shift between the upper and lower arm has to be chosen depending on whether there is an odd or an even number of submodules in each arm. When N is odd $\delta\theta_c = 0$ rad should be chosen since it implies that the first group of harmonics in $v_{u,l}$, around $N\omega_c$, will be eliminated in v_s, and the first harmonics will appear at twice this frequency. If N is instead an even number, the proper choice is $\delta\theta_c = \pi/N$ rad. If instead the dc-bus harmonics should

Table 5.1 Impact of carrier phase shift on the harmonic canceling within a modular multilevel converter phase leg. The two last columns tell at which multiples of the switching frequency pNf_c harmonics remain in the spectra seen from the ac and dc sides of the phase leg.

$\delta\theta_c$ [rad]	N	v_s	v_c
0	odd	p even	p odd
	even	All	None
π/N	odd	All	None
	even	p even	p odd
π	odd	All	None
	even	All	None

be reduced, $\delta\theta_c = \pi$ rad is a suitable choice regardless of the number of submodules. It will ensure that the submodule string voltages will be complementary so that there is (in theory) no harmonics at all in v_c.

Notably, the only harmonics in v_c that will appear on the dc side of the converter are those that are common to the phase legs. Differential mode harmonics, that sum to zero, will instead only drive currents circulating between the phase legs. Analogously, on the ac side generally only harmonics in v_s that are differential between the phases (positive or negative sequences) may cause distortion on the ac side of the converter. Common mode harmonics (zero sequence) cannot drive harmonic currents into the ac side if there is no other link between the ac and dc sides than through the converter. This will be the case if the dc bus is ungrounded, or if there is a transformer on the ac side that does not have a grounded midpoint.

5.4.1.2 Capacitor Voltage Balancing with Phase-Shifted Carrier Modulation

As discussed in the previous sections, PSC modulation can directly assign submodules for switching at constant and equal frequency. Thus, the switching pattern can be fully defined by the modulation scheme, and no additional submodule selector is required. With such a scheme it has to be ensured that energy balancing between the submodules will be upheld. This is particularly the case at the very low switching frequencies that are made possible by MMCs. To analyze this, an arbitrary submodule string in an MMC is considered. An upper arm (connected to the positive dc terminal) is considered, but for symmetry reasons the analysis will apply also to any lower arm. The energy balance for the phase leg as usual reads as follows:

$$\hat{v}_s \hat{i}_s \cos\varphi = \frac{2}{M} V_d I_d. \tag{5.46}$$

At the submodule level this translates to

$$\mathrm{Re}[V_{ul,1}^i I_{ul,1}^*] = \frac{1}{MN} V_d I_d \tag{5.47}$$

assuming that the capacitor direct voltages are maintained at V_d/N. The current components flowing through all the submodules in an arm will obviously be identical. Thus, to uphold the energy balance also at the submodule level, the alternating voltage components $V_{ul,1}^i$ have to be of the same magnitude and phase.

As clarified before, in addition to these dc and low-frequency ac components the spectrum will contain a number of harmonics located in groups around multiples of the carrier frequency. At the low switching frequencies made possible by MMCs the first of these groups will be close to the reference frequency. In particular, at an integer frequency ratio the lower sidebands of the first carrier group will coincide with the desired ac component in the reference waveform. This can be understood from Figure 5.3. When the carrier frequency is three times the reference frequency the first lower sideband ($m = 1, n = -2$) will overlap the reference. Likewise, at a frequency ratio of five, the fourth lower sideband will be at the reference frequency. Both these sidebands are generally of significant magnitude and will therefore modify the component at frequency ω_1 considerably. According to Equation (5.16), the phase displacement of the sidebands will be different in the submodules since it is a function of the carrier phase shift θ_c, which differs between the submodules. Thus, the component at the reference frequency will also differ between the submodules. Therefore, *there is no way by which the energy balance can be fulfilled in all submodules simultaneously at a low integer frequency ratio.* Instead, non-integer pulse numbers generally have to be used to allow for balancing of the capacitor voltages over time.

Notably, this reasoning has been simplified and a more complete analysis has to take several more effects into account. This concerns the impact of the harmonics in the capacitor voltages as well as the circulating currents. Such an analysis can be found in [3]. The study presented in this reference shows that the conclusion that a non-integer frequency ratio is a requirement holds also under more general conditions.

5.4.2 Level-Shifted Carriers

A second class of carrier-based methods works by using a number of carriers stacked on top of each other, dividing the available direct voltage range between them. In this way the voltage range that can be achieved by one phase of the multilevel converter is split into a number of subranges, each corresponding to a single two-level waveform. They are referred to as *level-shifted carrier methods*. As previously, for a converter that can provide $N + 1$ equidistant levels N carriers corresponding to the same number of two-level waveforms are required. The resulting multilevel waveform can also in this case be obtained as a sum of these two-level waveforms according to Equation (5.20).

With these methods, there is a single reference waveform that spans the entire direct voltage range and it is therefore not normalized by the number of levels, as in the case of PSC. A consequence is that the carrier frequency should be N times higher to produce the same pulse frequency. Considering that the switchings will be split among N submodules, the average switching frequency will be approximately the same as with PSC.

Several different ways of phase shifting the level-shifted carriers with respect to each other have been devised [4]. These are illustrated in Figure 5.7, and can be described as:

Phase disposition (PD) Here all carriers have the same phase angle and thus only differ in terms of a level offset.

Phase opposition disposition (POD) In this case the carriers above zero level are π rad phase shifted with respect to those below zero.

Alternative phase opposition disposition (APOD) By this method alternating phase shifts of zero and π rad are used so that adjacent carriers will be in anti-phase.

As evident from Figure 5.7, the different possibilities for phase shifting the carriers all result in multilevel waveforms with essentially constant pulse frequency. Comparing graph (c) of this figure with Figure 5.6(a), it is evident that the carrier pattern for APOD is the same as that of PSC. Thereby, the carrier-reference intersections will also be the same, resulting in the same switching instances. Thus, the resulting multilevel waveforms of these two methods will be the same, although the switchings associated with a certain carrier will be different in the two cases.

In a submodule-based converter, such as the MMC, it is theoretically possible to assign one of the level-shifted carriers to each submodule in a similar fashion as discussed for the phase-shifted carrier method in the previous sections. Thus, the switching function of the submodule would be defined by the intersections of this specific carrier and the reference. This would, however, imply that the switchings would be rather unevenly distributed between the submodules. It is sufficient to observe Figure 5.7 to realize that the submodules would only

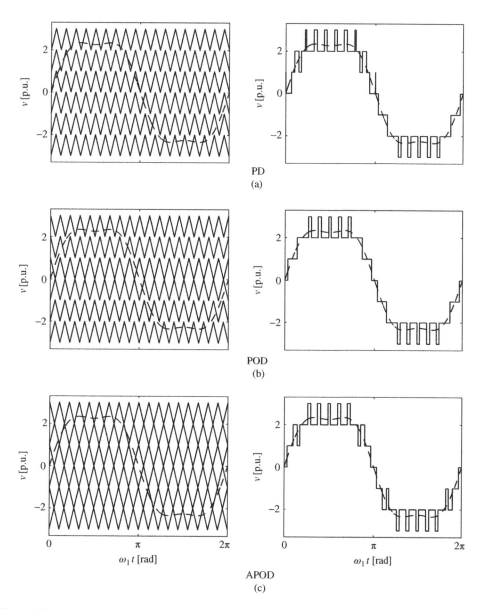

Figure 5.7 Multilevel modulation using level-shifted carriers with different carrier phase shift schemes. Left: carriers and reference, right: modulated waveform. The schemes are (a) phase disposition (PD), (b) phase opposition disposition (POD), and (c) alternative phase opposition disposition (APOD). All values are normalized by the voltage between two levels.

be switched during the intervals when reference passes through the concerned carrier bands. The switching patterns of the submodules would be very different, causing uneven loading and uneven distribution of switching losses. More importantly, however, is that this approach would not offer any possibility for capacitor energy balancing. This is because the dc and fundamental components of the output voltage of the submodules would differ considerably. Therefore, the power balances cannot be fulfilled simultaneously. For these reasons, assigning carriers to submodules is not an attractive method in this case. Instead, methods that decouple the waveform generation from the submodule selection are used in practice; see [5]. Submodule selection methods are described in Section 5.6.

The harmonic content of level-shifted-carrier modulated waveforms can be derived by double Fourier series methodology in a similar fashion as explained in Section 5.4.1. However, the calculations tend to be much more complex than in the two-level case, as shown in [1]. Also, the expressions for the Fourier coefficients are more complex and may not offer useful insights into the harmonic properties of the switched waveforms. This is particularly the case for higher numbers of levels, which are typical for MMCs in power transmission applications. The value of the analytical approach to the frequency analysis is therefore not obvious, and numerical methods for obtaining the spectra can be justified instead.

Numerically computed spectra of the multilevel voltage provided by the different alternatives are provided in Figure 5.8. A case representative for an MMC, i.e. with a large number of levels (31) and low switching frequency (three times the reference frequency), was chosen. On the left side in the figure phase voltages are shown, whereas on the right side the corresponding line-to-line voltage spectra in a three-phase system with a symmetric set of reference waveforms are displayed. Notably, the methods produce quite different spectra. In the literature PD has been suggested as the preferred choice since it concentrates much of the spectral density to the harmonic component the pulse frequency, in this case at harmonic order $3 \times 30 = 90$. This component will cancel out between the phases in a polyphase converter in a similar fashion as in a polyphase two-level converter, as discussed in Section 5.3.3. Evidently, this harmonic has high amplitude in the phase voltage, but is absent in the line-to-line spectrum. A drawback of the PD method is, however, that harmonics of significant amplitude stretch all the way down to very low frequencies. Also the spectra resulting from the POD scheme exhibit this weakness.

The spectrum obtained with APOD, on the other hand, is different. As explained, the waveform produced by APOD is identical to the one from the phase-shifted carrier scheme described previously. Therefore, obviously the spectra will also be the same. The analysis made in Section 5.4.1 predicts that the harmonics will be sidebands located at multiples of $N \times p$. Indeed, the harmonics seen all belong to the first such group, centered around the order 90. Evidently, the sidebands drop off relatively fast, meaning that the harmonics extending down to zero are much smaller than in the other two cases. The discussion above concerns the modulation of a single cell string. However, the methodology can obviously also be applied for the modulation of the different cell strings of an MMC. Also in this case certain harmonics can be made to cancel out between the upper and lower phase arms, thus improving the harmonic content on the ac or dc side of the converter [6].

5.5 Nearest-Level Control

Nearest-level control (NLC) is an alternative method to carrier-based PWM for modulation of multilevel converters, with the advantage of being simple to implement. It was originally introduced for large multilevel drives applications, using direct torque control (DTC) [7], but NLC can also be applied in MMC-HVDC applications.

The main idea is to sample the reference at high frequency f_s and to approximate it with the nearest available level. This approximation can be formulated mathematically by applying the

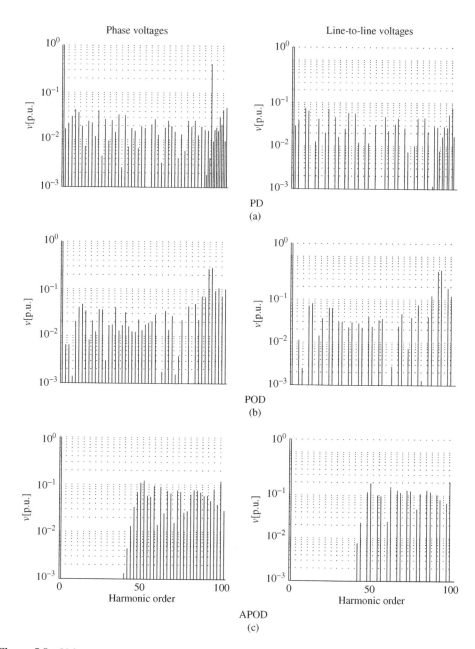

Figure 5.8 Voltage spectra obtained with level-shifted carrier modulation with different sets of carrier phase shifts. Left: phase voltages, right: line-to-line voltages in a three-phase system. The schemes are (a) phase disposition, (b) phase opposition disposition, and (c) alternative phase opposition disposition.

round(x) function, which is approximating the continuous argument to the closest integer as follows:

$$\text{round}(x) = \begin{cases} \text{floor}(x) & x < \text{floor}(x) + 0.5 \\ \text{ceil}(x) & x \geq \text{floor}(x) + 0.5 \end{cases} \tag{5.48}$$

where floor(x) is the largest integer lower than x and ceil(x) is the lowest integer higher than x. Thus, the reference waveform becomes a staircase, and here a challenge is that the lower levels will be used for longer than the higher ones, potentially leading to capacitor voltage unbalance. Therefore, NLC is unsuited for directly assigning the submodules to be inserted and bypassed. Instead, it requires a sorting algorithm as first described in [8] in order to ensure submodule-energy balance (see Section 5.6). In Figure 5.9, the concept of NLC is illustrated, and the sampling period $T_s = \dfrac{1}{f_s}$ is introduced.

In NLC, since there are no carriers, the harmonic content for a given number of submodules N and modulation index m_a can only be changed by the sampling frequency f_s. The sampling frequency has to be high enough to avoid steps in $N_{u,l}$ exceeding one level (e.g. a two levels step); otherwise, the content of lower order harmonics will increase, which is not desirable.

The impact of f_s on the harmonic distortion of the output voltage obtained with NLC has been studied in [9]. In order to fully utilize all submodules and always obtain a number of levels equal to $N + 1$ with all voltage steps equal to one level, the minimum required sampling frequency can be calculated as

$$f_{s,\min} = \pi N f_1 \tag{5.49}$$

where f_1 is the fundamental frequency of the grid. For a typical large number of submodules ($N = 40$) the sampling frequency should be higher than 6.3 kHz, for instance. Increasing the sampling frequency beyond this value will not further improve the harmonic performance (THD).

Assuming an infinitely high sampling frequency, and that all voltage levels are constant and equal to their mean value $\overline{v_{cu,l}^i}$, and an even N, the inserted voltage of the upper and lower arms generated by NLC can be calculated for unitary modulation index as a Fourier series [9]:

$$v_{u,l}(t) = \frac{4\overline{v_{cu,l}^i}}{\pi} \sum_{h=1,3,5\cdots}^{\infty} \left[\frac{1}{h} \sum_{i=1}^{N/2} \cos(h\theta_i) \sin(h\omega_1 t) \right] \tag{5.50}$$

where θ_i ($1 \leq i \leq N/2$) is the ith switching angle in a quarter of the period. The switching angles are the angles at which each level is starting to be imposed.

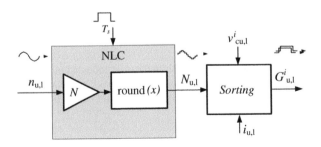

Figure 5.9   NLC concept.

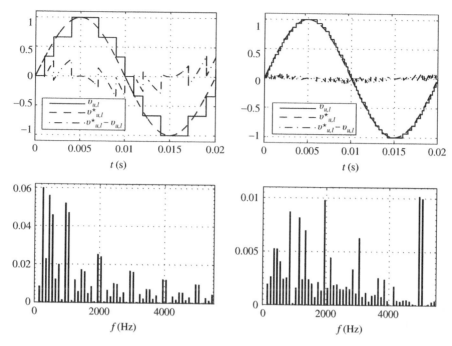

Figure 5.10 Output quality for nearest-level control with $N = 5, f_s = 1$ kHz, THD $= 13.8\%$ (left) and $N = 30, f_s = 5$ kHz, THD $= 1.9\%$ (right). Output, ideal reference, and tracking error (top) and calculated amplitude spectrum [computed by Discrete Fourier Transform (DFT)] (bottom). All values are normalized by $V_d/2$.

Considering that the switching angles will be identical in both arms, the ac-side emf of the MMC v_s, which amounts to the difference between the inserted arm voltages, will exhibit the same harmonic spectrum.

By using different sampling frequencies for the upper and lower arms, the switching angles will be different, and, as a consequence, the ac-side emf of the MMC will exhibit an *artificial* doubling of the number of the levels. This is a feature similar to carrier offset between arms in PSC, as described in Section 5.4.1.

Using NLC there will be a stationary tracking error, defined as the difference between the resulting staircase voltage and the reference. In order to minimize this error, a high number of submodules can be used. This error is comparable with the tracking error obtained in regularly sampled carrier modulation, and there are ways of compensating for it. Eventually there will be a residual error, which will be rejected by the internal current control loop if designed with appropriate bandwidth.

In Figure 5.10 the output voltage of an MMC, using NLC, together with the tracking error (with respect to the reference) and the associated numerically calculated amplitude spectra are displayed.

It is worth mentioning that, in contrast to PSC, the spectrum of NLC is non-characteristic, and contains more or less all harmonic orders, whose amplitude can be reduced only by increasing N. The sampling frequency integer multiples appears as sidebands (e.g. 1 kHz for the case $N = 5$). For large N, NLC and PSC modulation will exhibit similar harmonic performance when measured as THD. The spectrum of NLC will be more distributed, while PSC will shift harmonics to higher orders, resulting in lower WTHD and thereby lower filtering requirements.

The resulting switching frequency using NLC in the ideal case (assuming infinitely large submodule capacitance) is 50 Hz. However, owing to the action of the sorting algorithm, some submodules may be replaced by others without changing the output level. Thus, the number of switching events will increase leading to a higher *average* switching frequency.

5.6 Submodule Energy Balancing Methods

For switching schemes that separate the tasks of modulation and submodule selection there is a need for methods to balance the capacitor energies, and thus limit the voltage differences among the submodule capacitors. These methods generally require that the waveform generation is made in a dedicated modulator, i.e. they take as input a staircase signal:

$$N_{u,l}^{\Sigma} = \sum_{i=1}^{N} n_{u,l}^{i} \tag{5.51}$$

representing the number of submodules to be inserted at any instant. It thus assumes values between zero and N. Whenever this signal changes by $+1$ or -1 a submodule is to be inserted or bypassed. It is the task of the balancing scheme to identify the submodule to switch to carry out this command from the modulator. It should preferably be done in such a way as to bring the individual submodule capacitor voltages closer to the average. Notably, the average capacitor voltage of each submodule string will generally contain low-frequency harmonics (primarily at fundamental and second harmonic orders) that cannot be affected by modulation since they reflect the total energy exchange with the string. The basic conditions for balancing the submodule capacitor energies of a submodule string can be understood from the equation linking the arm current to the current injected into the capacitor:

$$i_{cu,l}^{i} = n_{u,l}^{i} i_{u,l}. \tag{5.52}$$

In inserted submodules the capacitor will be charged when the concerned arm current is positive and discharged when it is negative. In bypassed submodules the capacitor voltage will remain unchanged. Several schemes for submodule energy balancing have been devised and a few of these are reviewed in the coming sections.

5.6.1 Submodule Sorting

One of the simplest methods for submodule energy balancing is based on identifying the submodules with the highest and lowest instantaneous capacitor voltage. Whenever a submodule should be switched the choice is made such that the submodules with the most extreme capacitor voltages are switched if this counteracts their deviation from the average. The method was first outlined for MMC in [8]. Four basic cases can be identified:

1. A submodule should be inserted and the arm current is positive. The bypassed submodule with lowest capacitor voltage is chosen since this will cause the capacitor to be charged, whereby its voltage starts to approach the average.

2. A submodule should be inserted and the arm current is negative. The bypassed submodule with highest capacitor voltage is chosen since this will cause the capacitor to be discharged whereby its voltage starts to approach the average.
3. A submodule should be bypassed and the arm current is positive. The inserted submodule with highest capacitor voltage is chosen since this causes the charging of the capacitor to stop and thus the voltage will not depart further from the average.
4. A submodule should be bypassed and the arm current is negative. The inserted submodule with lowest capacitor voltage is chosen since this causes the discharging of the capacitor to stop and thus the voltage will not depart further from the average.

Examples of capacitor voltage waveforms obtained by using the basic sorting method are shown in Figure 5.11. The figures depict a simulated case where an MMC is operated at pure active power flow, with frequency ratio three (a) and six (b). The nominal total capacitor energy of the converter amounts to 60 J/kW, i.e. the energy transferred during three fundamental cycles. The average capacitor voltage fluctuation, related to the overall power exchange with the string, is in this case 8.3%, peak-to-peak, of the nominal capacitor voltage of 2400 V. In the case where the switching frequency is only three times the fundamental there is a significant variation around the average with individual peak voltages significantly exceeding the average. The interval the capacitor voltages are traversing is approximately 12% of the nominal, amounting to a 45% increase, that must be taken into account when dimensioning the submodules in terms of voltage. In Figure 5.11(b), where the switching frequency has been doubled, the added ripple that is due to the switching of the submodules is greatly reduced, and the maximum voltages only marginally stray out of the envelope defined by the average ripple. Evidently, doubling the frequency ratio leads to significant reduction in the ripple. This is not surprising given that increasing the switching frequency means that the corrective actions are taken more often, whereby the deviation from the average is reduced. Notably, however, this comes at a cost in the form of increased switching losses since each switching of a submodule incurs power loss.

A benefit of this method is its simplicity. The only mathematical operation that needs to be made is the comparison of capacitor voltages. It can therefore be implemented in simple

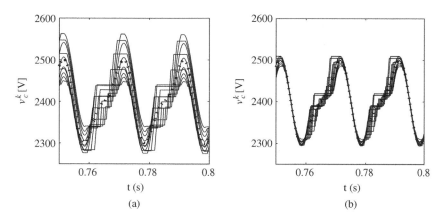

Figure 5.11 Simulated capacitor voltage waveforms using submodule sorting for balancing at pulse numbers (a) three and (b) six. The dotted lines show the average capacitor voltage.

logic hardware, such as field-programmable gate arrays (FPGAs). Such devices are useful for handling simple logic tasks when many digital I/O-signals must be handled. They are therefore well suited for providing the control signals for a large number of submodules that may be present in an MMC.

A more elaborate submodule sorting algorithm is shown in Figure 5.12, in conjunction with NLC. In the figure, $L[N] = L_1, L_2, \ldots, L_N$ is a list of all submodule numbers sorted according to the capacitor voltages, either in ascending or descending order. The flowchart is run through at a constant, high, sampling frequency. This sorting algorithm is able to balance the voltages well, and also reacts quickly to transient conditions since the submodule choice is always based on the instantaneous capacitor voltages. A fault case or a change of operating point can impose fast changes to the reference voltages and the arm currents. These may rapidly alter the capacitor voltages, and under such circumstances a method that responds quickly is valuable. However, the algorithm has the disadvantage that it sorts the capacitor voltages every sampling period, and every change in the ranking list will result in additional switchings. Often, these switchings are not really necessary as they only replace an SM with a lower voltage with another with higher voltage, with minimum impact on the output voltage.

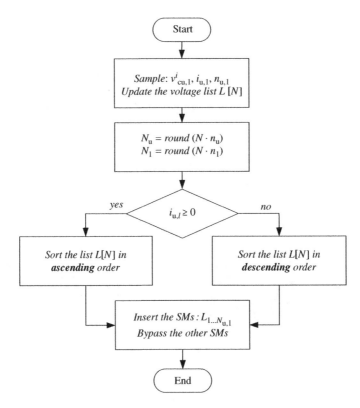

Figure 5.12 Flowchart of NLC with conventional sorting algorithm.

Note: SM = submodule.

In the following sections some more efficient sorting algorithms are described, with focus on reducing the switching losses.

5.6.2   Predictive Sorting

Recall that there is always an unavoidable sum-capacitor-voltage ripple. The ripple amplitude is determined by a number of parameters, as quantified in Chapter 3, but fundamentally it is inversely proportional to the submodule capacitance C. A smaller capacitance implies a lower capacitor cost, but results in a larger ripple amplitude and thus increased cost owing to higher voltage rating of the submodule capacitor and the semiconductors.

The minimum obtainable individual capacitor-voltage ripple is $1/N$ of the corresponding sum-capacitor-voltage ripple. If the capacitor voltages were perfectly balanced, so that the spread between them were negligible, this minimum ripple would be obtained. Unfortunately, for the desired low switching frequencies—which is one of the benefits of the MMC—a non-negligible spread is obtained. Yet, it is possible to constrain the peaks of all submodule-capacitor voltages to (or at least very close to) the same value, i.e.

$$\max_t v^i_{cu,l} = v^{T+}_{u,l}, \qquad i = 1, 2, \dots, N \tag{5.53}$$

where $v^{T+}_{u,l}$ are the targets for the peak values (which are discussed in more detail momentarily). In a similar fashion, the valleys of the submodule-capacitor voltages (i.e. the negative ripple peaks) can be constrained to the target values $v^{T-}_{u,l}$:

$$\min_t v^i_{cu,l} = v^{T-}_{u,l}, \qquad i = 1, 2, \dots, N. \tag{5.54}$$

Perfect balancing is thus obtained at the peaks and valleys, whereas elsewhere a large spread is accepted. The method is called *predictive sorting* [10], and it is now briefly described. The steps are illustrated by the flowchart shown in Figure 5.13.

Each fundamental frequency period is divided into a charging period and a discharging period. The charging period is the time interval during which the arm current is positive and the capacitor voltages are increasing. Conversely, the discharging period is the period during which the arm current is negative and the capacitor voltages are decreasing. At the beginning of each charging/discharging period, the target voltages are computed. This can be made either using the explicit formulas for the sum-capacitor-voltage ripples (scaled by $1/N$) that are derived in Chapter 3. Alternatively, the target voltages can be computed as the minimum and maximum of $v^\Sigma_{c,ul}/N$ evaluated over the preceding fundamental period. For each submodule, an online prediction of the individual peak/valley voltage $\hat{v}^i_{c,ul}$ is started. The details of this prediction are discussed in [10]. At each update instant (which normally occur at the sampling rate of the control system) during the charging/discharging period, the predictive sorting algorithm works as follows. First, the number of submodules per arm $N_{u,l}$ that should be inserted is determined as

$$N_{u,l} = \text{round}(Nn_{u,l}) \tag{5.55}$$

where the per-arm insertion indices $n_{u,l}$ are computed by the arm-balancing control stage, as shown in Chapter 3. The submodules that shall be inserted are stored in a list, where $L^i_{u,l} = 1$ indicates that the ith submodule shall be inserted. When all computations have been finished, the appropriate submodules are inserted by letting $n^i_{u,l} = L^i_{u,l}, i = 1, 2, \dots, N$.

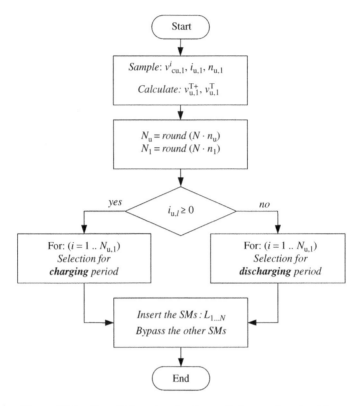

Figure 5.13   Overall flowchart for the predictive sorting algorithm.

Note: SM = submodule.

For the charging period, the following code is then executed at each update instant:

```
n_ins = 0                    % Counter for number of inserted submodules
for i = 1 : N                % Loop through all submodules
   L^i_u,l = 0               % Initialize submodule i as bypassed on the list
   if n_ins < N_u,l          % Desired number not reached?
      if v̂^i_u,l < v^T+_u,l   % Predicted peak voltage below target?
         L^i_u,l = 1         % Change submodule i to inserted on the list
         n_ins = n_ins + 1   % Increment counter
      end
   end
end
```

The desired number of submodules $N_{u,l}$ are this way inserted sequentially in their numbered order. A submodule that is predicted to reach the target voltage is bypassed, so that the target voltage will not be exceeded. Figure 5.14 shows an illustration of the code in a flowchart.

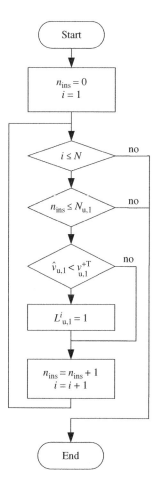

Figure 5.14 Flowchart for the charging period.

For the discharging period only one line of code has to be modified, that comparing the predicted voltage with the target voltage:

```
n_ins = 0            % Counter for number of inserted submodules
for i = 1 : N        % Loop through all submodules
   L^i_u,l = 0       % Initialize submodule i as bypassed on the list
   if n_ins < N_u,l  % Desired number not reached?
      if v̂^i_u,l > v^T-_u,l   % Predicted valley voltage above target?
         L^i_u,l = 1          % Change submodule i to inserted on the list
         n_ins = n_ins + 1    % Increment counter
      end
   end
end
```

It should be noted that this is just an embryo for the predictive sorting algorithm; in practice, additional lines of code have to be added. Particularly, as the code reads, not enough submodules will be inserted if all capacitor voltages have already reached the target voltage before n_{ins} has reached $N_{u,l}$. In that circumstance (which may occur at transients), the target voltage must be recomputed. Additionally, looping through the submodules in their numbered sequence will very likely result in uneven thermal stress. The sequence should be modified at the beginning of each charging/discharging period by an algorithm based on thermal monitoring.

5.6.2.1   Evaluation

The predictive sorting strategy is validated by simulation of an MMC with $N = 5$ submodules per arm. In order to illustrate the spread in the capacitor voltages that results from the use of conventional capacitor-voltage balancing techniques, the system is first simulated with the sorting algorithm described in [8], with 140 Hz switching frequency. The resulting capacitor voltages are shown in Figure 5.15. It is observed that the capacitor voltages are not well balanced when they reach their maximum values, which significantly increases the maximum peak voltage.

The proposed capacitor-voltage balancing strategy is evaluated under the same operating conditions. The resulting capacitor voltages are shown in Figure 5.16. It is observed that the predictive sorting successfully balances the capacitor voltages at their peaks and valleys. As a consequence, the peak-to-peak capacitor-voltage ripple is reduced by 29% as compared to Figure 5.15.

In many cases it is sufficient to constrain just the peaks. In such cases it may be sufficient to use the predictive sorting only during the charging period, whereas during the discharging

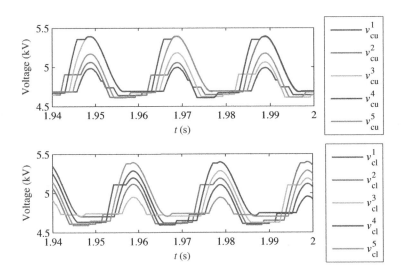

Figure 5.15   Capacitor voltages using conventional sorting.

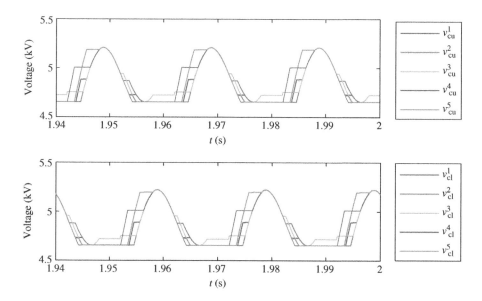

Figure 5.16   Capacitor voltages using predictive sorting.

period fundamental-frequency switching may be used. This results in the capacitor voltages shown in Figure 5.17. Even though the maximum peak value is much reduced as compared to Figure 5.15, the mean switching frequency is now merely 110 Hz.

5.6.3   Tolerance Band Methods

A number of ways of operating power electronic systems are based on implementing a direct link between the controlled variables and the switching of the power semiconductors. These are usually characterized by the fact that the switching actions are performed whenever a controlled variable falls outside a defined interval, labeled a *tolerance band*. This is made in such a way as to bring the variable back into the band. Thus, the distinction between control and modulation that characterize most other ways of operating power electronics is removed. An important general benefit of these methods is that switchings only occur when this is in some sense necessary. The switching frequency will therefore vary, which is different from, for instance, carrier-based methods, where the switching frequency is determined by the chosen carrier frequency. It will generally increase at disturbances and be reduced during steady-state conditions. These methods are therefore generally perceived as efficient with regard to the dynamic properties at a given average switching frequency. On the other hand, the varying switching frequency results in a non-characteristic frequency spectrum for the voltage and currents of the converter, which cannot be analytically described. This may make it more difficult to verify compliance with norms and standards regulating harmonic distortion.

An important example is the so-called *direct torque control* (DTC), which has achieved great popularity in drives applications [11]. In this case, tolerance bands are imposed for the torque and flux linkage of an electrical motor or generator. It has, in different varieties, been

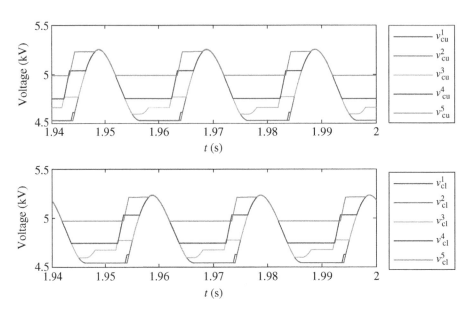

Figure 5.17 Capacitor voltages using predictive sorting for the charging period and fundamental switching for the discharging period.

employed in most of the motor drives offered by the manufacturer ABB in recent years. Tolerance band methods can also be applied in different ways to MMCs. The tolerance band approach can be utilized in the submodule selection process to maintain capacitor-voltage balancing between different submodules. Tolerance bands can be applied on either the submodule average voltage ripple (also known as an *average tolerance band* or ATB) or individual submodule voltage ripple (also known as *cell tolerance band* or CTB) [12]. These methods are generally compatible with any modulation method, for instance NLC. They are described in the following subsections.

5.6.3.1 Average Submodule Voltage Tolerance Band (ATB)

As explained in Chapter 1, the sum capacitor voltage ($v_{cu,1}^{\Sigma}$) of each converter arm ripples over time. A voltage-balancing method should ensure that this ripple is evenly distributed, so that each submodule has an average voltage ripple equal to $v_{cu,1}^{\Sigma}/N$. Therefore, this average submodule voltage can be used as a reference for balancing purposes. The ATB method sets a tolerance band, δ, around the $v_{cu,1}^{\Sigma}/N$ and performs a sorting action whenever any individual capacitor voltage hits the band boundaries. Figure 5.18 illustrates the sequence of the ATB method. In principle, this method monitors the deviation of capacitor voltages from the average voltage and keeps this deviation in a tolerance band with width δ.

    This method results in well-balanced capacitor voltages, not only during normal operation but also during transient conditions. However, it causes higher switching frequency in comparison to other tolerance band methods [12]. Figure 5.19 illustrates typical capacitor voltage waveforms for a converter utilizing the method.

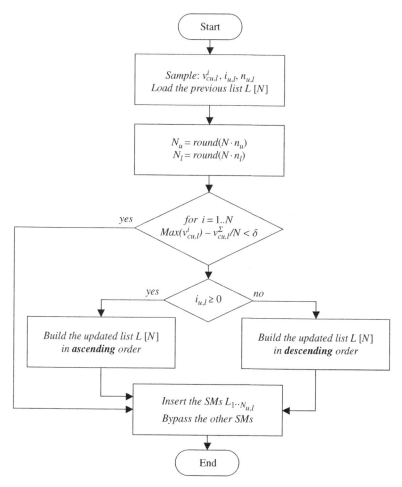

Figure 5.18 Flowchart of the average tolerance band (ATB) method, with used nearest-level control for the waveform synthesis part.

Note: SM = submodule.

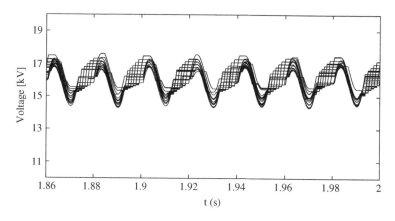

Figure 5.19 Simulated capacitor voltage waveforms using the ATB method.

5.6.3.2 Individual Submodule Voltage Tolerance Band (CTB)

Practically, in an MMC each submodule will be designed for a certain range of capacitor voltage ripple around the nominal value. Therefore, it is acceptable to charge the capacitors to a certain maximum value, and to discharge them to a certain minimum value during operation. These values are referred to as v_{min} and v_{max} and can be used as references for balancing purposes. The CTB method monitors the voltage of each individual capacitor and performs the sorting action at the time that any capacitor voltage violates the voltage boundaries (v_{min} or v_{max}). Thus, by utilizing the total available voltage range of each capacitor, this method minimizes the number of switchings made for the sake of balancing.

A flowchart explaining the CTB method is shown in Figure 5.20. As long as the list $L[N]$ is not re-sorted, the same list will be used, and no additional switchings are needed to maintain the voltage balancing. However, the inserted submodules will eventually reach the voltage boundaries v_{min} or v_{max}. When this happens, a new sorted list $L[N]$ will be generated and used

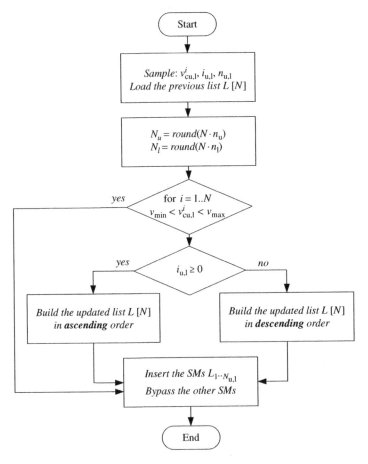

Figure 5.20   Flowchart of the CTB method, using NLC for the waveform synthesis part.

Note: SM = submodule.

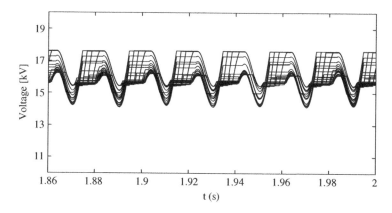

Figure 5.21   Simulated capacitor voltage waveforms using the CTB method.

for the next time steps, until another capacitor voltage reaches the boundary. The delay between the moment a capacitor voltage hits the maximum limit and the moment it is bypassed is only one time step (in the order of tens of microseconds). This is very fast in comparison with the capacitor charging time constant, resulting in a negligible tolerance band violation. Typical capacitor voltage waveforms for a converter using the CTB method are shown in Figure 5.21.

5.6.3.3   Optimized Submodule Tolerance Band (CTBoptimized)

The CTB method can be further improved with regard to efficiency, by optimizing the current level at switching instants, and also by concentrating sorting events around the zero-crossing of the current, as explained in [13], where the method is called *CTBoptimized*. Thus, most of the balancing switching events occur when the current is close to zero, and therefore cause minimal switching losses. Similar to CTB, CTBoptimized utilizes the individual maximum and minimum capacitor voltage limits, v_{min} and v_{max}, as a band for voltage variation. However, the CTBoptimized method assigns the submodules with high capacitor voltage (i.e. with a small margin to the upper limit) during periods with low arm current, in order to reserve the submodules with low capacitor voltage for high current intervals. This way, the number of switching events for balancing purposes at high current conditions is minimized.

Figure 5.22 shows how the CTBoptimized method can be implemented, again with NLC used for the waveform synthesis. The list $L[N]$ is re-sorted when the arm current is zero. The required number of submodules are inserted based on their index numbers in $L[N]$, while their indices will be reversed whenever one of the capacitor voltages reaches either of the voltage boundaries v_{min} or v_{max}. This method allows for high efficiency operation of the MMC, as investigated in [14]. Figure 5.23 shows typical capacitor voltage waveforms for a converter using the method.

5.6.3.4   Evaluation

Tolerance band methods have shown considerable benefits over phase-shifted carrier modulation where submodules are assigned to carriers, and higher efficiency is achievable for the

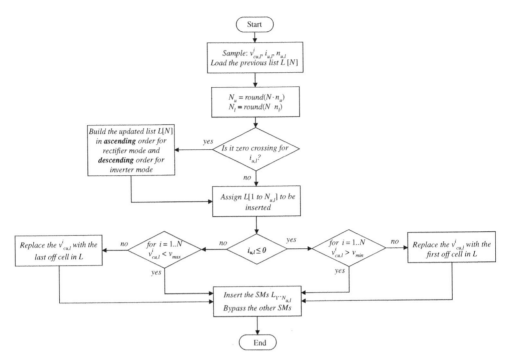

Figure 5.22 Flowchart of the CTBoptimized method, using NLC for the waveform synthesis part.

Note: SM = submodule.

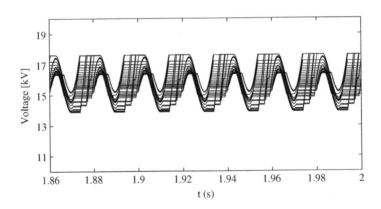

Figure 5.23 Simulated capacitor voltage waveforms using the CTBoptimized method.

same capacitor voltage ripple. The resulting average switching frequency depends on the tolerance band, and this dependency is illustrated in Figure 5.24. This figure also includes the phase-shifted PWM switching frequency for the same capacitor voltage ripple. For a typical tolerance band of $\pm10\%$, the average switching frequency for a 1-GW MMC HVDC station with $N = 40$ is found to be as low as 90 Hz [13]. However, the semiconductor losses are related

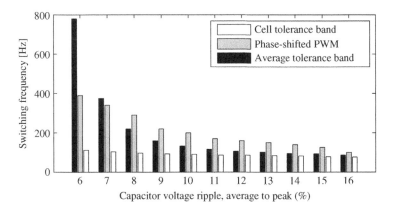

Figure 5.24 Resulting equivalent switching frequency as a function of tolerance band.

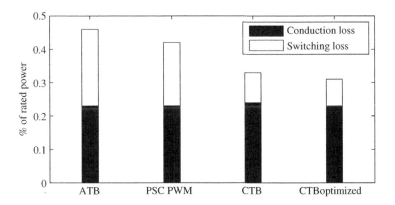

Figure 5.25 Comparison of semiconductor losses for tolerance band methods vs. phase-shifted PWM.

not only to the switching frequency but also to the current level at switching instants. Therefore, focusing on converter efficiency is more relevant for evaluating energy-balancing methods. A loss evaluation of different tolerance band methods is presented in [14] for a converter with 40 submodules/arm and 35 J/kW energy density. Figure 5.25 shows semiconductor losses for different tolerance band methods, and for phase-shifted carrier modulation.

Using average submodule voltage ripple as a control reference results in good voltage balancing, but higher semiconductor losses in comparison to other evaluated methods. On the other hand, using individual capacitor voltage ripple as a control reference allows more spread in capacitor voltages but lower semiconductor losses. As shown in Figure 5.25, the switching losses can be reduced by 60% compared to phase-shifted PWM.

5.6.4   Individual Submodule-Capacitor-Voltage Control

The arm-balancing control, as described in Chapter 3, ensures that sum capacitor voltage in each arm converges to the desired value, i.e. V_d, normally. That notwithstanding, when modulation using phase-shifted carriers is used, there is no inherent balancing function for the individual capacitor voltages. Such a function must be added to ensure that all capacitor voltages converge to the desired value, i.e. V_d/N, normally. This function can be implemented as follows. Starting from the per-arm insertion indices $n_{u,1}$, which are computed by the arm-balancing control, the individual insertion-index references are formed by adding an individual increment as

$$n_{u,1}^{refi} = n_{u,1} + G_{u,1}e_{u,1}^i. \tag{5.56}$$

In the added term, $e_{u,1}^i$ is the normalized deviation of the ith capacitor voltage $v_{cu,1}^i$ from its ideal value $v_{cu,1}^\Sigma/N$:

$$e_{u,1}^i = \frac{v_{cu,1}^\Sigma/N - v_{cu,1}^i}{v_{cu,1}^\Sigma/N} = 1 - \frac{Nv_{cu,1}^i}{v_{cu,1}^\Sigma} \tag{5.57}$$

and $G_{u,1}$ is a dimensionless gain. In order for the added control to give convergence of the individual capacitor voltages, $G_{u,1}$ must have the same sign as the corresponding arm current. One feasible selection is as normalized with the maximum allowed output-current peak value

$$G_{u,1} = \frac{G_0 i_{u,1}}{\hat{\imath}_{max}}. \tag{5.58}$$

The added term $G_{u,1}e_{u,1}^i$ needs to be small to avoid that disturbances are introduced in the modulation, i.e. $|G_{u,1}e_{u,1}^i| \ll 1$. An appropriate selection of the dimensionless constant G_0 is therefore $G_0 \ll 1$. (Tuning can be made iteratively using simulations.) This selection characteristically makes the individual capacitor-voltage control loops the slowest of all nested loops in the MMC control system.

5.7 Summary

Multilevel modulation both implies more voltage levels and a decoupling of the semiconductor switching frequency from the pulse frequency. This permits a reduction of the switching frequency while maintaining acceptable harmonic distortion in the external voltages and currents. The very large number of voltage levels (several hundreds) that can be implemented by an MMC makes these effects even more pronounced. Frequency ratios of 2–3 are feasible, which radically reduces the switching losses. This power loss reduction is one of the main factors behind the commercial success that MMC technology has enjoyed over the last decade.

However, the use of a switching frequency this low negatively impacts the capacitor voltage ripple, which does not benefit from the multilevel nature of the converter. The problem of submodule energy balancing is therefore critical and generally defines the lower limit for the switching frequency.

Two fundamental approaches to the modulation and submodule energy-balancing problem are reviewed in this chapter. The first relies on using a modulation method with phase-shifted carriers which also assigns submodules for insertion or bypassing. The other approach is based on separating the waveform generation from the submodule selection. Since both of the processes can be optimized separately, this alternative tends to produce better outcomes.

References

[1] D. G. Holmes and T. A. Lipo, *Pulse Width Modulation for Power Converters: Principles and Practice*. Wiley-IEEE Press, 2003.

[2] S. Bowes and B. Bird, "Novel approach to the analysis and synthesis of modulation processes in power convertors," *Proceedings of the Institution of Electrical Engineers*, vol. 122, no. 5, p. 507, 1975.

[3] K. Ilves, L. Harnefors, S. Norrga, and H.-P. Nee, "Analysis and operation of modular multilevel converters with phase-shifted carrier PWM," *IEEE Transactions on Power Electronics*, vol. 30, no. 1, pp. 268–283, 2015.

[4] G. Carrara, S. Gardella, M. Marchesoni, R. Salutari, and G. Sciutto, "A new multilevel PWM method: A theoretical analysis," *IEEE Transactions on Power Electronics*, vol. 7, no. 3, pp. 497–505, 1992.

[5] A. Hassanpoor, S. Norrga, H.-P. Nee, and L. Ängquist, "Evaluation of different carrier-based PWM methods for modular multilevel converters for HVDC application," *Proceedings of the 38th Annual Conference of the IEEE Industrial Electronics Society*, Montreal, Canada, 2012, pp. 388–393.

[6] B. P. McGrath, C. A. Teixeira and D. G. Holmes, "Optimised phase disposition (PD) modulation of a modular multilevel converter using a state machine decoder," *Energy Conversion Congress and Exposition*, Montreal, QC, pp. 6368–6375, 2015.

[7] S. Kouro, R. Bernal, H. Miranda, C. Silva, and J. Rodriguez, "High-performance torque and flux control for multilevel inverter fed induction motors," *IEEE Transactions on Power Electronics*, vol. 22, no. 6, pp. 2116–2123, 2007.

[8] A. Lesnicar and R. Marquardt, "An innovative modular multilevel converter topology suitable for a wide power range," *Proceedings of the IEEE Bologna Power Tech*, vol. 3, Bologna, Italy, 2003.

[9] Q. Tu and Z. Xu, "Impact of sampling frequency on harmonic distortion for modular multilevel converter," *IEEE Transactions on Power Delivery* vol. 26, no. 1, pp. 298–306, 2011.

[10] K. Ilves, L. Harnefors, S. Norrga, and H.-P. Nee, "Predictive sorting algorithm for modular multilevel converters minimizing the spread in the submodule capacitor voltages," *IEEE Transactions on Power Electronics*, vol. 30, no. 1, pp. 440–449, 2015.

[11] P. Vas, *Sensorless Vector and Direct Torque Control*. Oxford University Press, 1998.

[12] A. Hassanpoor, L. Ängquist, S. Norrga, K. Ilves, and H.-P. Nee, "Tolerance band modulation methods for modular multilevel converters," *IEEE Transactions on Power Electronics*, vol. 30, no. 1, pp. 311–326, 2015.

[13] A. Hassanpoor, A. Roostaei, S. Norrga, and M. Lindgren, "Optimization-based cell selection method for grid-connected modular multilevel converters," *IEEE Transactions on Power Electronics*, in press.

[14] A. Hassanpoor, S. Norrga, and A. Nami, "Loss evaluation for modular multilevel converters with different switching strategies," *Proceedings of the 9th International Conference on Power Electronics and ECCE Asia*, Seoul, Korea, pp. 1558–1563, June 2015.

6

Modeling and Simulation

This chapter intends to provide some guidelines for modeling and simulation of the complex dynamics of the MMC. Typical simulation studies range from power system integration to protection and loss estimation. Time-domain models with different levels of complexity ranging from detailed switching to aggregated averaging are described and useful guidelines for implementation in popular simulation packages are given.[1] A vectorization simulation technique for N cascaded electrical circuits and their modulators and control is introduced.

6.1 Introduction

Modeling and simulation of MMCs can be performed with different degrees of detail. When investigating voltage and current stress in critical main-circuit components during faults inside or immediately outside the converter, quantities of each individual submodule must be studied. Occasionally, the voltage or current of a certain switch may have to be followed closely during a transient. In such cases, detailed modeling of the MMC should be used. This may call for the modeling of parasitic inductances of bus bars and submodule capacitors. Additionally, dynamic modeling of the power semiconductor devices including adequate representations of the parasitic capacitances may have to be handled.

In contrast to such detailed modeling, the whole MMCs may be represented as adjustable voltage sources when investigating various power-system-related issues.

MMCs for HVDC transmission typically include several hundreds of submodules in each arm. If detailed simulation models of MMCs is used, hundreds or even thousands of switching elements would have to be represented. This results in an enormous computational effort and unreasonably long simulation times. In order to circumvent this problem, averaged or continuous simulation models, which provide similar dynamic properties as detailed simulation

---

[1] A collection of simulation models in Matlab/Simulink and PLECS is freely available for download from the book companion website.

Design, Control, and Application of Modular Multilevel Converters for HVDC Transmission Systems, First Edition.
Kamran Sharifabadi, Lennart Harnefors, Hans-Peter Nee, Staffan Norrga, and Remus Teodorescu.
© 2016 John Wiley & Sons, Ltd. Published 2016 by John Wiley & Sons, Ltd.
Companion Website URL: www.wiley.com/go/Sharifabadi/ModularConverters

models from a system perspective, are preferred. In [1], a simplified model of the MMC is proposed.

Several averaged or reduced-order simulation models of MMCs are proposed in [2–4]. These models are capable of representing most normal operating conditions, and in many cases this may be sufficient. An important shortcoming of these models is, however, that the blocked mode cannot be represented. This means that start-up and various operation scenarios during fault conditions are impossible to handle.

In [5] both a detailed reduced-order model and averaged models of a 401-level MMC are presented. The detailed model in [5] is capable of accurately representing the blocked mode, but the computational time required for high numbers of submodules is very high, according to an evaluation of the required processing time presented in [6].

The model presented in [7] is a continuous arm-based modeling approach, which is capable of representing both blocked and deblocked modes of operation. The model is shown to be able to model delays in the control system in a realistic way, and transients during blocking and deblocking are modeled accurately. In [8], the same model is used to study the start-up of an MMC as well as a pole-to-pole short-circuit conditions. The model was verified both by detailed circuit simulations and experiments on a 10 kVA prototype.

Later, also representation of the blocked-mode state of the MMC was presented in [9]. However, the models in [7–9] cannot be used to study the balancing control of the capacitor voltages and the operation of the MMC when redundant submodules are included in the arms. In [10], therefore, a detailed equivalent model capable of representing the balancing control of the capacitor voltages on the submodule level, along with blocking and deblocking, is presented.

The above overview indicates that significant research efforts have been carried out in the field of reduced-order modeling, and the handling of blocking and deblocking. Below, however, the presentation is based on the level of detail, and in what way different models can be implemented in available simulation platforms.

The following model types with different degrees of complexity are considered:

- leg-level averaged (LLA);
- arm-level averaged (ALA);
- submodule-level averaged (SLA);
- submodule-level switched (SLS).

In Table 6.1, a matrix of typical studies crossed with the suitable model types is shown.

Regarding the tools for computer simulation, there are generally two categories that can be used for MMC studies:

- mathematic equation solvers (Matlab, MathCad, Mathematica, Maple, etc.);
- circuit simulators (PLECS, Simulink/SimPowerSys, PSIM, PSCAD, PSpice, etc.).

Mathematic equation solvers are more abstract as they require everything (circuit, control, etc.) to be translated down to linear transfer functions (in the s- or z-domain) or to state-space form. The internal solver computes the chosen state variables for each time step. As these simulators are based on matrix representation, it is easy to vectorize the model of the circuit, i.e. N cascaded submodules, such as N becomes a parameter that easily can be changed.

Table 6.1 Classification of four modular multilevel converter model types and two study categories.

Power electronics	Typical time step	SLS	SLA	ALA	LLA
Hardware design and protection	0.1–10 μs	x			
PWM* switching transients	0.1–10 μs	x			
Semiconductor device losses	0.1–10 μs	x			
Harmonic analysis ($h < 40$)	10–100 μs	x			
Internal fault studies	10–100 μs	x	x		
Capacitor balancing	100 μs–1 ms	x	x		
Arm balancing	100 μs–1 ms		x	x	x
Outer control	1–10 ms			x	x
Power systems	**Typical time step**				
External fault studies	50–100 μs	x	x		x
Transient stability	10–100 ms			x	x
Power oscillation damping	10–100 ms			x	x
Voltage stability	10–100 ms			x	x
Frequency stability	10–100 ms			x	x
Load flow	10–100 ms			x	x

* PWM = pulse-width modulation.

Using time-domain circuit simulators is more intuitive as the user can define separately the circuit (using wires, devices, and passive components) and the control (using signals and controllers). This is conceptually closer to the real converter system. Typically, these simulators require more calculations than plain mathematical equations solvers. Therefore, the simulation time is longer. The challenge is to be able to vectorize the model for the parameter N, as explained later in the chapter.

In the following, some guidelines are given for both simulation techniques with relevance to MMC.

6.2 Leg-Level Averaged (LLA) Model

Consider the averaged MMC model described in Section 3.3.3, Equations (3.18) (right side) and (3.22), which represent the internal dynamics. With the state variables i_c, v_{cu}^Σ, and v_{cl}^Σ, this model can be expressed in state-space model form as

$$
\frac{d}{dt}\begin{bmatrix} i_c \\ v_{cu}^\Sigma \\ v_{cl}^\Sigma \end{bmatrix} = \begin{bmatrix} -\dfrac{R}{L} & -\dfrac{n_u}{2L} & -\dfrac{n_l}{2L} \\ \dfrac{Nn_u}{C} & 0 & 0 \\ \dfrac{Nn_l}{C} & 0 & 0 \end{bmatrix}\begin{bmatrix} i_c \\ v_{cu}^\Sigma \\ v_{cl}^\Sigma \end{bmatrix} + \frac{1}{2}\begin{bmatrix} \dfrac{v_d}{L} \\ \dfrac{Nn_u i_s}{C} \\ -\dfrac{Nn_l i_s}{C} \end{bmatrix}.
\tag{6.1}
$$

This model, first introduced in [8], is a computationally efficient continuous model of the internal dynamics of the MMC. It requires as input, besides the insertion indices $n_{u,l}$, the output current i_s and the direct voltage v_d. It reflects the interactions between the upper- and lower-arm sum capacitor voltages, circulating current, dc-bus voltage, output current, and (per-arm) insertion indices.

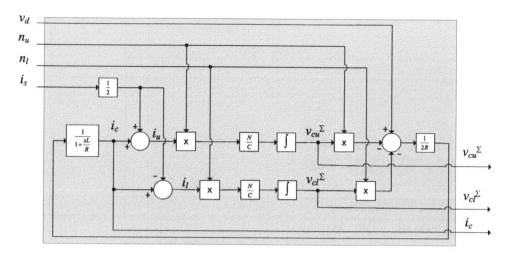

Figure 6.1 LLA continuous-time model.

In Figure 6.1, a block representation of the model is shown in a manner that makes it simple to implement it in the block-oriented or code-oriented simulators.

Assuming an even voltage distribution among the submodules in the arm, the state of the MMC can be described in terms of the sum capacitor voltages in each arm. In the LLA model, they are continuously updated depending on the arm current and the insertion indices. Describing the state of the MMC in terms of sum capacitor voltages greatly reduces its complexity even for large N. In order to avoid possible algebraic loops when closed-loop control is included, a computational delay equal to half of the sampling time can be added to the arm-insertion index input. In Figure 6.2, the LLA model is used to create a complete simulation model including the output-current and circulating-current controllers as well as arm-balancing control based using direct voltage control (direct modulation). This module can be easily implemented in simulators based on Laplace transfer functions and gives full dynamic response for an LLA model.

6.3 Arm-Level Averaged (ALA) Model

In this strategy, each converter arm (that consists of N submodules) is approximated by an aggregated ideal voltage source averaged over the switching period. The ALA model is described in Section 3.3.3 and is controlled by the arm insertion indices $n_{u,l}$ defined in Equation (3.116) for the direct voltage control arm-balancing strategy. The insertion indices are assumed continuous in the range [0, 1]. The expressions for the inserted voltages—Equation (3.17)—and the sum capacitor voltages—Equation (3.21)—are translated into a simple electrical circuit, as shown in Figure 6.3, where a controlled voltage source is used to model the inserted voltage for the arm and a controlled current source that feeds a capacitor is used to model the sum capacitor voltage.

As can be seen in Figure 6.3, this model is capable of representing the dynamics of the sum capacitor voltage with the typical fundamental and double-fundamental frequency components that result from the multiplication of two sinusoidal signals with same frequency and different phases ($n_{u,l}$ and $i_{u,l}$) to obtain the current through the capacitor. The ALA model is

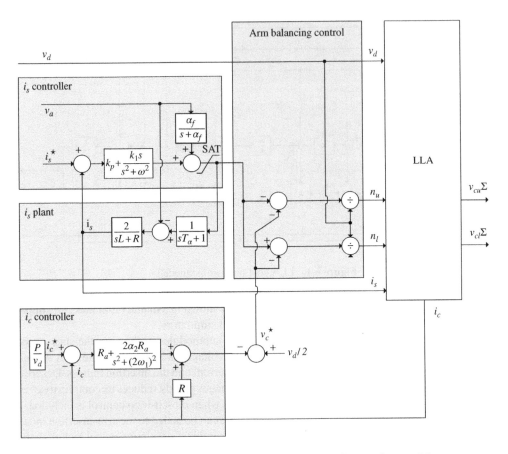

Figure 6.2   Complete (including controllers) LLA continuous-time model.

a fourth-order continuous model (state variables i_s, i_c, v_{cu}^{Σ}, and v_{cl}^{Σ}) and can be easily implemented in most of the popular circuit simulator programs. It can reflect the dynamics of both the ac and dc sides and is therefore suitable for control-design studies such as arm balancing or outer loops. The ignored aspects are:

- PWM dynamics;
- the discrete nature of the true insertion indices;
- capacitor voltage unbalance (individual variations) within the arm.

The biggest advantage of the ALA model is the low computational burden without dependency of N and the good suitability for control design with good capacitor voltage dynamics representation.

6.3.1   Arm-Level Averaged Model with Blocking Capability (ALA-BLK)

The continuous ALA model can be enhanced in order to represent the blocked state of the MMC, which is typical for fault conditions. The term *blocking* refers to the state when both

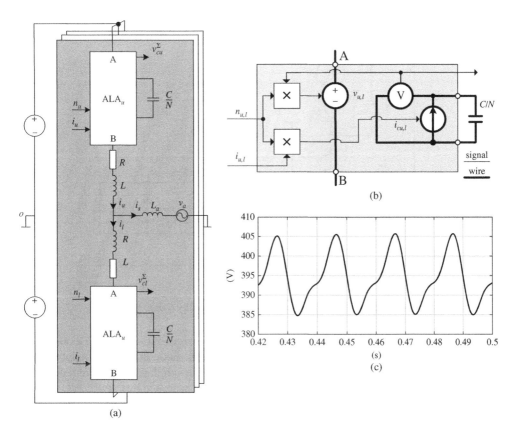

Figure 6.3 ALA model. (a) Electrical circuit. (b) ALA arm model. (c) Typical sum capacitor voltage for the upper arm.

switches in the half-bridge circuit are OFF at the same instant. The solution presented in [11] is shown in Figure 6.4.

The ALA model circuit has been enhanced by adding the following circuit elements:

- A normally closed switch S_{blk}, which is controlled by the blocking signal (BLK), is in turn produced by the protection circuit.
- A bypass diode D_{bp1} represents the anti-parallel diode of the upper switch in the submodule and is used to represent the reverse current capability in both normal and blocked states.
- A bypass diode D_{bp2} is introduced in parallel with S_{blk} in order to represent the anti-parallel diode of the lower switch in the submodule, allowing positive current to flow during a blocked state.

During the normal mode of operation, the switch S_{blk} remains closed and the bypass diode D_{bp1} remains reverse biased. Hence, both positive and negative current can pass through the submodule string. Blocking is made to protect the power transistors of the submodule switches from overcurrent. When the instantaneous value of the arm current reaches a given threshold, as detected by a protection algorithm, the BLK signal is activated. Switch S_{blk} is opened, forcing

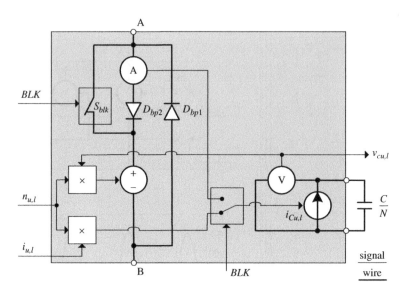

Figure 6.4   ALA model with blocking capability (ALA-BLK).

the insertion of the submodules in the arm for positive current (conducted through diode D_{bp2}), whereas for negative current all submodules are bypassed (as diode D_{bp1} then conducts).

The ALA-BLK model is a very simple and efficient way to simulate multi-terminal HVDC networks for dc fault studies, see, for example, the test circuit proposed by CIGRE WG B4 – 58 (Chapter 8, Section 8.2).

6.4   Submodule-Level Averaged (SLA) Model

For studies where the dynamics of each individual capacitor voltage are of concern, a model where each submodule is represented as an averaged voltage source can be used. This can be achieved by applying the concept of ALA particularized for $N = 1$ for each submodule and then cascading the N submodules in each arm in the so-called SLA model; see Figure 6.5(a). In contrast to ALA, SLA replaces the mathematical summation of the SM voltages by electrical summation (series-connection). As the arm insertion indices calculated by the arm-balancing control are in the range [0, 1], they can be used as the individual insertion index in the SLA as well. In addition, a resistance in parallel to the capacitor in each submodule can be introduced in order to account for internal losses.

As can be seen in Figure 6.5(b), the electrical circuit model is the same as for ALA with the difference that the order is increased to $2(N + 1)$, i.e. i_s, i_c, and $v_{cu,l}^i$ for $i = 1, 2, \dots, N$. In other words, this model can represent the capacitor voltage for each submodule. This is a very useful feature as, in practical conditions, the capacitor voltages will differ among submodules, owing to various factors, particularly

- capacitance tolerance, and
- variation of the internal losses of the submodules that are due to component tolerances and uneven insertion.

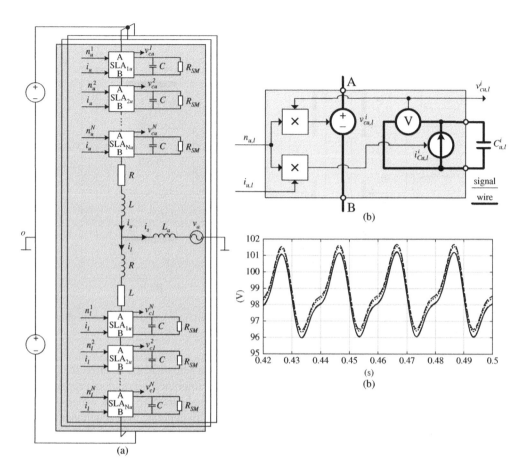

Figure 6.5 SLA model. (a) Electrical circuit. (b) SLA arm model. (c) Typical sum capacitor voltage for the upper arm for $n = 10$.

Owing to its simplicity, SLA is an efficient model that can achieve fast simulation speeds even for large systems. An inconvenience is, however, the fact that the model has to be redrawn every time N is changed, or in other words it is not vectorized.

6.4.1 Vectorized Simulation Models

An important limitation in simulating the MMC with a large number of submodules is that normally in circuit simulators all circuits have to be explicitly drawn. This is time consuming. Ideally, in a vectorized model, only one submodule circuit has to be drawn, which internally is able to clone itself $N - 1$ times for each arm. This is possible with some circuit simulators, such as PLECS, that allow vectorized simulations. By simply defining the size of the ports for the capacitor terminals as N, the circuit will be internally multiplexed N times, and consequently N physical capacitors will be internally represented. This allows the definition of the arm capacitance as a vector (C_i, $i = 1, 2, \ldots, N$) with N elements, so practical aspects such as parameter tolerance or capacitance reduction due to ageing can easily be introduced.

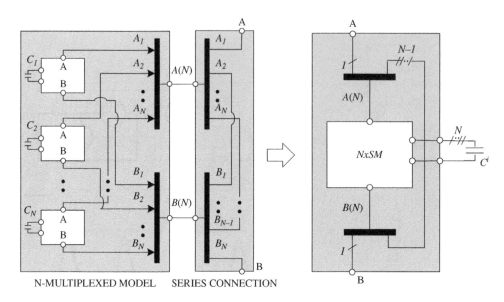

Figure 6.6 Left: Interconnection scheme for the vectorized circuit-simulation technique. Right: Equivalent vectorized model.

Random number generation is a useful method to create a statistical distribution for capacitance tolerances given a maximum range (as manufacturing tolerance is a random propriety). It can be used in a vectorized fashion in order to avoid individual parametrization, which—being a manual process—can be time consuming. A pseudo code to generate the capacitance vector with random tolerance within a given limit is shown as follows:

```
for  i = 1 : N
    C_i = C_rated[1 + tol_C · rand(1)]
end
```

where tol_C is the capacitance tolerance. The same concept can be applied to the shunt resistances that are used to represent the internal losses of the submodules.

The challenge with vectorized circuits is to connect the outputs in a proper way to reflect the real series connections of submodules within an arm. Basically, for vectorized circuits, the terminals in the form $(A_1, A_2, \ldots, A_N)$ and $(B_1, B_2, \ldots, B_N)$ respectively are multiplexed without interconnection. In order to interconnect these terminals in the desired way, i.e. $B_1 - A_2$, $B_2 - A_3$, etc., the interconnection scheme shown in Figure 6.6 can be used. Thus, an equivalent arm model based on vectorized SLA can be represented. With the vectorized SLA, the structure of an SLA simulation model for the MMC will be identical to the ALA model (Figure 6.3(a)) but with N as a parameter, i.e. changing N does not require circuit redrawing.

6.5   Submodule-Level Switched (SLS) Model

The SLS approach assumes the representation of the submodule using ideal switches (transistors and diodes). The equivalent circuit is shown in Figure 6.7(a), and a typical output

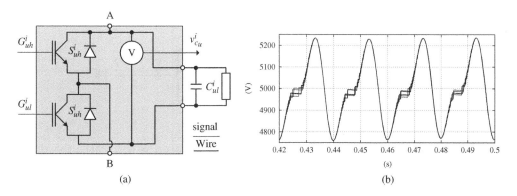

Figure 6.7   SLS model. (a) Electrical circuit. (b) Typical waveforms of the capacitor voltages.

waveform is shown in Figure 6.7(b). As can be observed, the bypassed periods are visible as flat lines.

The transistor and diode models that are used in circuit simulators are often ideal, i.e. they switch instantaneously but are able to represent the conduction losses, since the forward voltage is added to the model. The switching losses can be calculated based on 3D look-up tables based on the data sheets for the devices used.

The SLS model can be used for studies focusing on capacitor balancing methods and for loss evaluation. Similarly to SLA, SLS can be vectorized.

6.5.1   Multiple Phase-Shifted Carrier (PSC) Simulation

In phase-shifted carrier (PSC) modulation, N carriers with a certain individual phase shift have to be generated. In circuit simulators with vectorized processing capability, this can be realized using a common signal generator capable of generating triangular waveforms of certain frequency, amplitude, offset, and initial phase. By creating a vector with the desired initial phase values, the N carrier signals can be easily generated and multiplexed in one signal. Typically, the initial phase shift θ_c^i, $i = 1, 2, \ldots, N$ is chosen in a uniform way, as

```
for  i = 1 : N
    θⁱ_c = 2π(i−1)/(Nf_sw)
end
```

where f_{sw} is the switching frequency. In Figure 6.8, a typical implementation of the PSC pulse generation using vectorized carrier method is shown.

6.6   Summary

We present in this chapter the essentials of simulation of MMCs for various studies. Four different model types, with increasing level of complexity and accuracy, are introduced. In the averaged (aggregated) LLA and ALA models, only the sum capacitor voltage of each arm is

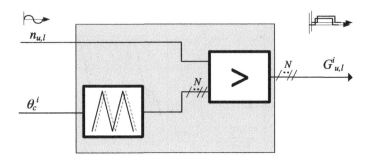

Figure 6.8   PSC modulation pulse generation using a vectorized carrier model.

modeled, allowing reduced complexity and shorter simulation times. The SLA and SLS models include each submodule capacitor voltage. Some simulation tools allow vectorization, which simplifies the task of setting up the model, particularly for a large number of submodules per arm N.

References

[1] S. Teeuwsen, "Simplified dynamic model of a voltage-sourced converter with modular multilevel converter design," *Proceedings of the IEEE/PES Power Systems Conference and Exposition*, pp. 1–6, Mar. 2009.

[2] U. Gnanarathna, A. Gole, and R. Jayasinghe, "Efficient modeling of modular multilevel HVDC converters (MMC) on electromagnetic transient simulation programs," *IEEE Transactions on Power Delivery*, vol. 26, no. 1, pp. 316–324, Jan. 2011.

[3] S. Rohner, J. Weber, and S. Bernet, "Continuous model of modular multilevel converter with experimental verification," *Proceedings of the 2011 IEEE Energy Conversion Congress and Exposition*, pp. 4021–4028, Sep. 2011.

[4] J. Xu, C. Zhao, W. Liu, and C. Guo, "Accelerated model of modular multilevel converters in PSCAD/EMTDC," *IEEE Transactions on Power Delivery*, vol. 28, no. 1, pp. 129–136, Jan. 2013.

[5] J. Peralta, H. Saad, S. Dennetiere, J. Mahseredjian, and S. Nguefeu, "Detailed and averaged models for a 401-level MMC-HVDC system," *IEEE Transactions on Power Delivery*, vol. 27, no. 3, pp. 1501–1508, Jul. 2012.

[6] H. Saad, J. Peralta, S. Dennetiere, J. Mahseredjian, J. Jatskevich, J. Martinez, A. Davoudi, M. Saeedifard, V. Sood, X. Wang, J. Cano, and A. Mehrizi-Sani, "Dynamic averaged and simplified models for MMC-based HVDC transmission systems," *IEEE Transactions on Power Delivery*, vol. 28, no. 3, pp. 1723–1730, Jul. 2013.

[7] N. Ahmed, L. Ängquist, S. Norrga, and H.-P. Nee, "Validation of the continuous model of the modular multilevel converter with blocking/deblocking capability," *Proceedings of the 10th IET International Conference on AC and DC Power Transmission*, pp. 1–6, Dec. 2012.

[8] N. Ahmed, L. Ängquist, S. Norrga, A. Antonopoulos, L. Harnefors, and H.-P. Nee, "A computationally efficient continuous model for the modular multilevel converter," *IEEE Journal of Emerging and Selected Topics in Power Electronics*, vol. 2, no. 4, pp. 1139–1148, Dec. 2014.

[9] H. Saad, S. Dennetiere, J. Mahseredjian, P. Delarue, X. Guillaud, J. Peralta, and S. Nguefeu, "Modular multilevel converter models for electromagnetic transients," *IEEE Transactions on Power Delivery*, vol. 29, no. 3, pp. 1481–1489, Jun. 2014.

[10] N. Ahmed, L. Ängquist, S. Mahmood, A. Antonopoulos, L. Harnefors, S. Norrga, and H.-P. Nee, "Efficient modeling of an MMC-based multiterminal dc system employing hybrid HVDC breakers," *IEEE Transactions on Power Delivery*, vol. 30, no. 4, pp. 1792–1801, Aug. 2015.

[11] W. Leterme, N. Ahmed, J. Beerten, L. Ängquist, D. Van Hertem, and S. Norrga, "A new HVDC grid test system for HVDC grid dynamics and protection studies in EMT-type software," *Proceedings of the 11th IET International Conference on AC and DC Power Transmission*, pp. 1–7, Feb. 2015.

7

Design and Optimization of MMC-HVDC Schemes for Offshore Wind-Power Plant Application

7.1 Introduction

In recent years the first offshore wind-power plants (WPPs) utilizing MMC-HVDC transmission technology has been connected to the mainland ac networks [1]. The German TSO TenneT, the responsible entity for development of the offshore grid in the German sector of the North Sea, has procured eight modular multilevel converter high-voltage direct current (MMC-HVDC) schemes for the export of energy from offshore wind farms to the mainland. At the moment of writing some of these schemes have been commissioned and are in operation. The last contracted scheme, the 900 MW BorWin Gamma, is planned for commissioning in 2019. See Table 7.1.

The experience from German offshore HVDC projects indicates that housing the HVDC converters in an offshore platform is more challenging than anticipated. Installing the HVDC converter equipment on a platform for use in offshore environments imposes new requirements and restrictions, affecting project execution and design as well as the manufacturing, installation, commissioning, operation, and maintenance of that equipment.

The integration of HVDC converters to offshore platforms is still in its infancy, as there are not many relevant standards, guidelines, and recommendations available. Consequently the immature nature of offshore HVDC projects causes uncertainties and exposure to higher risks for stakeholders.

Offshore HVDC equipment is subjected to harsh marine environmental conditions, such as high levels of humidity combined with the corrosiveness of the saline air. The offshore HVDC equipment is also subjected to mechanical loading from platform accelerations and vibrations. In addition, weather and sea conditions impact transportation, installation, operation, maintenance, and health and safety procedures.

Table 7.1   List of TenneT's offshore MMC-HVDC schemes for the export of wind energy to mainland network.

Project	Capacity	DC-link voltage
BorWin Beta	800 MW	300 kV
BorWin Gamma	900 MW	320 kV
DolWin Alpha	800 MW	320 kV
DolWin Beta	916 MW	320 kV
DolWin Gamma	900 MW	320 kV
HelWin Alpha	576 MW	250 kV
HelWin Beta	690 MW	320 kV
SylWin Alpha	864 MW	320 kV

The weight and volume of the offshore HVDC equipment, including the topside platform housing the equipment, influence its design, manufacturing, transportation, and installation. As an example, the center of gravity and large masses and volume might introduce challenges for the mechanical structure of the platform. At the same time the platform design may impose constraints on the layout of the converter components and the auxiliary systems. The remote location—and the concomitant access options (via boat or helicopter) that entails—requires sufficient design margins and redundancy options to be considered during the design phase of the offshore converter. The industry has more than a decade of experience with development of voltage-source-converter high-voltage direct current (VSC-HVDC) schemes onshore; however, the development and integration of HVDC equipment to offshore platforms is still under development and the industry continues to work on optimization and cost reduction.

Offshore HVDC projects for the export of offshore wind energy to the mainland have to comply with various requirements imposed by different stakeholders, such as the grid codes, regulatory frameworks, marine and shipping regulations, aviation, and offshore health and safety rules and regulations [2]. Different stakeholder requirements impact the design and increase the complexity of offshore HVDC projects [3]. These requirements may create a number of contradicting options and challenges that normally are not encountered with onshore MMC-HVDC scheme projects.

This chapter provides an overview on how different European regulatory frameworks impact the design optimization and ownership of HVDC schemes used for the export of offshore wind energy to the mainland. Main components of the offshore and onshore MMC-HVDC converters are presented with a brief introduction to various offshore platform technologies. Finally, the minimum set of system studies that are required for the design, development, and manufacture of these HVDC schemes system is presented.

7.2   The Influence of Regulatory Frameworks on the Development Strategies for Offshore HVDC Schemes

In Europe the third legislative package for the internal EU gas and electricity market released by the European Council requires the unbundling of ownership between transmission assets and generation assets. The implementation and adaption of this regulatory framework has resulted in two different European models for the construction and ownership of offshore generation and transmission assets. As an example, Germany and UK have developed different national regulatory frameworks and consequently two different practices and policies for the development and ownership of offshore WPPs and offshore transmission infrastructures [4]. National regulatory frameworks impact the development strategies and ownership of offshore HVDC assets [5, 6]. Recently ENTSO-E has published new sets of grid codes. It is expected that these set of new grid codes will harmonize the requirements across the European countries, and will be implemented in the internal electricity market of EU member countries by 2017.

7.2.1   UK's Regulatory Framework for Offshore Transmission Assets

The UK governmental departments, the Office of Gas and Electricity Markets (Ofgem) and the Department of Energy and Climate Change (DECC) oversee the new regulatory framework for offshore electricity generation and transmission. These departments are responsible for the integration of renewable offshore generation with the onshore electricity network. In the UK, separate offshore transmission owners (OFTOs), which are neither the wind farm project developers/owners nor the onshore transmission system operators (TSOs), take the ownership of the offshore transmission assets under a long-term license. This license guarantees revenues over the lifetime of the assets subject to certain conditions, such as satisfying performance obligations. Consequently, the OFTO is responsible for the design, building, financing, and operation and maintenance of the offshore transmission assets [7].

Offshore transmission assets are seen as an investment opportunity for infrastructure investors. For each offshore transmission asset, the owner of the asset is selected competitively. Ofgem screens the bidders for financial, management, operational, and technical capability. Bidders that satisfy screening criteria will then submit the 20-year tariff they will need to receive in order to undertake the OFTO role. This tariff is fixed and not subject to regulatory reviews. Offshore wind developers have a choice of constructing the export transmission assets themselves, known as *generator build*, or to handover the responsibility to an OFTO to undertake the development, construction, and integration of the offshore transmission system, known as *OFTO build*.

If the offshore wind project developer decides to construct the assets themselves (*generator build*), after the integration and trial test period, the HVDC scheme, including any offshore assets above 132 kV ac, must be transferred to an OFTO, according to strict regulatory frameworks.

The key driver of the OFTO is to deliver cheaper and timely offshore grid connections through innovation and competition. This may present an opportunity for innovative system design and focus to reduce the cost of energy produced by offshore wind. However, the transmission asset has to comply with requirements of the grid code specific to offshore networks.

The commercial relationship between the offshore WPP and the TSO is governed by the Connection and Use of System Code (CUSC). The CUSC forms the contractual basis for the connection and grid interface with the onshore ac transmission grid. The CUSC contains model agreements, including the Bilateral Connection Agreement (BCA), which details the more generic requirements of the grid codes. The BCA is negotiated and signed at the conceptual stage of an offshore wind generation project—between the National Grid, the TSO, the prospective developer, and the investor—and could be subjected to revisions as the project develops and technical details of the project are established. For an offshore HVDC converter station, the UK grid codes allow certain freedom (compared to an onshore converter with interface to the ac transmission grid) for the parties to negotiate the requirements on electrical performance. OFTOs may apply for derogations from the TSO's technical specifications. The TSO will only serve to check the onshore grid code compliance of these assets.

This allows the OFTO to be innovative and flexible in regard to levels of redundancy, rating of assets (export cables, transformers), and technologies applied.

The UK's regulatory framework opens up for optimizing the radial connections of export links from offshore wind farms to the mainland. Existing frameworks provide no incentives for the optimization of offshore platforms (e.g. interconnections between different wind farms and export assets).

7.2.2 Germany's Regulatory Framework for Offshore Transmission Assets

After the Fukushima nuclear power plant incident in Japan, Germany decided to speed up the integration process of renewable energy sources and to shut down its nuclear power plants by 2020. This processes, called the *German Energy Transition*, or *Energiewende*, gained international attention. The implementation of the Energy Transition Act required accelerating the infrastructure planning procedures (*Infrastrukturplanungsbeschleunigungsgesetz 2006*). Consequently, the German TSOs were instructed to develop, reinforce, and manage required grid connections to offshore wind farms in the North Sea and the Baltic Sea. Hence the TSOs were instructed to develop a long-term grid development plan (*Netzentwicklungsplan*) for the connection of renewable energy sources to the mainland grid. Based on these plans, TSOs have to facilitate, plan, design, construct, and operate the offshore ac or dc collector hubs for the export of offshore wind energy to the mainland networks.

Based on the enforced regulations, when an offshore WPP project receives the necessary permits for construction, and the grid connection agreement from the TSO, the required offshore transmission link to shore must be in place according to the agreement between the stakeholders. The German grid code for the connection of offshore wind does not differentiate between ac- or dc-connected wind farms. The WPPs have to comply with the very strict set of grid code requirements, such as voltage and reactive power control, frequency control, and fault ride-through (FRT) capability at the point of common coupling (PCC) offshore. The WPP developer is responsible for the development and operation of the offshore WPP, including the offshore high-voltage alternating current (HVAC) collector grid and substation and the ac export cables connecting the WPP to the offshore HVDC terminal. The HVDC link is developed, owned, and operated by the TSO.

7.3 Impact of Regulatory Frameworks on the Functional Requirements and Design of Offshore HVDC Terminals

The UK regulation requires that the offshore transmission assets of 132 kV and above are defined as offshore transmission assets and must comply with the GB grid code at the connection point between the offshore and onshore transmission system.

However, German regulation defines the technical requirements for offshore WPPs at the point of connection offshore for both ac and dc export assets.

For a split development and ownership model, like in Germany, the export asset and wind farm projects are developed separately with different timelines. This division of tasks, responsibilities, and ownerships has been driven by the national regulatory framework. The rationale of splitting between the ac collector substation platform and the HVDC platform is not based on technical or functional requirements. Consequently, it does not provide any incentives for optimization or innovative solutions that could reduce the project development costs for both the wind farm and the export assets. In the German model, the TSO is responsible for the planning and construction of the necessary offshore transmission export infrastructures in the planned and designated areas. In this case the offshore WPP has to design and develop the required infrastructure and fulfill the necessary grid code requirements at the point of connection offshore. The offshore wind farm and the offshore HVDC export link project evolve separately.

At the PCC (offshore) between the generation asset and the offshore HVDC terminal, both parties have to comply with grid code requirements. These requirements are universal and reflect a wide range of technical requirements, which do not take account of any technical optimization of the assets. The integration of separately designed and developed assets may require additional technical efforts. The split of ownership may impose major implications on the final technical features of the ac collector system and hence on its design (e.g. on the reactive and active power control and protection requirements imposed on the offshore ac system).

In the case of an integrated project development like in the UK, there are greater opportunities to design and optimize the assets, because the whole asset (wind farm, HVAC substation and the HVDC export scheme) is designed and developed as an integrated system. This approach creates opportunities for technical optimization in the early stages of the project, resulting in reduced investment and ownership costs for the offshore assets. This creates opportunities to optimize the design of the interconnected offshore system as a whole.

As an example, the reactive power compensation requirements offshore and at the point of connection onshore can be divided between the HVDC converters and the wind turbine generators (WTGs). From a systems perspective, considering the interconnected offshore system including the offshore HVDC terminal, the inherent reactive power capability of the WTG and the HVDC converters can be utilized to meet the most stringent reactive-power-balancing requirements. As a result, the offshore shunt reactors between the offshore wind farm ac-collector system and the offshore HVDC converter can be eliminated. In an optimized design both the HVDC converters and the wind turbines contribute to the reactive-power and voltage-control requirements both off- and onshore. In this way, the HVDC converter's MVA rating is not over-dimensioned, and in addition the contribution of the WTGs can be actively controlled.

There are a number of opportunities for design and integration refinements, which have not been explored yet. For example, it is possible to remove the ac collector substation and thereby connect the wind farm's 66 kV array cables directly to the primary side of the offshore HVDC converter transformer. Removing the intermediate ac substation platform will result in significant investment savings and will reduce overall maintenance costs, as there are fewer components to fail.

7.4   Components of an Offshore MMC-HVDC Converter

The most common collector grid voltage of offshore WPPs is in the range of 30–33 kV. Recently, WTG manufactures have announced the availability of 66 kV WTGs. The higher inter-array voltage (66 kV) reduces the electrical losses and allows more WTGs to be connected to a single feeder of the offshore substation [8]. With higher voltage levels, the inter-arrays collector grid can spread over a larger distance without unacceptable voltage profiles. This is an important factor for very large WPPs, where there is a strong desire to

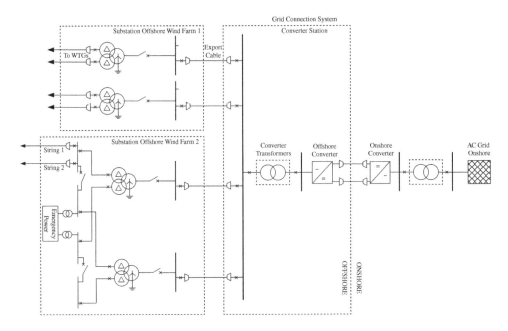

Figure 7.1 Offshore wind farms with HVAC substations connected to a HVDC export link.

reduce the number of ac collector platforms, and thereby capital investment requirements. Figure 7.1 shows a typical design for two offshore ac collector substations integrated to an HVDC export link [5].

The offshore HVAC collector substation platform can be located inside or outside the wind farm perimeters [9]. The choice of HVAC platform location is influenced by several factors:

- requirements from shipping and aviation authorities;
- water depth and seabed condition;
- ac cable voltage profile and cable losses (cable length).

The offshore HVDC terminal should be designed with the black-start capability. The off-shore HVDC converter is normally energized from the dc side and should be able to energize the islanded offshore ac network (wind farm ac collector grid) and supply the required auxiliary power for start-up of the WTGs. It should be noticed that some designs include auxiliary power generators on the offshore ac substation that are used for energizing the offshore ac network. These diesel generators may require installation of reactive power compensation equipment offshore to compensate the reactive power generated by the subsea collector grid cables. As a rule of thumb, cross-linked polyethylene 33 kV cables generate 100–150 kVAr and 132 kV XLPE cables generate approximately 1000 kVAr, reactive power per kilometer. During nor-mal steady-state operation of the offshore WPP, the offshore HVDC converter operates as the "slack bus" and controls the offshore collector grid frequency, while the onshore MMC-HVDC converter controls the dc link voltage.

Grid codes imposed by utilities regulate the features and operation of MMC-HVDC schemes which connect offshore wind farms to mainland networks. FRT specifications require the adoption of specific measures to avoid the trip of the HVDC link during ac network faults (e.g. installation of braker equipment to absorb and dissipate the excessive energy under fault conditions). In addition, the onshore HVDC terminal must provide the required ancillary services at the PCC to the onshore ac network. Consequently, the control and protection requirements of the MMC-HVDC terminals connected to the islanded offshore WPPs become more complex and sophisticated. The aforementioned issues require the development of common technical requirements for communication and the integration of control and protection systems between the WPP and the HVDC scheme.

7.4.1   Offshore HVDC Converter Transformer

The offshore HVDC station adapts its converter voltage according to the varying power flow and reactive power supply/absorption requirements. The basic principle for the voltage control at the islanded offshore power system is to keep the ac network voltage constant.

The converter transformer provides galvanic isolation between the converter and the offshore ac network.

The converter transformer adds additional impedance to the overall circuit, which limits the initial rate of rise of possible fault currents.

Normally, the main circuit arrangement on the large MMC-HVDC platforms is based on two parallel connected three-phase power transformers. One reason is that there are no transformer units available on the market that can handle the full power throughput of approximately 1 GW. It is not unusual to implement two identical transformer units, each with 75% of the total capacity for the offshore HVDC platform. If one of the transformers is not in operation or out of order, the parallel transformer configuration will allow the operation of the HVDC scheme with at least 75% of the export capacity. Under such operating conditions, the single transformer can be overloaded periodically, as the generated power from the offshore wind farm will have natural fluctuations. It is foreseen that the average offshore wind farm capacity factor is between 45 and 60%. Hence the offshore HVDC terminal can, theoretically, operate at 100% of the designed capacity for a number of months without degrading the remaining transformer in operation. But this overload will not impact the lifetime of, or degrade, the remaining transformer. Different operation scenarios provide the functional requirements which must be considered during the design phase:

- **normal operation**: with parallel operation of two transformers each 70–75% of nominal scheme rating, naturally cooled;
- **emergency operation**: with only one transformer in operation, overloaded with possibly forced cooling, which equates to 100% of the nominal scheme rating.

Potentially, the operation modes of the abovementioned configuration may impose limitations on the power transmission capability and reactive power exchange capability of the converter. For large HVDC schemes (e.g. 1 GW), the manufacturing requirements imposed on the interface transformers could potentially push the design to exceed the limits of manufacturing capabilities. The potential size and weight of such transformers impose transport limitations from the transformer factory to the dockyard or assembly site.

It is generally recognized that severe transformer failures, which require transport back to the factory for repair, are very rare events. The converter transformers are the single heaviest equipment items on the platform. Normally, the platform cranes lift-up capacities are between 40 and 80 tons and cannot be utilized for the lift-up of offshore converter transformers. The dimensions for a three-phase 154/400 kV, 650 MVA transformer are approximately 13000 L, 45000 W, 47000 H with an approximate weight of 700 tons. The removal of the converter transformer from the platform and shipping it to shore requires a heavy-lift vessel, and the lifting process is restricted to limited weather windows. During the design phase, practical limitations on maximum lift weight offshore with special vessels must be considered. Designs options with single phase transformers may reduce the constrains imposed by weight and volume of three-phase transformers.

In order to take full advantage of operation with a stand-alone transformer unit while the other unit is out of operation, the system must be designed with adequate busbars, circuit breakers, disconnector switches, measuring transformers, surge arresters, and any other equipment required to operate the offshore HVDC converter with one transformer.

Parallel connection of offshore transformers adds to the complexity of the design because of the design of the converter phase reactors. When designing the main parameters of the converter phase reactors, reactance of the transformers is considered and included. The total converter phase reactance will change when the system is operated only with one transformer.

Offshore HVDC terminals with black-start functionality require transformer designs with a tertiary winding. The tertiary winding of the offshore HVDC transformer provides the necessary auxiliary power for the offshore HVDC terminal.

7.4.2   Phase Reactors and DC Pole Reactors

The three-phase modules of the MMC-HVDC converter are connected in parallel at the dc side as shown in Figure 9.4. As the three generated direct voltages of the phase modules are not exactly equal, balancing currents occur between the phase units. The converter phase units are equipped with air-core phase reactors that suppress these circulating currents and make them controllable by means of appropriate methods. In addition, the reactors limit the rate of rise of possible fault currents. The sizing and selection of reactors impact the overall design of the converter. The actual phase reactor value at maximum and minimum and steady-state dc current have to be determined by detailed system design studies in cooperation with the reactor manufacture. It should be noted that the phase impedance is a combination of the transformer's impedance and the phase reactor's impedance. The air-insulated phase reactors are located in a separate room on the platform close to the valve-hall room. The reactor room is among the largest rooms on the offshore platform as the reactors must keep the required electrical clearance between different phases and between the phase and the ground. In addition, the magnetic field of reactors can cause circulating currents in the nearby platform steel structures. Protective measures should be taken in order to limit the magnetic fields of the reactors to surrounding environment. The reactor room dimensions for 75 mH reactors employed in a 900 MW +/−320 kV MMC-HVDC scheme is approximately 25 m × 15 m × 10 m (reactors and disconnectors). With higher dc-link voltage, the required clearance distances becomes larger, and therefore a larger volume is needed on the reactor room and consequently on the offshore platform. Apart from electrical and magnetic clearances, it is necessary to consider the space required around the equipment for maintenance purposes. Based on the operation and maintenance experiences from the ac reactors, these components are exposed to higher failure rates (e.g. insulation damage and faults caused by hotspots). Therefore, the design

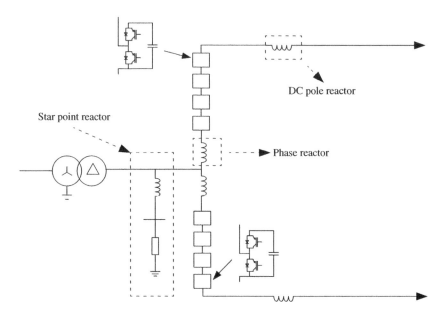

Figure 7.2   Star-point reactor between the converter transformer and the valves.

must allow easy access and the ability to remove and replace the reactor. Material handling of especially bulky and/or heavy materials offshore plays an important part in the design of the platform.

In some converter designs, midpoints of MMC-HVDC phase reactors are connected to a capacitor branch which suppresses the circulating second harmonic currents. However, the undesired second harmonic can be suppressed by control as well, thus eliminating the need for a suppressing circuit.

The industry is working on oil-encapsulated reactors in order to reduce the required insulation distances.

Some designs include a dc pole reactor. DC pole reactors are connected in series with each converter pole. The main function of the dc reactor is to smooth the dc ripple and harmonics and limit the peak magnitude and rate of rise of the dc fault current in the event of a dc fault.

The sizing of the dc pole reactor is a compromise between dc fault current handling requirements, electrical losses, reactor volume, and cost.

As MMC dc capacitors are distributed among the converter submodules, the converter requires a different grounding strategy compared with the two-level VSC-HVDC. One manufacturer has patented a star-point reactor. This special grounding device is installed between the converter transformer secondary side and ac side of converter arms to provide a reference to ground in one station. The star-point reactor provides the zero potential point for the dc side of the converter. This device consists of three star-connected inductors with their neutral connected to the ground, as shown in Figure 7.2. This reactor provides a low impedance path to the ground for dc from the ac side of the converter without requiring dc current flowing through the transformer windings [10]. Due to the space limitations at the offshore platform, normally the star-point reactor is installed at the onshore converter station.

7.4.3   Converter Valve Hall

The half-bridge MMC-HVDC terminal consists of two separated valve halls on the offshore platform.

Each hall accommodates the submodule structures corresponding to three arms of the converters. The converter arms are physically made of stacking the submodules to a tower construction, which can be shaped in different manners, regarding height and length without affecting its functionality. Three rows of towers are arranged in each of the valve-halls. One example of a valve-hall is shown in Figure 7.3. Each arm is equipped with surge arresters to protect the components from possible overvoltages. The stacks of submodules, depending on the manufacture's design, can be suspended from the ceiling or installed on isolator columns. The submodule stacks should resist any mechanical resonances and accelerations sourced from the platform's movements during transportation or installation at the offshore environment. In bad weather conditions, platforms can have displacement movements of up to 1 m (subject to the platform and jacket design).

Each submodule contains the insulated-gate bipolar transistor (IGBT) devices with the anti-parallel diodes, dc capacitors, driver, and protection circuits. The large dc capacitor in each submodule can act as an energy source for the auxiliary power circuit to supply the gate drives and the communication interface electronics. Voltage quality, converter harmonics, and total converter losses are influenced by the total number of submodules in each arm. Optimization of the total number of submodules with regard to acceptable levels of harmonic distortion and converter loss must be considered carefully during the design phase.

The converter valve hall is another large room on the platform. The converter valve hall dimension reflects the electrical insulation safety clearances required for direct-voltage insulation within the valve hall. Increasing the dc voltage requires a corresponding increase in the safety clearances, and consequently the volume of the valve hall will increase, which

Figure 7.3   A 24-MW demonstrator facility valve and valve reactor hall (Courtesy of Alstom Grid – © Alstom Grid UK Ltd).

increases the platform's weight and volume. As an example, average valve-hall dimensions for a 900 MW, +/−320 kV MMC-HVDC converter is about 40 m × 40 m × 12 m. It is expected that with emerging MMC topologies the valve hall volume will shrink in future.

The IGBTs are force cooled, using de-ionized water as the primary cooling medium. The heat from the valves is rejected into the platform freshwater system which is, in turn, cooled using a seawater cooling system or with external forced air cooling. Extensive studies are required to design adequate cooling system capacities. Water cooling pumps are the components with the highest failure rates. An adequate number of pumps and monitoring systems should be included in the design in order to provide redundancy.

7.4.4   Control and Protection Systems

Special designed rooms on the platform are designated for housing the control, protection, and communication equipment. In addition, there will be dedicated rooms for the onsite control and monitoring of the converter with human–machine interface (HMI) equipment. These rooms may impose special design requirements for easy access, firefighting, ventilation, and shielding against electromagnetic interference (EMI) and electromagnetic compatibility (EMC) sourced from the converter and high-voltage equipment.

7.4.5   AC and DC Switchyards

HVDC platforms are equipped with specially designed ac and dc switchyards. Normally a combination of ac and dc GIS or AIS busbars and switchgears are used for offshore converters. However, in order to reduce the required volume, the trend is to utilize only GIS components. Recently, some manufacturers have launched dc GIS busbars and 320 kV dc compact switchgears (dc disconnectors). The new offshore design trends employ only GIS components.

7.4.6   Auxiliary Systems

7.4.6.1   Emergency or Essential Generator

The offshore HVDC platforms must be equipped with a secondary source of auxiliary power, such as diesel power generators. The diesel generator is necessary to provide the back-up power in the event of a loss of power. Emergency power is needed when the HVDC converters are not in operation. In addition, the onboard generators can be utilized to supply the required power during commissioning and maintenance. During the normal operation of the offshore HVDC converter, required auxiliary power is tapped from the tertiary winding of the converter transformer. The platform should contain adequate fuel storage for several days of continued generator operation without the need for refueling, in case of bad weather conditions. Fuel storage will add to the weight and volume of the platform, and in addition it will have an impact on the required firefighting capacity on the platform. Several key points drive the insurance policy costs; one of them is believed to be the presence of fuel in the platform topsides. The generator capacity and rating can be optimized by careful equipment selection and only supplying the essential loads.

7.4.6.2   Seawater Cooling Systems and Heat, Ventilation Systems

Seawater cooling and closed heat exchangers are required for the forced cooling of de-ionized water and other equipment with forced cooling (e.g. converter transformers designed with a forced cooling system). Water-purification equipment will provide the necessary freshwater on the platform. Sewage and sludge tanks must be provided on platforms. The sewage treatment unit will be provided to treat the waste before it is discharged overboard.

The platform must be equipped with adequate heating, ventilation, and air-conditioning (H/V/AC). Heat dissipation of HVDC equipment influences the sizing and design of heating, ventilating, and air-conditioning equipment. The H/V/AC system maintains specified levels of temperature, air change rate, humidity, fresh-air quality, and pressure inside the platform. The system should be designed to provide air-cooling of the converter rooms. Air-conditioning systems for the valve-hall must fulfill the strict particle filtering and humidity requirements given by the HVDC manufacture. Failure of the H/V/AC system affects the overall operation of the installation and may potentially cause a system shutdown.

7.4.6.3   Firefighting Systems

National regulations define the minimum firefighting requirements for the platforms, or in most cases the DNV-OS-D301 specification is used in tenders. Normally, the transformer room will be provided with high-pressure water deluge systems, while the protection and control rooms will be provided with inert gaseous suppression systems. A freshwater system with fog sprinklers will be required inside the living quarters. However, it should be noted that the firefighting strategies and equipment impact the total weight, volume, and design of the platform.

7.4.6.4   Uninterruptible Power Supplies and Battery Rooms

Uninterruptible power supplies (UPSs) with a dc battery system are required for protection, control, and communication systems. A backup power supply with UPS is essential for the reliable operation of the HVDC terminal. These systems must be designed with adequate redundancy in order to manage in-service failures. It is important to ensure the repairs can be made on-line without having to switch off the whole platform and ensure that any single power supply component failure will not cause the shutdown of the HVDC terminal.

7.5   Offshore Platform Concepts

There are different types of platforms and jackets that can be utilized for housing the offshore HVDC equipment:

- conventional jackets (with lift- or float-over topside installation);
- gravity-based substructures (with lift-over or integrated topsides); (see Figure 7.4)
- self-installing platforms.

The selection of platform concept is normally influenced by many factors such as water depth and jacket design, and the quantity/size of cables which have to be fed into the platform. Cables should not enter the platform from all sides as this could limit the access to the platform

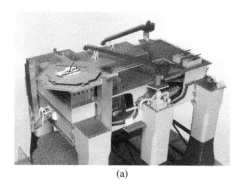
(a)

(b)

Figure 7.4   Gravity-based platform DolWin Beta (Reproduced with permission of Aibel).

via boats, or may lead to cables being damaged during future works should a jack-up vessel operate close to the installation. Normally, one side of the platform should be free from any cable entries. The platform concept influences the layout of rooms and HVDC components.

7.5.1 Accommodation Offshore

Unmanned platforms ("not normally manned") must be equipped with emergency overnight accommodation (EOA) in case the onboard personnel cannot leave the platform because of bad weather conditions. Temporary accommodation provides refuge in the event a crew becomes trapped on the platform in bad weather conditions. The possible need of an accommodation module or EOA should be based on analyses of manning requirements, the frequency and duration of platform visits, etc. The unmanned MMC-HVDC platforms can be designed with EOA for up to 12-18 persons, used mainly during the maintenance periods. The EOA should include necessary technical rooms for maintenance and workshop areas on the platform.

If the platform is designed with EOA, during commissioning and major maintenance periods, a ship with necessary accommodation (flotel) can be utilized. The flotel can anchor nearby the HVDC platform and the personnel can be transported between the flotel and the HVDC platform with special transport vessels or via special temporary bridges connecting the flotel to the platform. Platforms can be equipped with a helideck or a heli-hoist to accommodate the transportation or evacuation of the personnel with helicopters.

The layout of equipment and rooms should consider which locations contain hazardous materials and those which could be used during normal operations. For example, the situation of the control room adjacent to transformer rooms may require blast and firewall considerations. The layout must also allow for the fundamental principle that escape routes are designed to ensure that personnel can leave by at least one safe route to a designated evacuation area.

The platform size and cost is influenced by the choice of onboard accommodation (permanent or EOA) and the choice of helideck or heli-hoist.

7.6 Onshore HVDC Converter

The layout and construction of onshore HVDC terminals is not as complicated a task as it is with offshore terminals. Figure 7.5 shows the INELFE MMC-HVDC terminal.

Figure 7.5 INELFE interconnector, HVDC PLUS 2 × 1000 MW terminal (Reproduced with permission of Siemens AG).

The 2 × 1000 MW, 65 km underground HVDC scheme connects the Spain and France power systems. The HVDC terminal equipment, such as the valve hall and control and protection systems, are indoor installations, while the transformers, phase reactors, and ac switchyard are normally outdoor installations.

The design and integration of an onshore HVDC inverter to the onshore ac grid has to comply with the project's technical requirements and satisfy the grid code requirements imposed by the ac grid operator. The onshore HVDC terminal normally should comply with requirements such as FRT capability undervoltage and frequency contingencies, harmonic content and flicker limits, active and reactive power exchange control, etc. The grid operator may define the magnitude and time profile of active power recovery after temporary fault conditions. Grid codes define the technical requirements and features that must be fulfilled with the VSC-HVDC schemes.

7.6.1   Onshore DC Choppers/Dynamic Brakers

The offshore WPP represents an isolated electrical system when connected via a MMC-HVDC transmission scheme to the onshore ac network.

The grid codes normally require that the onshore HVDC inverter shall sustain any single-phase or three-phase faults in the ac network. During onshore ac network faults, the ability of the onshore converter to export the offshore-generated power into the ac onshore network is reduced and, depending on the severity of the fault (e.g. single-phase or three-phase fault), the power injection to the onshore network will be disturbed. During onshore network faults, the offshore wind turbines keep generating power and feed the offshore HVDC terminal. The excess power (difference of power, fed from the WPP and the reduced export capacity to the onshore grid) causes a rise of the dc link voltage and will activate the converter overvoltage

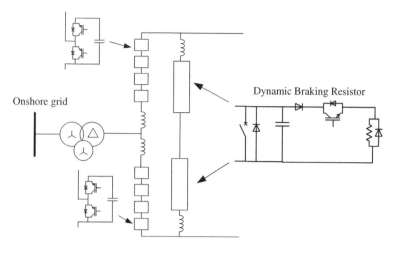

Figure 7.6   Onshore dc chopper or dynamic braking resistor.

protection system and trip the HVDC converters, if additional protective measures are not implemented. In order to limit the propagation of onshore ac faults to the other side of the dc system, and offshore ac collector grid, a dedicated dc chopper is installed between the dc plus and minus pole at the onshore converter station as shown in Figure 7.6. The dc chopper or dynamic braking system is capable of dissipating the generated power from the WPP during temporary onshore grid faults. The grid operators require installation of a dynamic braking system (DBS), dc chopper, or braking resistor for the HVDC links being built today for the export of offshore wind power to shore. The functionality of a DBS is to dissipate the rated power from the wind farm for a short period, normally longer than the onshore auto-recloser dead-time. This period should be long enough to clear the fault on the ac network connected to the HVDC inverter [11].

It is important to mention that the need for dynamic braking equipment has been questioned and in many cases assumed to be an unnecessary requirement imposed by onshore grid operators. The modern wind turbines utilized for offshore wind farms include a local DBS within the nacelle, rated to absorb only the power output of the turbine itself. In case of any ac grid contingencies, the WTG dynamic resistor and the WTG blade pitch control can reduce the impact of such temporary fault conditions.

The DBS can be integrated to the dc side of the offshore HVDC rectifier. However, this approach adds to the complexity, weight, and volume of the offshore installation and is not recommended. Different HVDC equipment manufacturers have different designs and configurations for such DBS or dc choppers.

7.6.2   Inrush Current Limiter Resistors

During the energizing process of an onshore HVDC converter station, it should not impose any large dynamic voltage distortions on the connecting ac network. Therefore, most of the onshore converters are equipped with a pre-insertion resistor bank between the converter transformer secondary side and the ac side of converter, as shown in Figure 7.7. This resistor bank is

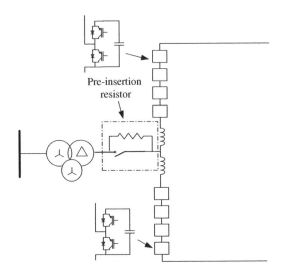

Figure 7.7   Pre-insertion resistor at the secondary side of the ac side of a converter.

equipped with a parallel bypassing breaker. The resistor limits the impact of the potential voltage distortion caused by the charging process of the dc side capacitors from the ac side. The resistor is only in service for 0.5–1.0 s during the initial energizing process of the scheme and thereafter is shorted out by a bypass-switch.

7.7   Recommended System Studies for the Development and Integration of an Offshore HVDC Link to a WPP

In order to bring together all the relevant aspects of the complex process of designing large offshore WPP with HVDC export link to shore, several system studies must be carried out through different stages of the design, manufacturing, and commissioning of the project. Simulation software and real-time simulators (RTSs) are important tools for the reliable design and analysis of offshore wind farms connected to MMC-HVDC schemes.

The studies provide the necessary information throughout the lifecycle of an HVDC project for all parties involved in the HVDC project. The system studies provide the necessary input for the design, development, manufacturing, commissioning, and integration of HVDC equipment to the power system.

The system studies carried out with simulation tools rely on the availability of accurate models from the ac networks, the WPP, the HVDC converters, and the cables [12]. Modeling of MMCs requires special attention and simplification of models as they contain large amount of nonlinear switching devices.

The design and development of a WPP with offshore ac collector grid connected to an MMC-HVDC terminal requires the accurate modeling and simulation of both systems in time and frequency domains.

There is no single universal simulation tool with appropriate models, which can present the behavior of the interconnected system or equipment for all types of studies [13]. Different simulation tools and models are required for different types of studies. These studies will be performed during the different stages of a project's lifecycle. The listed studies in the following subsection are just an indicative proposal for such studies. Different projects, stakeholders, and vendors may require different sets of studies at different stages of the project. Different European grid codes define a set of studies which must be performed in order to document and ensure the grid code compliance of the project. It is not unusual to verify the simulations and the simulation results with a third party.

The minimum set of simulation studies required for the integration of a WPP to offshore HVDC converter include:

- feasibility studies (pre-project studies);
- technical pre-specification studies for the tendering process;
- detailed design studies for manufacturing, front-end engineering design (FEED), including equipment rating studies (main circuit parameters);
- system integration studies;
- steady-state load flow studies;
- grid code compliance studies (defined and required by the grid operators or TSO);
- dynamic load flow studies;
- transient stability studies;
- harmonic analysis studies;
- insulation coordination studies.

CIGRE brochure 563 "Modelling and simulation studies to be performed during the lifecycle of HVDC systems" provides a detail overview of studies required during the lifecycle of HVDC schemes. Table 7.2 provides an overview for such studies [14].

7.7.1   Conceptual and Feasibility Studies with Steady-State Load Flow

All projects start with a conceptual design and analysis stage. The design and integration of offshore wind farms to an offshore HVDC converter requires a number of feasibility and concept development studies. These studies are performed by the project owners or third parties to determine a conceptual design, feasibility, and possible constraints of the project. These studies require simplified associated models from the offshore wind farm including the WTGs, ac collector grid, the ac substation, and the offshore HVDC terminal. At this stage the project can identify the feasibility of the design and provide a high level of project evaluation of the alternatives and constrains. Steady-state load flow analyses are needed to determine the active and reactive power flow, current, and voltage distribution of the entire offshore power grid, and the onshore network. The modeling tool used would normally be load flow analysis programs with models of the components and systems.

The minimum main objectives of studies should be:

- grid-code compliance requirements (steady state);
- to determine the reactive power capability of the interconnected system;
- to determine the voltage and current distribution in the offshore ac system (cable cross-sections, transformer, and tap changer ratings, etc.);
- to calculate active power losses throughout the interconnected system.

Table 7.2 Study proposal required during the lifecycle of an HVDC scheme [14] (Reproduced with permission of CIGRE).

Project lifecycle phase	Study	Tool	Responsibility
Pre-specification	System adequacy	Power flow	HVDC client
	Short-circuit calculation	Short-circuit	
	Dynamic stability	Transient stability/EMT/ small signal stability	
	Harmonic studies	Frequency domain/EMT	
Bid	Main scheme parameters	Typically vendor specific in-house HVDC design scripts and simulation tools	HVDC OEM
	Reactive power		
	AC harmonic filtering		
	DC harmonic filtering		
	Insulation coordination		
	Dynamic performance study	EMT	
Post award	Main circuit design	Typically vendor specific in-house HVDC design scripts and simulation tools	HVDC OEM
	Reactive power		
	AC harmonic filtering		
	DC harmonic filtering		
	Insulation coordination	EMT	
	Dynamic performance study	EMT	
	Transient stability study	Transient stability program	
	HVDC sub-synchronous torsional interaction study	EMT/Real-time simulator	
	Dynamic performance study	Real-time simulator	
Commissioning	Energizing of the components	EMT	HVDC client + TSO, ac grid operator + HVDC OEM
	Commutation failure test	EMT/transient stability	
	Performance of high level controls	EMT/transient stability	
	AC and dc system faults	EMT/transient stability	
Post-commissioning	Planning of transmission network and operational planning	Any tool applied by the owner/utility as integral part of its system planning and operations	HVDC client + ac network operator
	Post-disturbance analysis		
	Pre-specification studies of new equipment		

7.7.2   Short-Circuit Analysis

The main goal of the study is to verify the correct sizing of the high-voltage and medium-voltage circuit breakers in terms of short-circuit breaking capacity and for the evaluation of the short-circuit conditions to which the main electrical components cables, busbars, and transformers will be exposed to.

The study must cover symmetrical and asymmetrical fault conditions in the offshore collector grid. The influence of the grounding arrangements is of importance and must be included in the study.

The short-circuit analysis is an important input used for the design of the overall ac protection scheme. However, it should be noted that the accuracy of WTG and HVDC converter models will have a major impact on the results of the short-circuit calculations as the WTG and the HVDC converter control their short-circuit current contribution to the system under ac fault conditions.

7.7.3   Dynamic System Performance Analysis

The objective of the dynamic performance study (DPS) is to determine that the interconnected system will have stable operation and good dynamic performance during different operational conditions including post fault recovery. The DPS will provide the basis for the evaluation of the stable operation of the system (e.g. during start-up, shutdown, different loadings and power generation variations sourced from the wind variations, FRT capabilities, and stable post-fault operation and recovery). The DPS provides an overview of parameter variations during different operation stages and scenarios of the system. The analysis provides, among others, parameter variations of voltage, current, and reactive power under fault and post recovery. Verification of the FRT behavior of the system and compliance with the offshore and onshore grid code is of high importance. It is also necessary to simulate onshore ac network faults in order to verify the size of the dc dynamic braking resistor or the onshore dc chopper of the HVDC scheme. The onshore fault simulation can also be used to investigate the oscillation/fluctuation in the HVDC link and in the offshore power grid during and after the onshore fault.

7.7.4   Transient Stability Analysis

A transient stability analysis determines the values of voltage and current during the fault conditions and immediately after the recovery from a fault condition. The control systems of the HVDC rectifier and the WTG play a major role in the control and shape of the fault currents in offshore ac networks. Both the WTG's converter and the HVDC converters are designed to control their contribution to the fault current. The transient system analysis covers a timescale of some hundreds of microseconds to a few milliseconds from the occurrence of the fault to the few milliseconds after recovering from the fault. The transient system analysis requires accurate and adequate models of the HVDC converters, WTG and associated control systems, and the passive components such as cables and transformers in the positive, negative, and zero sequences.

7.7.5   Harmonic Analysis

Harmonics are created when nonlinear loads draw nonsinusoidal current from a sinusoidal voltage source. In isolated offshore ac networks a major source of harmonics are WTG converters. Sometimes sinusoidal voltage components generated by the WTGs voltage source converters can be of a different frequency than the power system's fundamental frequency. WTG converters generate harmonics because of the pulse-width modulation (PWM) of the semiconductors. The harmonic components are affected by both the power system fundamental frequency and the PWM carrier signal fundamental frequency. The consequence is that the harmonic components are not integer multiple of the power system fundamental frequency. It should be noted that different WTGs from different manufacturers can introduce different harmonic spectrums as they apply different modulation techniques in the WTG converters. These components are identified as *inter-harmonics*. Harmonic components generated by VSCs can be integer multiple of grid frequency, PWM carrier frequency, or a mixture of both components. Harmonic currents and voltages generated in a wind farm have a negative effect on the interconnected system. Voltage and current harmonics can cause electrical losses and overheating of power cables and transformers. In extreme cases it may cause an undesired trip of protection relays and measurement errors in voltage and current transformers. The evaluation of the harmonic current emissions of wind farms is based on the summation of individual WTG harmonic currents, included in their power quality (PQ) test certificates applying the quadratic summation rule recommended in IEC 61000-3-6. Limits for harmonic emissions are often given only for harmonic currents, not for harmonic voltages. This will imply that the harmonic voltages must be calculated from the harmonic current emission of the WTG, while the frequency-dependent short-circuit impedance may vary as the WPP is connected to an HVDC rectifier. The variation of offshore grid impedance with frequency makes these calculations very complicated and difficult.

Modern offshore WTGs significantly contribute to overall harmonic emissions in the large offshore wind farms. The harmonic contribution at the PCC of a wind farm is the sum of the harmonic generated by each WTG with characteristic harmonics and non-characteristic harmonics. The characteristic harmonic emissions are determined by the converter topology and the switching pattern. Non-characteristic harmonics are determined by the control strategy of the WTG and the control strategy of the wind farm and operating point of the WTGs.

It has been observed that some of the HVDC-connected offshore wind farms were subjected to operation disturbances, assumed to originate from harmonic and resonance phenomena amplifying the undesired harmonics to unacceptable levels. Preliminary measurements indicate that the constant current model used for harmonic analysis did not represent the actual harmonic emissions for the interconnected designs.

The IEC standard 61400-21 recommends modeling of WTGs as a harmonic current source. This approach may cause wrong assumptions on wind turbine harmonic behavior. Modern offshore wind turbines are equipped with VSCs, hence modeling these converters as a harmonic voltage source will provide more accurate results.

7.7.6   Ferroresonance

Ferroresonance is a special case of resonance involving nonlinear inductances that mainly affects the functionality of transformers. Ferroresonance is characterized by high sustained overvoltages and overcurrents with maintained levels of current and voltage waveform distortion causing irreversible damage to a converter's components [15].

When the magnetic flux in a core (e.g. transformer) exceeds a certain value, resonance conditions may occur. The phenomenon occurs when the magnetic core of an inductive device is saturated, making its current-flux characteristic nonlinear. Because of this nonlinearity, resonance can occur at various frequencies. In practice, ferroresonant oscillations are initiated by momentary saturation of the core of the inductive element as a result of switching operations. Ferroresonance can also be initiated in other situations, such as capacitive coupling between parallel lines, ferroresonance between the VT and the power transformer's internal capacitance, or single-phase disconnection in grounded networks.

The effects of such resonance are further aggravated if damping is insufficient. Such occurrences are frequent in the voltage transformers (VTs) that transform high and medium voltage into low voltage for instrumentation or protection purposes. The rated power of VTs is usually very low, because of their metering capability. Nominal primary currents in the transformer winding are typically in the order of a few milliamps, while the voltage is up to tens of kilovolts. As an example the ferroresonance phenomenon can occur when VTs are connected between phases and ground in a system with no direct neutral grounding. Currents can occur that exceed nominal values by orders of magnitude, risking damage to the VTs. The serial and parallel resonance of circuits containing an inductance and a capacitance is a well-known physical phenomenon and the resonance pulsation exists in both parallel and series circuits. At and near to the resonance frequency in the series circuit, voltages across the capacitor and the inductance can reach values that exceed the source voltage significantly. In the parallel circuit, it is the currents through these components that are similarly amplified. Such extreme values can damage e.g. VT, converter surge arresters, as well as insulation damage to cables and transformers, if no remedial action is taken.

It is recommended to perform EMT system design studies to assess how and when ferroresonance may occur in the offshore grid, and to analyze which design alternatives or components are required to mitigate the possible ferroresonance conditions in the system.

7.8 Summary

An offshore environment imposes different sets of requirements on the design and application of HVDC equipment for the export of wind energy to mainland networks. Regulatory frameworks influence the ownership and options for optimization of offshore HVDC assets utilized for offshore WPPs. Integrating with and housing the HVDC converters on an offshore platform is more challenging compared with onshore installations. Different platform technologies can be applied for housing offshore HVDC converters. These options influence the economy of the projects. Components of the HVDC converters and the minimum set of studies which are required for the integration of HVDC terminals with offshore WPPs are presented in this chapter.

References

[1] CIGRE Technical Brochure 370, Working Group B4.39, "Integration of large scale wind generation using HVDC and power electronics."

[2] DNV-OS-J201, "Offshore substations for wind farms," https://rules.dnvgl.com/docs/pdf/DNV/codes/docs/2013-11/OS-J201.pdf, accessed Mar. 30, 2016.

[3] CIGRE Technical Brochure 492, Working Group B4.46, "Voltage source converter (VSC) HVDC for power transmission: Economic aspects and comparison with other AC and DC technologies."

[4] CIGRE Technical Brochure 450, Working Group C6.08, "Grid integration of wind generation."

[5] CIGRE Technical Brochure 619, Working Group B4.55, "HVDC connection of offshore wind power plants."

[6] CIGRE Technical Brochure 370,Working Group B4.39, "Integration of large scale wind generation using HVDC and power electronics."

[7] Ofgem, "Our role in offshore transmission," https://www.ofgem.gov.uk/electricity/transmission networks/offshore-transmission/our-role-offshore-transmission, 2016, accessed Mar. 30, 2016.

[8] A. Rygg Årdal and K. Sharifabadi, "Grid integration of offshore wind power plants with oil & gas installations," *Proceedings of the 12th Wind Integration Workshop.*

[9] CIGRE Technical Brochure 483, Working Group B3.26, "Guidelines for the design and construction of ac offshore substations for wind power plants."

[10] V. Hussennether, D. Woorthington, and M. Siebert, "Projects BorWin2 and HelWin1: Large scale multilevel voltage sourced converter technology for bundling of offshore windpower," *B4 Proceedings: CIGRE*, Paris, 2012.

[11] K. Friedrich, "Modern HVDC PLUS application of VSC in modular multilevel converter technology," *Industrial Electronics (ISIE), 2010 IEEE International Symposium*, pp. 3807–3810, Jul. 4–7, 2010.

[12] J. Glasdam, J. Hjerrild, and L. Kocewiak, "Review on multilevel voltage source converter based HVDC technologies for grid connection of large offshore wind farms," *Power System Technology (POWERCON), 2012 IEEE International Conference*, pp. 1–6, Oct. 30–Nov. 2, 2012.

[13] CIGRE Technical Brochure 604, , Working Group B4.57, "Guide for development of models for HVDC converters in a HVDC grid."

[14] CIGRE Technical Brochure 563, Working Group B4.38, "Modelling and simulation studies to be performed during the life cycle of HVDC systems."

[15] CIGRE Technical Brochure 569, Working Group C4.307, "Resonance and Ferroresonance in Power Networks."

8

MMC-HVDC Standards and Commissioning Procedures

8.1 Introduction

Most of the modular multilevel converter high-voltage direct current (MMC-HVDC) projects are planned and developed as point-to-point transmission scheme projects, where the HVDC equipment are designed, manufactured, and installed by the same HVDC original equipment manufacturer (OEM).

The HVDC project customers define the functional requirements, including the grid-code requirements that must be fulfilled by the procured equipment. During the bidding process, the HVDC OEM normally will document the compliance of the tendered HVDC scheme with the customer requirements through adequate system studies and simulations. At the moment of writing, there are no unified requirements or grid codes available, or any standards covering the MMC-HVDC transmission technology. Different grid codes impose different sets of technical requirements which must be fulfilled by the voltage-source converter HVDC (VSC-HVDC) equipment. There are ongoing efforts by European Network of Transmission System Operators for Electricity (ENTSO-E) to harmonize the European grid codes.

The MMC-HVDC market has been growing at an unprecedentedly high rate during the last few years. This growth has been driven by the increasing demand for the connection of offshore wind-power plants (WPPs) to onshore ac networks and MMC-HVDC interconnector projects in Europe. The rapid development of MMC-HVDC technology has made the above-mentioned applications feasible, and often is the most economic option. Owing to the rapid development speed of MMC-HVDC technology, the level of standardization efforts has been very limited—while the line-commutated converter HVDC (LCC-HVDC) technology and equipment is covered by a substantial number of standards and test procedure recommendations [1],[2],[3].

MMC-HVDC transmission technology has become an attractive option compared to LCC-HVDC transmission technology. MMC-HVDC schemes with transmission capacity of above 1 GW from a single converter are being offered. Many HVDC project developers

Design, Control, and Application of Modular Multilevel Converters for HVDC Transmission Systems, First Edition.
Kamran Sharifabadi, Lennart Harnefors, Hans-Peter Nee, Staffan Norrga, and Remus Teodorescu.
© 2016 John Wiley & Sons, Ltd. Published 2016 by John Wiley & Sons, Ltd.
Companion Website URL: www.wiley.com/go/Sharifabadi/ModularConverters

are considering the option of interconnecting the existing point-to-point HVDC schemes in order to create a multi-terminal direct current (MTDC) transmission network. These projects range from offshore WPPs, interconnectors between power markets and countries, and embedded schemes to relieve the congestion issues in the ac networks. At the heart of these concepts is the idea of an MTDC network that can provide the same functionality as the existing ac networks but also bring additional features which are distinct to an HVDC transmission system. An important next aspect in developing multi-terminal HVDC grids will be the inter-operability between different individual projects and technologies of different manufacturers. Interoperability requires standardization of the basic principles of design, testing procedures, and operation of HVDC grids. If the dc-link voltage, protection principles, load-flow control principles, and strategies are not standardized, it might not be possible to interconnect the existing HVDC schemes and create an MTDC grid. Not ensuring interoperability will have serious consequences for the future extendibility of MTDC grids. Interoperability between systems and equipment enhances competition and eases the introduction of innovative technologies. Lack of HVDC equipment standardization is seen as the main obstacle to the successful development of multi-terminal VSC-HVDC networks. However, there is a general concern that any unnecessary standardization of systems and equipment may diminish their development and constrain technical innovation, which is particularly undesirable for a technology that is relatively new. Ideally, any specifications developed for HVDC networks should be functional in nature and should not constrain new designs and innovative technologies.

Worldwide, various professional associations have taken initiatives to facilitate and develop the necessary recommendations and standards. Professional associations such as IEC, CIGRE, CENELEC, and IEEE are promoting and facilitating activities toward the standardization of MMC-HVDC technology worldwide.

This chapter provides an overview of standards applicable to HVDC equipment developed by professional associations in the field of MMC-HVDC technology. In addition, best practice recommendations for factory and site acceptance tests are presented.

8.2 CIGRE and IEC Activities for the Standardization of MMC-HVDC Technology

CIGRE, the Council on Large Electric Systems, was founded in 1921 and is an international non-profit association for promoting collaboration with experts from around the world by sharing knowledge and joining forces to improve the electric power systems of today and tomorrow. CIGRE boasts more than 3500 experts from across the globe working actively together in structured work programs coordinated by CIGRE's 16 Study Committees. Their main objectives are to design and deploy power systems for the future, optimize existing equipment and power systems, respect the environment, and facilitate access to technical information, recommendations, and standards. CIGRE Study Committee B4 (SC B4) addresses all the relevant target groups in the power industry interested in different aspects of power electronics, HVDC, and FACTS (flexible alternating current transmission system) [4].

The Study Committee's activities include:

- **HVDC**: economics of HVDC, applications, planning aspects, design, performance, control, protection, and control and testing of converter stations, i.e. the converting equipment itself and the equipment associated with HVDC links;

- **Power electronics for AC systems and power quality improvement**: economics, applications, planning, design, performance, control, protection, construction, and testing;
- **Advanced power electronics**: development of new converter technologies including controls, use of new semiconductor devices, applications of these technologies in HVDC, power electronics for ac systems, and power quality improvement.

SC B4 focuses on the provision of unbiased and up-to-date application guides for the implementation of HVDC and FACTS projects. SC B4 also provides such technical information to support the standardization required for the testing and application of new technology solutions.

SC B4 has published several technical brochures focusing on HVDC equipment and testing procedure recommendations, as well as application guidelines for the planning and operation of HVDC schemes, including descriptions and analysis of the interaction between ac networks and HVDC schemes. Many of these technical brochures have formed the foundations on which IEC HVDC technology standards have been developed.

CIGRE has set up many working groups to look in more detail at aspects of the VSC-HVDC and HVDC grid technologies. An MTDC test circuit has been proposed by CIGRE WG B4-58, shown in Figure 8.1 (WG B4-58 is a CIGRE working group tasked with writing a technical

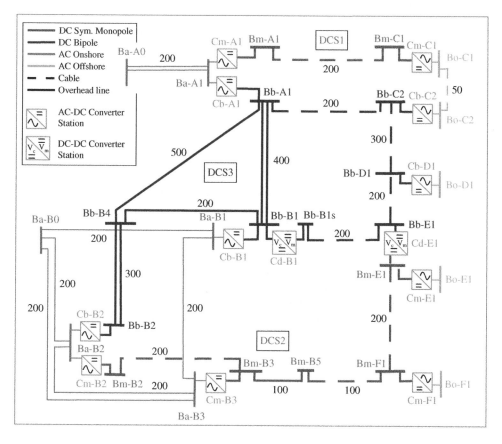

Figure 8.1 CIGRE MTDC test grid proposal.

brochure entitled *Devices for Load flow Control and Methodologies for Direct Voltage Control in a Meshed HVDC Grid*). This MTDC test circuit proposal has been adopted by many of the working groups as a common basis for DC grid evaluation. Details of the MTDC test circuit can be found at [5].

The following is a list of CIGRE's technical brochures and technical working committees which focus on dc grid technologies:

- CIGRE Technical Brochure 447: "Components testing of VSC system for HVDC applications"
- CIGRE Technical Brochure 496: "Recommendations for testing dc extruded cable systems for power transmission at a rated voltage up to 500 kV"
- CIGRE Technical Brochure 612: "Special considerations for ac collector systems and substations associated with HVDC- connected wind power plants"
- CIGRE Technical Brochure 604: WG B4.57: "Guide for the development of models for HVDC converters in a HVDC grid." The purpose and aim of the WG was to facilitate the development of generic models that would allow a high-level study of the HVDC grid and ac network interactions to be performed.
- CIGRE Technical Brochure 563: "Modelling and simulation studies to be performed during the lifecycle of HVDC systems"
- WG B4.63: "Commissioning of VSC HVDC schemes"
- CIGRE Technical Brochure 657: "Guidelines for the preparation of 'connection agreements' or 'grid codes' for HVDC grids"
- WG B4.57: "Guide for the development of models for HVDC converters in a HVDC grid"
- WG B4.58: "Devices for load flow control and methodologies for direct voltage control in a meshed HVDC grid"
- Joint WG B4/B5.59: "Control and protection of HVDC grids"
- WG B4.62: "Connection of wind farms to weak ac networks"
- WG B4.64: "Impact of ac system characteristics on the performance of HVDC schemes"
- Joint WG B4/C1.65: "Recommended voltages for HVDC grids"
- WG B4.67: "Harmonic aspects of VSC HVDC, and appropriate harmonic limits"
- WG B4.69: "Minimizing loss of transmitted power by VSC during overhead line faults"
- WG B4.70: "Guide for electromagnetic transient studies involving VSCs"
- WG B4.71: "Application guide for the insulation coordination of VSC HVDC stations."

CIGRE technical brochures and guidelines are available at the website: www.e-cigre.org.

The International Electrotechnical Commission (IEC) is a non-profit, non-governmental organization founded in 1906. The IEC's members are drawn from industry, government bodies, associations, and academia to participate in technical work and standardization activities. The IEC has published a few technical requirements focusing on VSC-HVDC transmission technology, including:

- IEC 62747, 2014: "Terminology for voltage-sourced converters for high-voltage direct current (HVDC) systems"
- IEC 62501, 2014: "Voltage sourced converter valves for high-voltage direct current power transmission: Electrical testing"

- IEC 62751-1, 2014: "Power losses in voltage sourced converter valves for high-voltage direct current (HVDC) systems: Part 1: General requirements"
- IEC 62751-2, 2014: "Power losses in voltage sourced converter valves for high-voltage direct current (HVDC) systems: Part 2: Modular multilevel converters."

CENELEC is the European Committee for Electrotechnical Standardization and is responsible for standardization in the electrotechnical engineering field. CENELEC prepares voluntary standards to facilitate trade between countries, create new markets, cut compliance costs, and support the development of a single European market. CENELEC creates market access at the European level but also at an international level, adopting international standards wherever possible, through its close collaboration with the International Electrotechnical Commission (IEC). CENELEC is focusing on standardization in the field of HVDC transmission technology above 100 kV. The task includes HVDC system oriented standards as design aspects, technical requirements, construction and commissioning, reliability and availability, and operation and maintenance.

The European HVDC Grid Study Group was established in September 2010, based on an initiative by the German Commission for Electrical, Electronic & Information Technologies (DKE). The Study Group has been aiming to develop "technical guidelines for first HVDC grids" with the following main objectives:

- To elaborate the basic technical principles of HVDC grids.
- To prepare functional specifications for the major power components of HVDC grids.
- To propose new work item proposals for standardization work to CENELEC TC 8X.

8.2.1 Hierarchy of Available and Applicable Codes, Standards and Best Practice Recommendations for MMC-HVDC Projects

Table 8.1 presents a list of the available international standards and recommendations applicable to the design, engineering, procurement, manufacturing, installation, and commissioning of HVDC equipment. During the contractual phase of HVDC projects, the customer and the HVDC manufacturer must agree on which hierarchy of codes and standards should be applied and included in the agreement.

8.3 MMC-HVDC Commissioning and Factory and Site Acceptance Tests

The commissioning process consists of factory acceptance tests (FATs), or offsite tests, and site acceptance tests (SATs), or onsite tests. During the commissioning process, the HVDC manufacturer, together with the representatives of the customer, verifies and demonstrates the full functionality, operation, and compliance of the procured and installed equipment according to the functional requirements and specifications. Commissioning processes allows the owners, developers, and/or end-user of the HVDC system to witness and document the equipment compliance with the specified contracted technical requirements. System testing is defined as

Table 8.1   Hierarchy of codes and standards applicable to VSC-HVDC equipment.

IEC	60044-1	Instrument Transformers: Part 1: Current Transformers
IEC	60071	Insulation Coordination
IEC	60076	Power Transformers
IEC	60099-4	Gapless Metal Oxide Surge Arresters for AC Systems
IEC	60099-5	Surge Arresters: Selection and Application Recommendations
IEC	60137	Insulated Bushings for AC Voltages above 1000 V
IEC	62501	Voltage Sourced Converter (VSC) Valves for High Voltage Direct Current (HVDC) Power Transmission, Electrical Testing
IEC TR	62543	High-voltage direct current (HVDC) power transmission using voltage sourced converters (VSC)
IEC	60146	Semiconductor Converters: General Requirements and Line Commutated Converters
IEC	60076-6	Reactors
IEC	60358	Coupling Capacitors and Capacitor Dividers
IEC	60633	Terminology for HVDC Transmission
IEC	60694	Common Specifications for High-Voltage Switchgear and Controlgear Standards
IEC	61071-1	Power Electronic Capacitors: General
IEC	61803	Determination of Power Losses in HVDC Converter Stations
IEC	61954	Power Electronics for Electrical Transmission and Distribution Systems: Testing of Thyristors Valves for SVCs
IEC	60815	Guide for Selection of Insulators in respect of Polluted Conditions
IEC	60273	Characteristics of Indoor and Outdoor Post Insulators for Systems with Nominal Voltages greater than 1000 V
IEC	60282-2	High-Voltage Fuses
IEC	60439	Low-Voltage Switchgear and Controlgear Assemblies
IEC	60715	Dimensions of Low-Voltage Switchgear and Controlgear
IEC	60129	AC Disconnectors and Earthing Switches
IEC	60265	High-Voltage Switches
IEC	61000-4-3: 2002	Radiated RF Immunity
IEC	61000-4-6: 2004	Conducted RF Immunity
IEC	60255-22-4: 2002	Fast Transients Immunity
IEC	60255-22-1: 2005	Damped Oscillatory Wave Immunity
IEC	60225-5: 2000	Dielectric Strength Immunity
IEC	60225-5: 2000	Impulse Voltage Withstand Immunity
IEC	61000-4-2: 2001	Electrostatic Discharge Immunity
IEC	60643-1	Surge Protection
IEC	60643-12	Surge Protection
ISO	9001	Quality Systems Model for Quality Assurance in Final Design, Development, Production, Installation and Servicing
NEMA	CC-1	Electric Power Connectors for Substations
IEEE	837	Standard for Qualifying Permanent Connections Used in Substation Grounding

the start-up and testing of the complete HVDC scheme in operation. The system tests should follow the structure of the HVDC system, starting from the smallest, least complex operational unit, and end with the total system in operation.

The HVDC project contract between the stakeholders may define and outline the required offsite (FAT) and onsite (SAT) test procedures and the test acceptance criteria. Normally, the required test procedures are defined and agreed in advance between the stakeholders and may include the following test programs:

- offsite testing (e.g. components and subsystem tests including control and protection tests, auxiliary equipment tests, dynamic performance tests);
- verification of onsite installation;
- onsite HVDC terminal energizing tests;
- operation of the HVDC terminal in STATCOM mode;
- controlled start-up and shutdown tests and emergency shutdown tests;
- steady-state power transmission tests, including power ramping;
- power quality and interference tests;
- black-start operation and loss of auxiliary (disturbance) tests.
- end-to-end system tests;
- AC network interaction tests (e.g. grid-code compliance tests, staged faults, reactive power and voltage control tests, and run-back and protection function tests).

Offsite testing provides the opportunity, among other things, to discover and correct possible hardware and software errors in the control and protection systems. Such faults are easier to find and correct during offsite testing than during the onsite testing. Correcting such faults reduces the probability of disturbing the power system during onsite commissioning.

8.3.1   Pre-Commissioning

The commissioning process requires thorough planning and cooperation between the involved parties.

During the project development phase, the parties usually agree on the required tests, test plan, test procedures and processes, test acceptance criteria, and finally the test documentation and protocols. The following is indicative of offsite and onsite test stages and procedures. Each project will have its unique test stages and procedures agreed between the parties.

The offsite commissioning starts with the pre-commissioning procedure. This consists of the inspection and documentation of manufactured equipment and subsystem test preparations. Subsystem tests include electrical and mechanical tests and simple functional tests confined to a subsystem. The objectives of these tests are to inspect the equipment's condition and verify the manufacturing, installation, and operation according to the technical requirements and standards.

Some projects may require routine inspections during the manufacturing phase of converter components in order to perform routine tests of components (e.g. the converter transformer, converter reactors, and converter valves). There are IEC test procedures and CIGRE test recommendations for component testing of VSC-HVDC equipment. CIGRE Technical Brochure

447 "Components Testing of VSC System for HVDC Applications" and WG B4.63: "Commissioning of VSC HVDC schemes" provide best practices and guidelines for the VSC-HVDC component testing.

8.3.2 Offsite Commissioning Tests or Factory Acceptance Tests

Offsite commissioning tests (FATs) include end-to-end tests and verification of all subsystems, including complete HVDC control and protection systems. Normally at this stage, the high-voltage equipment and the ac networks are modeled and simulated with a real-time simulation (RTS) tool. A digital RTS is a hardware-based simulation tool that provides an environment for fast, reliable, accurate, and cost-effective studies of power systems with complex HVAC and HVDC networks. The RTS is a fully digital electromagnetic transient power system simulator that operates in real time.

The RTS software containing actual models from the HVDC converter and interfaced ac networks is capable of simulating in real time the electrical characteristics of the interconnected HVDC and connected HVAC systems. The studies performed during the design phase of the HVDC converters evidently provide a reference for the expected outcome of the commissioning test(s).

For test purposes, the actual control and protection cubicles are wired to the RTS and auxiliary systems, including test and measurement equipment such as fault recorders. The objective of these tests is to determine the correct behaviors of the control and protection systems in terms of the expected input and output signals. The tests normally cover all modes of operation such as energizing, de-blocking the converter, standby and steady-state operation, ac and dc fault conditions, converter blocking cases, and trip of converter and start-up and shutdown scenarios. All operation modes of the HVDC terminals are simulated and the predicted behavior of the system is verified. In addition, it is important to verify the proper operation and signaling of auxiliary systems such as converter cooling system and firefighting systems.

The FATs should cover the steady-state operation and dynamic performance behavior of the HVDC converter during ac and dc network fault conditions. For these tests, different ac and dc network fault conditions are simulated, and the HVDC control and protection functionalities are verified and documented. The possible deviations between the results from the design phase studies and the recordings from the commissioning tests always require careful analysis. The analysis may lead to changes in the implemented RTS models, or in certain cases it will require the implemented control and protection functions of the HVDC equipment to be tuned.

During the design phase of HVDC converters, manufacturers utilize electromagnetic transient (EMT) simulation tools to model the HVDC converters. The behavior of the system is then verified through dynamic simulations in the time domain. The EMT software is typically an electromagnetic transient simulation tool (e.g. PSCAD). The EMT model provides nearly identical results to the actual converter controls and is a very important tool during the design phase. During the offsite testing, the EMT simulation results are useful to verify and control outcomes of the FATs. For test purposes, it is important that the EMT model is validated against the real-time analysis software model. The EMT converter model represents the actual controls of the HVDC scheme as the EMT converter model is used during the initial design and tuning of the HVDC system. Normally, the settings determined in the EMT simulation are directly implemented into the manufacturer's control systems. The EMT converter model is a useful tool to troubleshoot any issues that may arise during commissioning.

Dynamic performance tests focuses on the dynamic behavior of the system (interconnected HVDC terminals and the ac networks). These tests are conducted with computer simulation tools using RTS software with project-specific control and protection hardware. Results from the tests are important because they can demonstrate that the equipment complies with basic and grid-code functional and technical requirements.

The FAT procedures should also include tests to verify system redundancy, changeover functionalities, communication, and human–machine interface (HMI) functions.

8.3.3   Onsite Testing and Site Acceptance Tests

When all the HVDC station equipment, dc cables or overhead powerlines, and the connecting ac yard are installed at the project site, the SATs can be initiated. SATs normally consist of the following stages:

- mechanical and electrical verification of installed equipment;
- individual subsystem tests;
- single terminal energizing tests;
- operation in STATCOM mode;
- grid-code compliance tests;
- end-to-end power flow tests;
- heat run and the trial operation test.

At the first stage, normally the HVDC scheme client verifies that the equipment and the installations are in accordance with the agreed technical specifications, guidelines, and standards. Verification of physical and mechanical installation and completion of the equipment is verified at the pre-commissioning phase of the SAT. The completeness and correctness of the overall system installation at the site should be visually verified, with the aid of checklists, diagrams, drawings, and installation instructions.

The next stage may include verification and acceptance tests of subsystems such as cooling systems, ac switchyard, dc yard apparatus and components, control and protection cubicles, IGBT valves and valve halls, transformers, reactors and station auxiliary systems.

Before energizing the subsystems, all possible piping and hard wirings between the cubicles and the equipment must be verified and documented. The verification includes testing of cabling termination and identification of each power, control, and communication cable within the converter station. Normally during this procedure, all cabling and connections between the subsystems are identified using schematic diagrams and marking off each cable on these documents after the cable is confirmed to originate and terminate at the correct terminals.

The signaling and communication between the submodules, IGBTs, and the control system must be verified. The aim is to ensure that the submodules and, specifically, the IGBTs are connected to the correct location in the valve control and monitoring systems. It should be noticed that in an MMC-HVDC converter a few thousand fiber-optic wires connect the IGBTs and submodules to the converter control system. It is not unusual to discover some mistakes of cabling during this stage. The discrepancies should be documented, corrected, and verified once more.

After verification of the system's cabling and signaling, the next stage may consist of starting-up, energizing, and shutting down each subsystem.

As an example, the onsite test procedures for the valve-cooling subsystem can consist of the following stages:

- visual inspection of cooling system and piping: checking that the piping and interfaces are in accordance with the technical specification and standards and carrying out a leakage test of pipes;
- checking all valve cooling measurement devices and transducers: this involves checking the signals from flow meters, conductivity meters, pressure transducers, and temperature transducers back to the converter control and protection equipment;
- functional test of pumps: checking the operation of valve cooling pumps, including control signals from the converter control and protection systems and the operation of pump protections. This will include operation of the changeover from active to standby pumps;
- functional test of cooling fans and heat exchangers: checking the operation of each cooling fan and heat exchanger, and the control of these fans from the converter control and protection systems. The tests should include operation of any motor-operated valves, water temperature measurement, and the correct starting/stopping of cooling fans;
- functional test of other cooling system elements: including any water pressure systems, makeup/filling and storage systems, deionization systems, including the conductivity measurement equipment, oxygen management systems, heaters, expansion tanks, etc.;
- overall check of cooling system from control and protection systems: coordinated operation test of pump and fans to ensure correct water regulation and checking valve cooling protection, including flow, pressure, water level, leakage detection, conductivity alarms, audible sound, and trips.

During the SAT, start-up and shutdown of all interfaced computers, including HMI systems, supervisory control and data acquisition (SCADA) and telecommunication systems must be verified and documented.

Changeover from standby and the redundancy of all subsystems, especially the control and protection systems, must be tested and documented.

It is important that, during SATs, as a minimum, the normal operation of all subsystems and the signaling between them are controlled, monitored, and documented, before high-voltage energizing tests are initiated.

8.3.4   Onsite Energizing Tests

After the approval and signoff of the onsite subsystem acceptance tests, the energizing of the HVDC terminal can be initiated. It is of the upmost importance that, prior to the energizing procedures and tests, all safety procedures are established, and the relevant staff are accustomed and acquainted with health and safety procedures.

The energizing tests are normally initiated by energizing the ac switchyard and the substation transformers.

During ac yard energizing, local and remote operation and the interlocking functions of the circuit breakers, disconnectors, and earth switches should be tested and verified. In addition, the proper operation of the main transformer, tap changer, and current and voltage transformers must be verified. After the acceptance and approval of the energized ac switchyard, the converter transformer can be energized and tested. At this stage, the converter valves are blocked and possibly grounded. During the next stage, the converter's dc components, such as the converter valves, can be energized and tested. The converter station is still in the blocked mode. The proper function of all subsystems at this stage must be verified. After approval of all subsystems including the auxiliary systems and the control and protection system, the HVDC converter is ready to be deblocked. The HVDC converter terminal tests are performed on individual terminals while they are disconnected from the opposite converter terminal and the dc cable or line. The HVDC converter energizing procedures have usually been developed by the HVDC equipment manufacturer according to project and customer requirements.

The converter terminal tests constitute a set of verifications of the ac-bus-voltage control and reactive power control modes (STATCOM mode of operation) that are to be conducted when the converter terminal is connected to the adjacent ac network for the first time.

Energizing and subsequent deblocking of a converter terminal for the first time is one of the most critical stages in the operation of an HVDC system. At this time, the converter terminal is isolated from the dc transmission cable/line. During the terminal tests, the two converters can be operated and tested independently without any interface between the two stations. The onsite HVDC terminal tests should cover the tests and verification of energizing, standby operation, and deblocking and blocking sequences and procedures. Different modes of operations must be verified as standalone and independent systems.

However, the SAT and energizing procedures for offshore HVDC converters connected to wind farms can be different when compared to onshore HVDC terminals connected to an ac network. The energizing of offshore HVDC terminals requires the availability of an auxiliary power source, such as diesel generators.

The converter energizing procedures may vary depending on the agreed procedures between the involved stakeholders and the equipment manufacturer.

Finally, the transmission system tests are performed with the HVDC terminals connected to the dc transmission link. These tests are also called *end-to-end tests* where the transmission of dc-side power is verified.

The transmission system tests constitute a set of verifications of transmission operation between the connected HVDC terminals. During these tests, the active and reactive power capability of the converters is verified. In addition, step responses, and verifications of the active power control modes in conjunction with the reactive power capabilities of each individual converter terminal, are verified.

During the end-to-end tests, the following control modes should be verified:

- AC-bus-voltage control at both terminals;
- reactive power control at both terminals;
- DC-bus-voltage control at both terminals;
- DC power flow and step control response;
- independent control of active (P) and reactive (Q) powers at both terminals;
- system redundancy and changeover tests;
- trip tests;
- communication and operator control tests (e.g. transfer of HMI station and system functions between individual control locations, including local station control at both terminals).

During various P and Q commissioning tests (e.g. operation at the extremes of the PQ curve), the HVDC converter station will be required to generate or absorb relatively high levels of reactive power. A converter trip during these tests may result in unacceptable voltage levels at the ac busbar and cause stability issues for the interfaced ac network. This incident can be of concern to the system and grid operators. Therefore, it is important that the SATs are communicated and coordinated with the related ac network asset owners and operators. Any form of converter trip tests may require the availability of rotating reserves, depending on the strength of the connected ac networks. The agreements with the responsible authorities and stakeholders must be clarified and agreed on in advance, before such onsite testes can be performed.

After the completion of SATs, projects may require *heat test runs*. The intention of these tests is to prove that all components of the project are capable of continuous operation at the maximum designed load levels guaranteed in the contract. This will be achieved by carrying out the heat run test and calculating the components steady-state temperatures from the measurement results. The transmission capacity of the HVDC link will be brought to the agreed test value or the maximum available capacity and held constant for a longer period (agreed between the parties). During this period, all relevant measurements will be performed and recorded. Additionally, it is required that the surface temperatures of dc reactors, breakers, and bushings are recorded by means of a thermal-imaging camera. The temperature trends versus time curves will be developed and evaluated in order to verify the maximum load performance of the HVDC scheme and components.

Following the completion and approval of the SATs and commissioning, a trial operation can be initiated. The duration, purpose, and requirements of the trial operation period are normally defined and agreed between the HVDC equipment manufacturer and the customer during the procurement phase and detailed in the contract.

Each project defines the objectives of the trial operation and its acceptance criteria. In some cases, the trial operation objective is to demonstrate the contracted availability, reliability, and performance requirements of the HVDC scheme (e.g. maximum number of forced outages or trips during the trial period). In other cases, the trial period may be defined as the operation period where no forced outages are allowed and an availability target of 100% is required. In either case, the contract or an agreement before commencement of the trial operation will need to define what happens when the test does not achieve the acceptance criteria (e.g. the extension of the trial operation period until the acceptance criteria are achieved).

After the successful trial operation period, the equipment may enter the guarantee period. Terms and conditions of the guarantee period are normally defined in the project contract between the stakeholders.

However, each and every HVDC scheme will have its own unique challenges. As each HVDC scheme project is designed and operated under different technical requirements, national grid codes and regulatory frameworks imposed by the various stakeholders.

The commissioning planning process must ensure that the project functional requirements, grid-code requirements, and regulatory and power market requirements are identified early and planned for through their inclusion in the overall commissioning program. The requirement for the tests, their objectives, preconditions, procedures, and acceptance criteria must be identified in the early stages of the project and preferably agreed upon during the contractual phase of the project.

8.4 Summary

This chapter presents factory and site acceptance test procedures for HVDC equipment. The hierarchy of codes and standards applicable to VSC-HVDC equipment are reviewed.

References

[1] IEC 61975, "HVDC system test."
[2] Cigre Technical Brochure 447, "Components Testing of VSC System for HVDC Applications."
[3] IEEE 1378-1997, "Guide for Commissioning High-Voltage Direct-Current (HVDC) Converter Stations and Associated Transmission Systems."
[4] http://b4.cigre.org/.
[5] CIGRE, "Documents related to the development of HVDC Grids," http://b4.cigre.org/Publications/Documents-related-to-the-development-of-HVDC-Grids, accessed Mar. 30, 2016.

9

Control and Protection of MMC-HVDC under AC and DC Network Fault Contingencies

9.1 Introduction

The modular multilevel converter (MMC) has become the most attractive voltage source converter (VSC) topology for high-power applications.

AC or dc short-circuit fault conditions cannot be avoided in high-voltage direct current (HVDC) schemes, independent of the converter and the transmission link technology (e.g. cables or overhead lines). The probability, frequency, and characteristics of these faults depend on the employed transmission technology, converter technology (e.g. two-level or MMC), and the grounding topology (e.g. direct, resistive, or high-impedance).

Generally, the fault conditions for an HVDC scheme can be divided into four main scenarios:

- AC network faults at point of common coupling (PCC);
- AC faults inside the converter station;
- DC faults, e.g. pole-to-pole or pole-to-ground faults;
- converter internal faults, e.g. submodule faults, modulation and control faults, phase-reactor faults.

MMC-HVDC terminals can withstand symmetrical and asymmetrical ac network fault conditions. However, the half-bridge MMC-HVDC transmission technology is vulnerable to dc-pole fault conditions. One of the major challenges associated with the half-bridge MMC-HVDC technology is the lack of dc-side short-circuit fault-handling capability.

Currently, all VSC-HVDC installations are point-to-point connections. Protecting these installations against dc faults relies on operating the ac circuit breakers and isolating the converters from the ac network. DC fault detection and protection is seen as one of the enabling technologies for the realization of HVDC multi-terminal networks. MMC topologies

Design, Control, and Application of Modular Multilevel Converters for HVDC Transmission Systems, First Edition.
Kamran Sharifabadi, Lennart Harnefors, Hans-Peter Nee, Staffan Norrga, and Remus Teodorescu.
© 2016 John Wiley & Sons, Ltd. Published 2016 by John Wiley & Sons, Ltd.
Companion Website URL: www.wiley.com/go/Sharifabadi/ModularConverters

with the inherent capability of blocking the dc fault current composed with alternative submodule topologies (e.g. full-bridge) present an alternative solution for managing the dc fault current.

This chapter presents and compares the two-level VSC-HVDC and MMC-HVDC fault characteristics and behaviors under ac and dc fault conditions. The impacts of different component failures are discussed, and related control and protection techniques are presented.

9.2 Two-Level VSC-HVDC Fault Characteristics under Unbalanced AC Network Contingency

The grid code is a document normally written and governed by the transmission system operators. This document defines functional and technical requirements that must be fulfilled by the equipment connected to the ac network. Most European grid codes define functional and performance requirements for VSC-HVDC equipment. Low-voltage ride-through (LVRT) capability is defined as the ability of the HVDC converter to withstand balanced and unbalanced grid faults for a limited amount of time. Typically, the HVDC converter is required to remain connected to the ac system in case of a nearby ac network fault for up to 140–200 milliseconds. In addition, the converter should be able to support the ac network by injecting positive sequence reactive power proportional to the positive sequence voltage drop caused by the fault, and, once ac network conditions become balanced again, the converter must be able to immediately support the network by ramping up its active power, as shown in Figure 9.1. Moreover, the presence of the fault should not be indicated on the other side of the HVDC transmission link.

VSC-HVDC terminals are designed to fulfill FRT requirements and remain synchronously connected to the network and provide the required ancillary services during the unbalanced ac

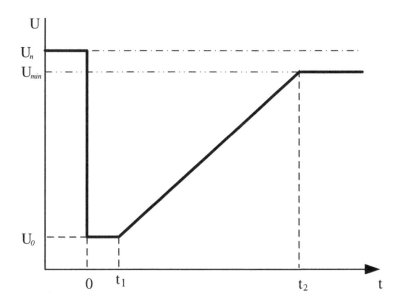

Figure 9.1 Typical grid code FRT requirement for VSC-HVDC stations (converter ac-bus voltage).

fault conditions. VSCs have limited overloading capabilities, and during such faults they can only supply the fault current up to their designed rated current. DC transmission systems based on VSCs can comply with FRT requirements with minimum current and voltage stresses on the converter switching devices.

Normally, the converter control objective during unbalanced conditions is to eliminate the double-frequency oscillations in active power flow usually associated with the unbalanced three-phase system, and keep the dc link voltage stable. In conventional two-level VSCs, the converter currents can be controlled to achieve various objectives for the control of active and reactive power flow at PCC. When an asymmetrical fault occurs at the ac network, there will be negative and zero-sequence components in the grid voltages. In two-level converters with centralized dc capacitors, only positive- and negative-sequence current components are fully controllable. As in most configurations the converter transformer is a star/delta configuration, and the zero-sequence components are isolated from the secondary side of the transformer. However, some zero-sequence components on the converter side will appear and are unavoidable. To deal with the unbalanced grid conditions, a negative-sequence current controller is implemented in the two-level VSC converters. The controller eliminates the negative-sequence current and keeps the ac current balanced under unbalanced grid voltages [1]. Therefore, the positive- and negative-sequence components of the power ripple in each phase, which are circulating between three phases, have a negligible negative effect on the three-phase real power and the dc link voltage. Depending on the implemented converter control strategy, it may happen that the dc link voltage contains a relatively small second-harmonic dc voltage and current. DC harmonics are damped through the long dc cables and thus have a negligible negative effect on the dc-bus voltage [2–4].

The basic performance of the converter under unbalanced ac fault conditions can be summarized as: the converter controller detects the low ac-bus voltage, and increases the converter's capacitive current output, in order to supply reactive power to the ac system. As the ac-bus voltage drops, the converter current increases up to its maximum current limit, until the fault is cleared by the ac protection system.

After the fault is cleared, the converter controls the ac-bus voltage during the system recovery. The converter ramps up its reactive power and reduces its reactive power injection.

Based on the operation of various VSC-HVDC schemes, it has been concluded that recovery from ac network disturbances is highly system-design-dependent. It can be observed that the two-level VSC-HVDC has a satisfactory behavior and fulfills the FRT requirements.

However, the two-level VSC- HVDC converters are very vulnerable to ac faults, when the fault location is on the secondary side of the converter transformer. Internal ac bus faults are usually caused by insulation failure or the failure of devices connected to the ac bus (e.g. ac filters) between the secondary side of the transformer and the converter valves. The faults in the ac bus between the transformer and converter shall be considered as permanent faults. A short-circuit fault of the converter ac bus to ground is much severer; as the midpoint of dc capacitors in a two-level converter is normally grounded in order to provide a reference point for the dc voltage [5, 6].

Figure 9.2 shows the single-phase to ground fault on the two-level VSC converter terminal bus. The dc-link capacitors discharge through the short-circuit loop rapidly and cause destructive overcurrents to the converter semiconductors. The generated negative-sequence component causes second-harmonic dc ripple voltages and currents on the dc bus. The ground potential of the dc-side neutral point will float and the dc-bus voltage will contain

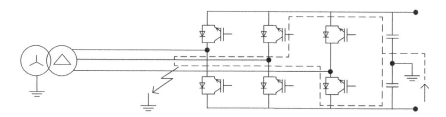

Figure 9.2   Converter internal ac bus short-circuit and the fault current path loop.

high-frequency oscillations. At this point, as a result of the charge and discharge of the dc-link capacitors through the short-circuit path loop, the dc-link voltage can reach a peak overvoltage of about two times its nominal voltage.

Isolation coordination studies must cover such severe overvoltage conditions and design the converter surge arresters accordingly. Consequently, the converter must be blocked and disconnected from the connected ac network immediately.

9.2.1   Two-Level VSC-HVDC Fault Characteristics under DC Fault Contingency

Two-level VSC-HVDC converters are very sensitive to dc fault conditions. The dc-side short-circuit faults can cause serious damages to the converter valves. When a dc fault occurs, the dc link capacitors discharge through the fault location and cause a large dc fault current, a rapid decrease in the dc-bus voltage, and a substantial increase of the ac current flow to the converter valves [7]. Grounding strategy and the grounding impedance influence the dynamics of the short-circuit current. Once the control and protection of the converters detect a dc fault, the converters are blocked immediately and converters are isolated from the ac networks from both sides by initiating the trip signals to the converter ac breakers. The VSC cannot control the dc side-fault current even when the insulated-gate bipolar transistors (IGBTs) are blocked, as the anti-parallel diodes will act as an uncontrollable rectifier and feed the dc short-circuit with the current from the ac-side network.

DC bus fault condition has three distinctive stages:

- Stage 1: Capacitor discharge phase: Immediately after occurrence of fault the converter's large dc capacitors are discharged through the fault location. The equivalent circuit, shown in Figure 9.3, can be presented as a second-order differential equation which has an oscillatory behavior if $R < 2\sqrt{L/C}$. The magnitude and steepness of the first current peak is determined by the dc link voltage and the cable resistance and impedance.
- Stage 2: Freewheel diode phase: At this stage, the converter capacitor voltage has decayed to zero and the cable capacitance discharges through the anti-parallel diodes in the converter bridge. Depending on the dc cable design and the cable length, the cable capacitance discharge current can be very large.
- Stage 3: AC grid feeding the fault current: Finally, in the third phase, the fault is fed from the ac grid through the anti-parallel diodes acting as an uncontrollable rectifier. At this stage, the fault current peak depends on the short-circuit capacity of the ac grid, the converter filter impedance, the dc-cable impedance, and the fault resistance.

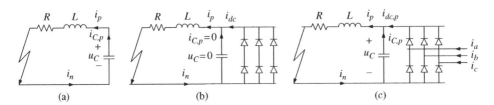

Figure 9.3 Equivalent circuits for fault on dc-side during (a) capacitor discharge phase, (b) freewheel diode phase, (c) grid feeding phase.

Under a pole-to-ground fault, the faulty pole capacitor will discharge and the capacitor voltage will collapse, while the voltage at the healthy pole will rises up to 1.5–2 p.u. and impose overvoltage on the healthy pole capacitor. Faulty pole capacitors will be imposed to overcurrent caused by rapid discharging through the short-circuit loop.

During a pole-to-pole fault, both dc link capacitors will rapidly discharge through the fault loop, and contribute to the fault current flowing through the short-circuit path. Simultaneously, the ac network will be subjected to a three-phase short-circuit condition. Pole-to-pole faults are more severe compared with the pole-to-ground fault scenarios and demand that both side HVDC converters are blocked and disconnected from the ac-side networks.

Pole-to-pole faults are rare with submarine cables, while pole-to-ground faults are seen as more likely since they are related to cable insulation deterioration.

DC pole-to-ground or pole-to-pole faults are considered the most severe fault conditions for two-level VSC-HVDC schemes.

9.3 MMC-HVDC Fault Characteristics under Unbalanced AC Network Contingency

MMCs are robust against the unbalanced ac network fault conditions and can fulfill the LVRT requirements defined in the grid codes.

Normally, the MMC-HVDC converter transformers adapt the star/delta (Y/Δ) configuration with the delta connection on the converter side designed with high zero-sequence impedance. Furthermore, with this design the converter output voltage capabilities can be increased by implementing the third-harmonic reinjection, as shown in Section 3.7.4. During unbalanced ac network conditions, the zero-sequence current is almost blocked by the delta connection of the transformer in the converter side, and thus only positive- and negative-sequence currents must be controlled. However, with an MMC, the distributed location of energy storage capacitors permits the independent control of each phase. Thus, a zero-sequence current control becomes possible in addition to the positive- and negative-sequence current control [8–11].

During the unbalanced ac fault conditions, the converter will be subjected to double frequency active power oscillations at the PCC. The active power oscillation increases the amplitude of voltage ripples at the submodule capacitors. However, the higher submodule voltage ripples will not affect the voltage balance between the converter submodules, as it affects equally all the submodules in the converter arms. As the dc voltage ripple of the submodule capacitors increases, the converter dc voltage will decrease.

Even by applying the star/delta winding vector group, the currents between the transformer and converter valves may not be symmetrical and the energy balance between the converter's phase units will become imbalanced. Correction of submodule's capacitor voltages will generally require an exchange of power between the three phases. The converter's inner current control loop will act against the imbalanced charges in the phase units, and initiate limited circulating currents within the converter. As the dc current is not equally shared among the converter's three-phase units, the initiated circulating currents compensate and rebalance the charges among the converter phase units [12].

During unbalanced ac network conditions in the primary side of the star/delta transformer, negative-sequence voltage and current will flow to the converter side. The converters are equipped with positive- and negative-sequence current controllers, controlling the converter's line currents under such conditions [13–15].

Under asymmetrical ac fault conditions each converter phase can operate independently without causing voltage distortion across the other phases, because there is no common dc capacitor shared by all phases. Consequently, the MMC can operate with continuous grid voltage imbalances. Under single phase to ground fault conditions the two healthy phases can continue the full power transmission, limiting the net energy transfer deficit to about one-third of the maximum value. This is particularly beneficial when connected to weak ac networks.

The impact of ac faults on converters is largely reduced by dedicated converter fault management control systems. The control system switches the converter control objectives from normal operating conditions to objectives which are more appropriate under such fault conditions. AC faults are normally handled by the proper design of the converter control systems and by engaging the ac fault protection equipment to remove the ac short-circuit fault [16]. MMC-HVDC converters are designed with inherent FRT capability in case of any ac fault conditions at the primary side of the converter transformer and are robust when it comes to the unbalanced ac network conditions, owing to the individual voltage and energy flow control in each phase [17, 18].

Unbalanced ac grid conditions cause ac-bus-voltage fluctuations and phase shifts. AC voltage phase shifts, combined with voltage fluctuations, may cause a change in the converter's active power flow. The active power controller is capable of correcting the imposed changes immediately. When an ac fault occurs at the HVDC terminal operating in inverter mode, if the rectifier terminal does not reduce the real power injected into the dc grid, and if this real power is higher than the maximum power that can be exported to the ac grid by the inverter terminal during the fault condition, it is necessary to drain the power surplus into dissipative devices (e.g. dynamic braking resistors). Therefore, in such applications, it is necessary to install a dc breaking chopper that is aimed to compensate the real power mismatch between the two converters connected to the HVDC scheme.

Temporary asymmetrical grid conditions are not seen as harmful to the converter itself. The converter recovers quickly when the ac fault is removed and the converter switches are not exposed to overvoltage or current stresses since the modular converter cell's capacitor is not compromised [19].

9.3.1 Internal AC Bus Fault Conditions at the Secondary Side of the Converter Transformer

The dynamics and fault behavior of VSC-HVDCs is influenced by the converter transformer winding group configuration and grounding philosophy. Most VSC-HVDC schemes adapt a

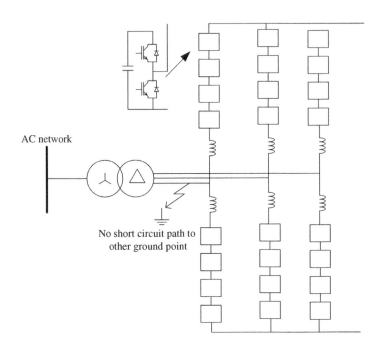

Figure 9.4   MMC-HVDC internal ac bus single phase-to-ground short-circuit.

star/delta winding group and, consequently, there is no path for the zero-sequence components flowing from the primary side ac network to the converter, as shown in Figure 9.4.

For any ac bus faults inside the converter station between the secondary side of the converter transformer and the converter, the MMC-HVDC terminal should be tripped immediately to avoid component damages, as usually the internal ac bus faults are permanent. Overvoltages will be limited by the arrester scheme until the trip is completed.

Under single phase to ground fault at secondary side of the transformer with the star/delta transformer, the faulty phase voltage sags to zero and the non-faulty phase rises to the $\sqrt{3}$ times of nominal value. Either insulation levels must allow for this voltage increase or (normally) arresters come into action to hold the voltages down. With the delta winding of the transformer in the secondary side, no zero-sequence current from the ac network can flow into the converter. At this stage, the voltage of the healthy phases at the converter bus will be higher than the dc pole voltage. As a consequence the dc cable is charged through the submodule diodes. Under such conditions, the converter poles can be subjected to overvoltages, as the anti-parallel diodes in the healthy phases charge both poles to the peak value of the phase-to-phase ac voltage. The charging current is limited by the impedance of the converter transformer and the converter phase reactors. The converter surge arresters limit the imposed overvoltage, and if the HVDC scheme is equipped with dynamic braking resistors, the dc bus overvoltage can be controlled by activating the dynamic braking resistors [19].

During single phase to ground fault, the negative-sequence voltage in the primary side of the transformer can flow to the secondary side and result in negative-sequence current in the converter side if there is no negative-sequence controller is implemented.

It should be noticed that, unlike the two-level VSC-HVDC, there is no pathway for the zero-sequence components and the discharge of converter capacitors, as discussed in the previous section. Therefore, the converter semiconductors will not be compromised under single phase to ground conditions.

9.4 DC Pole-to-Ground Short-Circuit Fault Characteristics of the Half-Bridge MMC-HVDC

In symmetric monopole MMC-HVDC schemes, under a dc pole-to-ground short-circuit fault condition a voltage displacement between the poles occurs. The faulty pole voltage decreases to a lower level, given by the fault resistance, while the healthy pole voltage increases considerably, and may rise up to 2 p.u. of the pre-fault voltage [20]. It is important that the dc cable's maximum insulation voltage is designed to sustain such temporary overvoltage. The converter surge arrester protects the components from any abnormal overvoltages. It should be noticed that the pole-to-pole voltage U_{dc} keeps constant but with a displacement between the poles. In contrast to the two-level VSC-HVDC, the dc-link voltage will keep constant because there isn't a charging path between the capacitor and the grounding point [21–24].

For minimizing the overvoltage stress on the healthy pole it is important that the dc pole-to-ground fault is cleared as fast as possible. The overvoltage and the voltage unbalance are detected independently by the dc voltage protection in both HVDC terminals. This protection will block the converters.

9.4.1 DC Pole-to-Pole Short-Circuit Fault Characteristics of the Half-Bridge MMC-HVDC

HVDC schemes can be subjected to dc fault conditions due to the deterioration of cable isolations caused by aging processes or physical damage, and overhead lines may experience short-circuit conditions sourced from lightning or extreme weather conditions. After detection of the dc pole-to-pole fault, the control and protection systems block the converters immediately. When the converter switches are blocked, the anti-parallel diodes act as an uncontrolled rectifier, and a voltage with the opposite polarity of the ac network voltage will be fed into the converter arms as the natural commutation processes of the anti-parallel diodes and the dc short-circuit current is fed from the ac-side network, as shown in Figure 9.5. To interrupt the dc fault current supplied from the ac network in a half-bridge converter topology, the ac circuit breakers associated with the converter must be tripped [25].

In the absence of any feasible commercial dc breakers, the pole-to-pole short-circuit fault is the most severe fault condition for MMC-HVDC schemes, and the converters must be tripped and isolated from the connecting ac networks. Control and protection systems of the HVDC converter are designed to detect such fault conditions and immediately block the HVDC converters on both sides of the scheme.

The half-bridge converter topology, just like the conventional two-level converters, will lose control of its currents in case of a stiff pole-to-pole dc short-circuit fault. In general, the dc fault characteristics of MMC-HVDC schemes are very different compared with the two-level VSC-HVDC schemes. Surge current contributions sourced from the discharge of dc capacitors are limited as the dc capacitors are distributed in the MMC. The converter phase reactors along with the converter transformer reactance limit the amplitude and the rise rate of the dc fault current fed from the ac side. Clearing the fault by opening the ac circuit

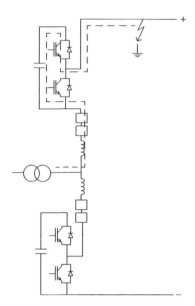

Figure 9.5   Short-circuit current path through the MMC during a pole-to-pole short-circuit fault.

breaker will typically take three to four ac system cycles. The half-bridge MMC topology is not able to block the dc fault current fed from the ac network through the anti-parallel diodes. Alternative submodule topologies (e.g. the full-bridge topology) are capable of quenching the dc fault current in the case of dc pole fault conditions, hence an offload disconnector can be utilized to isolate the faulty section of the dc grid and reconfigure the dc network in a no-current condition. This topology is seen as the preferred topology for HVDC schemes with overhead dc transmission lines.

A dc pole-to-pole short-circuit can be divided into three distinct stages.

Stage 1: Discharge of Cable and Submodule Capacitors

This stage refers to the first few milliseconds after occurrence of a dc pole-to-pole short-circuit fault, before the converters are blocked. When the dc bus overcurrent or dc-bus undervoltage is detected, the controller system blocks the converter. One approach to detect the dc fault is based on measuring the instantaneous converter's dc current, I_{dc}, and comparing it with a predefined threshold current, I_{fault}. If the measured deviations fulfill the predefined conditions, then the control system will block the converter valves. During this short period, the cable and the distributed submodule capacitors connected in the circuit will discharge and contribute to a rise of the dc fault current. However, owing to the small amount of distributed dc capacitors inserted in the circuit, the maximum short-circuit current is expected to be much lower for the MMCs compared with the two-level VSCs. The converter phase reactors limit the rate of rise of the fault current di/dt and protect the free-wheeling diodes from surge currents. The component sizing and design of the converter components should not only optimize the converters for the steady-state operation but also ensure the converter operates within safety margins during the ac- and dc-side fault conditions.

Stage 2: Anti-Parallel Diodes Act as Uncountable Rectifiers and Feed the Fault Current from the AC Side Network

The second stage starts after the IGBTs are blocked and the short-circuit fault current is fed through the anti-parallel diodes. This stage is characterized by the short-circuit capacity level of the connected ac networks. The anti-parallel diodes utilized in the industrial IGBT packages have a limited ability to withstand a surge overcurrent without sustaining permanent damage. MMC-HVDC designs utilizing industrial IGBT packages employ a protective bypass, press-pack thyristor connected in parallel with each submodule. The thyristor is switched on in the event of a dc pole fault condition, hence most of the fault current flows through the thyristor and not through the anti-parallel diodes.

Stage 3: Activation of Converter AC Breakers and Isolation of HVDC Converters from the AC Network

This stage will last for at least three to four fundamental ac cycles, until the ac breakers are activated and the converters are isolated from the ac network. However, it is observed that after disconnection of the HVDC converters from the ac network, a limited amount of current from the converter will flow out, to the short-circuit location. The source of the remaining current is the stored energy at the converter phase reactors and the stray cable capacitances.

A dc pole-to-pole fault condition for half-bridge MMC-HVDC schemes is seen as the most severe fault condition. The converters must be blocked and isolated from the connection ac networks.

9.5 MMC-HVDC Component Failures

During the lifetime of an MMC-HVDC terminal, its components—such as the converter transformer, submodule components, phase reactors, and control and protection systems—could be subjected to irreversible permanent damage. MMC-HVDC schemes are normally designed with a high level of component and subsystem redundancies. Most usual redundant components are submodules, control and protection systems, and submodule cooling pumps. In offshore converter designs, normally a redundant transformer is included in the scheme. Any repair and possible replacement of the heavy offshore transformer is dependent on good weather conditions. With redundant main converter components, in the event of a single failure, the HVDC scheme can continue its normal operation until the planned maintenance period. Some other converter components such as phase reactors are not redundant. In the event of reactor failures the converter must be taken out of operation until the reactor is repaired or replaced.

9.5.1 Submodule Semiconductor Failures

The converter submodule redundancy enables the uninterrupted operation of the HVDC scheme in case of submodule failures. The amount of redundant submodules is a merit of the availability targets and the functional requirements, valve-hall size, and the cost of adding redundant submodules to the converters. Normally, HVDC clients require availability targets higher than 98%. This requirement can be interpreted as meaning the scheme operates for one year without any need to interrupt the power transmission in order to replace any damaged submodules. Consequently, the converters must be designed with adequate reserve or redundant submodules in the circuit. During the annual planned maintenance period, the defective or damaged submodules are replaced.

A submodule semiconductor failure must not result in an open circuit fault condition of the respective submodule, as the converter arm must continue operating with the remaining healthy submodules. Instead, the faulty semiconductor must enter into a short-circuit failure mode (SCFM) or activate the submodule bypass switch. The shorted submodule must be capable of conducting current until it can be removed and replaced during the scheduled maintenance period.

ABB utilizes press-packed IGBT modules, while Siemens and Alstom utilize industrial-packed IGBTs for submodule semiconductors. Press-pack IGBT modules are designed so that the faulty semiconductor and the submodule enter into a safe SCFM.

Industrial-packed IGBT devices are constructed with multi-chips connected through wire-bonding. In case of surge fault currents, the wire-bonding will melt and the semiconductor chip will fail. The open-mode of semiconductor wire-bonding may cause internal flashover, which can damage the IGBT package and/or other materials outside the IGBT packages. It is therefore necessary to take measures to avoid cascading failures of neighboring submodules, and detect the open-mode failure before the severe damage occurs. In case of submodule semiconductor failures the converter control and monitoring system will detect the faulty submodule, energize the high-speed bypass switch and short-circuit the submodule, and exclude the faulty submodule from the circuit, as shown in Figure 9.6. At the same time, the converter controller includes and inserts the redundant submodule to the arm. This control action is made seamless without disturbance to the normal operation of the converter. For details of this control function, please refer to Chapter 3 and Section 3.9.

9.5.2 Submodule Capacitor Failure

A failure of a submodule dc capacitor is limited to the respective submodule and is therefore included in the redundancy of submodule levels. Failures of the capacitor usually lead to abnormal voltage waveforms. Submodule capacitor voltage and current are monitored continuously

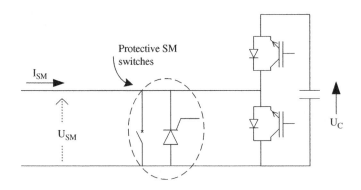

Figure 9.6 Submodule failure bypass switch.

by the control and protection systems of the converter. Abnormal values will initiate a subsequent bypass of the respective submodule from the circuit.

9.5.3 Phase Reactor Failure

The reactors used in MMC-HVDC converters are of dry-type air-core reactors. Therefore, the only possible failures for the phase reactors are flashovers to ground or flashovers across the reactor, which may be partial. Usually the reactor damages are caused by the hotspots and abnormal reactor heating. In fact, most of the reactor failures start as turn-to-turn failures, which usually develop into a complete reactor flashover in later stages. The reactor failure impacts the converter circulating currents and normally results in higher capacitor voltage ripples, distortion of three-phase current, and oscillation of dc pole voltages. The control and protection systems of the converters are able to detect the above-mentioned changes and initiate an alarm or trip the converter in severe cases. The converter protection system is not able to detect the reactor turn-to-turn flashover in the initial stage. If the reactors are located indoors, an arc detector, heat detector, and smoke detector can detect any flashovers at an early stage. Damaged reactors must be repaired or replaced in order to restart the normal operation of the HVDC scheme.

9.5.4 Converter Transformer Failure

The MMC-HVDC transformers are ordinary single- or three-phase power transformers which normally adapt the star/delta winding group. The transformer design may include a tap changer on the primary side. Transformers of HVDC schemes with black-start capability are normally equipped with tertiary winding. The tertiary winding feeds act as the station auxiliary power system and supply the auxiliary systems of the converter required for the converter start-up (e.g. valve cooling system pumps, control and protection cubicles).

Based on the available operational experience of symmetrical monopole VSC-HVDC schemes, and transformer reliability analysis, it is predicted that one transformer failure may occur during 40 years operation of the HVDC scheme. However, in the event of transformer failure the repair time will be long if no reserve transformer is available. Offshore HVDC platforms are normally designed with two parallel transformers each with 75% full load capacity. In case one transformer fails, the HVDC scheme can operate with a reduced transmission capacity.

Transformers in symmetrical configurations are not subjected to dc component stresses, as is the case with asymmetrical HVDC configurations. Hence, the failure rates are very low. The converter transformers are equipped with typical sets of transformer protection equipment (e.g. differential protections and gas relays).

9.6 MMC-HVDC Protection Systems

The objective of an HVDC protection system is to isolate the faulted component or subsystem from the service in order to protect the components of the HVDC scheme from any severe damage caused by fault conditions. A protection system must be set up to operate reliably and shall therefore also include self-supervision. Any detected failures in protection hardware

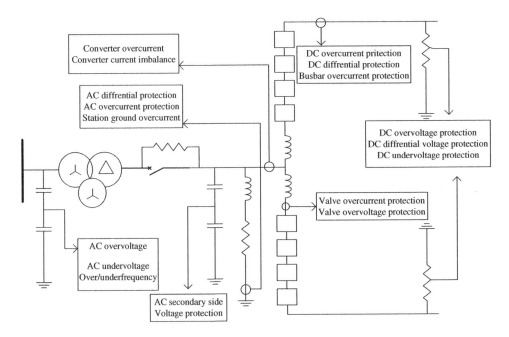

Figure 9.7   Typical protection system for MMC-HVDC.

shall result in appropriate action being taken, such as the shutdown of the faulty protection and switchover to redundant or standby protection system. Redundant protection systems shall preferably be equipped with separate measurement transducers and circuits and separate or secondary backup tripping circuits.

MMC-HVDC converters are equipped with hierarchical control and protections systems. Normally, the protection hierarchy is divided between the following areas (see also Figure 9.7):

- converter ac bus
- converter dc bus
- converter transformer
- submodule or valve protection
- phase reactor protection.

During the design phase of the HVDC equipment, the HVDC manufacturer and the client or end-user agree on the details of the *protection coordination philosophy* and surrounding ac network protection equipment.

It is expected that the following requirements are fulfilled by the protection coordination philosophy:

- All equipment shall be encompassed by at least one protection zone.
- Zones of protection shall overlap when possible.
- Faulty equipment should be tripped and isolated as quickly as possible.
- In case of redundant subsystems, minimum disturbance to the system should be obtained with the switchover to the redundant components or subsystem.

- Backup protection should operate if the primary protection fails.
- Each converter shall have its own independent set of protections and should be self-protected.
- All faults within a converter should be detected by two different protections (main protection equipment and back-up protection with separate transducers as far as practicable).

When a fault is detected, and depending on its location, the following protective actions are executed:

- change of converter control mode;
- temporary block of the converter;
- permanent block of converter followed by trip of ac breakers.

When a trip signal is sent to the main converter ac breaker, at the same time normally an order is also sent to start the breaker failure protection. If the main converter breaker fails to open, the breaker failure protection initiates the trip of the next breaker or breakers in line.

When the converter is isolated from the ac side network, the dc bus will be disconnected from the dc line/cable. Converter dc bus isolation can be executed manually, or as part of the protective actions during the converter shutdown process.

The basic protection philosophy is based on ac and dc-bus undervoltage, overvoltage, and overcurrent measurement. The overcurrent protection is normally based on a differential protection method.

9.6.1 AC-Side Protections

The protective zone is the converter's ac feeder with the primary objective of protecting the ac bus against phase-to-phase or phase-to-ground faults. The protection includes ac bus overcurrent and overvoltage, ac bus differential protection, and abnormal frequency detection. All currents flowing into the protective zone are compared phase by phase and the protection is completely phase segregated. The protection system is only sensitive to the fundamental current and designed to handle current transformer (CT) saturation or open CT circuit conditions. The protection operates if the vector sum of measured values differs from zero.

9.6.2 DC-Side Protections

The protective zone is the converter dc bus and the converter phase reactors. The primary objective is to protect the dc-side components against overcurrent conditions.

9.6.3 DC-Bus Undervoltage, Overvoltage Protection

The dc-bus undervoltage protection system only monitors the dc voltage in the rectifier terminal. If the dc-bus voltage persists below a threshold level for a predefined period, the converter is blocked.

The derivative protection monitors the dc-pole voltage. An earth fault on the dc line is characterized by the fact that the voltage collapses to a low level with a high voltage derivative rate.

High-derivative dc-bus voltage variations can trigger the protection, and a converter blocking signal will be initiated.

A reduced dc-bus voltage level in combination with overcurrent condition might be used as backup protection. It is recommended to implement interlocking schemes between different dc-bus voltage protection systems in such a way that an inadvertent protection operation is avoided if the deviations are caused by normal switching operations (e.g. start or stop of HVDC converters, energizing converter transformers).

9.6.4   DC-Bus Voltage Unbalance Protection

The protection compares the positive- and negative-pole voltages. The pole-to-ground voltage is measured by a resistive voltage divider on the positive and negative poles. The primary objective is to protect the converter from overvoltage, especially at high current, and prevent switching outside the safe switching operating area (SSOA) in the event of pole-to-ground fault conditions.

If the difference is greater than a threshold level for a predefined time, the converter is blocked.

9.6.5   DC-Bus Overcurrent Protection

The converter overcurrent protection protects the converter components (e.g. submodules and phase reactors) if abnormal overcurrent conditions are detected. The protection monitors segregated phase currents in and out of the converter. If the current is greater than a threshold level for a predefined time, the converter is blocked.

9.6.6   DC Bus Differential Protection

The primary objective of the converter dc bus differential protection is to protect the inner converter equipment dc fault conditions. The incoming currents entering and leaving the dc bus are measured and compared. A large deviation between the currents indicates a ground fault within the converter, and so the converter must be blocked immediately.

For the dc transmission link protection, the dc link current in both ends of the dc transmission line/cable are measured and compared. The protection is to be coordinated with possible time deviations owing to the communication between the stations in regard to start/stop sequences and power ramping. The protection requires telecommunication between the two ends of the transmission line.

9.6.7   Valve and Submodule Protection

VSC-HVDC converters are equipped with many different types of control and protection systems (e.g. valve overcurrent protection, valve overvoltage and undervoltage protection, abnormal dc capacitor voltage). When an abnormal condition is detected by the valve protection systems, normally the valves will be blocked immediately and removed from the circuit. In addition, valves are equipped with thermal overload protection which prevents the valves from overheating. An algorithm is used to estimate the IGBT junction temperature. If the temperature exceeds the predefined limit, the IGBTs will be blocked.

9.6.8   Transformer Protection

The primary objectives of the transformer protection are to protect the converter transformer from internal ground fault, phase-to-phase faults, and winding turn-to-turn faults.

The converter transformers are equipped with various protection systems (e.g. overcurrent protection, differential protection between the primary and secondary windings, high temperature protection, and transformer gas detection and protection).

Typical protection is based on vectorial comparison of the primary side transformer currents with the secondary side currents and with respect to the transformer's turn ratio and the tap changer position. The protection detects a difference in the fundamental currents, and operates if the equal ampere-turn criterion is not fulfilled. The equal ampere turn criterion is only valid under steady-state operation.

The transformer earth fault protection monitors the difference between the fundamental frequency zero-sequence current on the high-voltage side and the grounded side of the converter transformer. Under transformer fault conditions, a significant rise and difference in the zero-sequence currents can be detected by the protection system.

In addition, other protection systems for converter transformers are equipped with typical protective relays, such as:

- Buchholz relay
- transformer winding temperature
- tap changer pressure
- transformer and tap changer low oil detector
- transformer cooling system.

In case of any transformer failures or serious alarms, the ac breakers will be tripped and will isolate the converter transformer.

9.6.9   Primary Converter AC Breaker Failure Protection

This protection is a secondary protection system in case the converter ac breakers fail to operate or fail to break the converter's feeding current.

After the trip signal to the primary ac breakers has been initiated, the protection system monitors the current in each three phases of the ac circuit breakers. If the breaker current is not reduced to a certain level then the trip signal to the next ac breaker is initiated.

9.7   Summary

This chapter presents the fault behaviors of two-level and MMC-HVDC terminals under ac and dc fault conditions. MMC-HVDC terminals comply with the most stringent low-voltage fault ride-through (LVFRT) requirements and can ride through the ac fault conditions and support the ac network with required ancillary services. The MMC-HVDC transmission technology is vulnerable to dc pole-to-ground and pole-to-pole fault conditions. Under dc fault conditions the converter control systems block the converter and isolate them from the connecting ac networks in order to protect the converter components from possible surge currents and permanent damages.

Converter protection is normally divided between the ac side and the dc side of the converter. Different techniques are applied to detect and protect the converter components from ac and dc fault conditions.

References

[1] J. Hu and Y. He, "Modeling and control of grid-connected voltage sourced converters under generalized unbalanced operation conditions," *IEEE Transactions. Energy Conversion*, vol. 23, no. 3, pp. 903–913, Sep. 2008.

[2] L. Xu, B. Andersen, and P. Cartwright, "VSC transmission operating under unbalanced AC conditions: Analysis and control design," *IEEE Transactions on Power Delivery*, vol. 20, no. 1, pp. 427–434, Jan. 2005.

[3] A. Yazdaniname and R. Iravani, "A unified dynamic model and control for the voltage-sourced converter under unbalanced grid conditions," *IEEE Transactions on Power Delivery*, vol. 21, no. 3, pp. 1620–1629, Jul. 2006.

[4] L. Tang and B. Ooi, "Managing zero sequence in voltage source converters," *Industry Applications Conference, 37th IAS Annual Meeting, conference record*, vol. 2, pp. 795–802 XXXX.

[5] X. Shi, Z. Wang, Y. Liu, and L. M. Tolbert, "Characteristic investigation and control of a modular multilevel converter-based HVDC system under single-line-to-ground fault conditions," *IEEE Transactions on Power Electronics*, vol. 30, no. 1, pp. 408–421, 2015.

[6] L. Xiao, S. Huang, and K. Lu, "DC-bus voltage control of grid-connected voltage source converter by using space vector modulated direct power control under unbalanced network conditions," *IET Power Electronics*, vol. 6, no. 5, pp. 925–934, 2013.

[7] J. Rafferty, L. Xu, and D. J. Morrow, "DC fault analysis of VSC based multi-terminal HVDC systems," *AC and DC Power Transmission, 10th IET International Conference*, pp. 1–6, 2012.

[8] P. Bordignon, M. Marchesoni, G. Parodi, and L. Vaccaro, "Modular multilevel converter in HVDC systems under fault conditions," DOI: 10.1109/EPE.2013.6634467, 2013.

[9] S. Norrga, L. Ängquist, and K. Sharifabadi, "*Power electronics for HVDC grids: An overview*," CIGRE HVDC Symposium, Lund, Sweden, 2015.

[10] G. P. Adam, G. Kalcon, S. Finney, D. Holliday, O. Anaya-Lara, and B. Williams, "HVDC Network: DC fault ride-through improvement," *CIGRE Canada Conference on Power Systems*, Halifax, September 6–8, 2011.

[11] S. Cui, Su. Kim, J.-J. Jung, and S.-K. Sul, "Principle, control and comparison of modular multilevel converters (MMCs) with dc short-circuit fault ride-through capability," *Applied Power Electronics Conference and Exposition (APEC), Twenty-Ninth Annual IEEE*, pp. 610–616, 2014.

[12] X. Chen and C. Zhao, "Research on the fault characteristics of HVDC based on modular multilevel converter," *Electrical Power and Energy Conference (EPEC)*, pp. 91–96, 2011.

[13] J. Rafferty, L. Xu, and D. J. Morrow, "Analysis of VSC-based HVDC under DC line-to-earth fault," *Industrial Electronics Society, 39th Annual Conference of the IEEE*, pp. 459–464, 2013.

[14] A. Nami, J. Liang, F. Dijkhuizen, and P. Lundberg, "Analysis of modular multilevel converters with dc short-circuit fault blocking capability in bipolar HVDC transmission systems," *Power Electronics and Applications*, pp. 1–10, 2015.

[15] M. K. Bucher and C. M. Franck, "Analysis of transient fault currents in multi-terminal HVDC networks during pole-to-ground faults," *Proceedings of the International Conference on Power Electronics*, Trondheim, 2013.

[16] G. P. Adam, K. H. Ahmed, S. J. Finney, and B. W. Williams, "AC fault ride-through capability of a VSC-HVDC transmission systems," *Proceedings of the IEEE Energy Conversion*, pp. 3739–3745, 2010.

[17] A. Antonopoulos, L. Ängquist, and H.-P. Nee, "On dynamics and voltage control of the modular multilevel converter," *Power Electronics and Applications*, pp. 1–10, 2009.

[18] I. Wu, S. Zhang, R. Hou, and D. Xu, "DC fault analysis and protection design in modular multilevel based HVDC systems," *Electrical Power and Energy Conference*, pp. 91–96, 2011.

[19] V. Hussennether, D. Worthington, B. Hühnerbein, M. Siebert, A. Barth, and M. Rapetti, "*Projects BorWin2 and HelWin1: Large scale multilevel voltage-sourced converter technology for bundling of offshore wind power*," CIGRE Session, B4 proceeding, Paris, 2012.

[20] M. Saeedifard and R. Iravani, "Dynamic performance of a modular multilevel back-to-back HVDC system," *IEEE Transactions on Power Delivery*, vol. 25, no. 4, pp. 2903–2912, Oct. 2010.

[21] Y. Zhou, D. Jiang, J. Guo, P. Hu, and Y. Liang, "Analysis and control of modular multilevel converters under unbalanced conditions," *IEEE Transactions on Power Delivery*, vol. 28, no. 4, pp. 1986–1995, Oct. 2013.

[22] X. Chen, C. Zhao, and C. Cao, "Research on the fault characteristics of HVDC based on modular multilevel converter," *Electrical Power and Energy Conference*, pp. 91–96. 2011.

[23] M. Guan and Z. Xu, "Modelling and control of a modular multilevel converter-based HVDC system under unbalanced grid conditions," *IEEE Transactions on Power Electronics*, vol. 27, no. 12, pp. 4858–4867, Dec. 2012.

[24] Q. Tu, Z. Xu, Y. Chang, and L. Guan, "Suppressing DC voltage ripples of MMC-HVDC under unbalanced grid conditions," *IEEE Transactions on Power Delivery*, vol. 27, no. 3, pp. 1332–1338, Jul. 2012.

[25] J. Candelaria and J.-D. Park, "VSC-HVDC system protection: A review of current methods," *Power Systems Conference and Exposition*, pp. 1–7, 2011.

10

MMC-HVDC Transmission Technology and MTDC Networks

10.1 Introduction

Line-commutated converter high-voltage direct current (LCC-HVDC) transmission technology is widely recognized as being advantageous for bulk power transmission over long distances, asynchronous interconnections of ac power systems, and interconnectors with long submarine cable. In addition, the LCC-HVDC technology is utilized as firewall in back-to-back configurations.

New modular multilevel converter (MMC) designs have broadened the potential range of HVDC transmission to include application of HVDC schemes in weak ac networks (e.g. integration of remote offshore wind-power plants, or WPPs, to mainland ac networks). MMC-HVDC technology is foreseen as the most suitable technology for development of multi-terminal direct current (MTDC) networks. This chapter presents different HVDC scheme configurations and MTDC configurations. Various dc grid control and protection strategies, including dc breaker technologies, are presented.

10.2 LCC-HVDC Transmission Technology

The first commercial LCC-HVDC scheme was commissioned in 1954.

The LCC technology is a mature dc transmission technology with many years of operation experience and more than 100 installations worldwide. These installations are HVDC configurations with overhead lines, submarine cable links, or back-to-back schemes. The total rating of LCC-HVDC schemes in service, at the time of writing, is about 100 GW and growing fast. UHVDC schemes with ratings of up to 8000 MW and with dc transmission voltage of up to ± 800 kV are in service.

LCCs utilize thyristor semiconductors, which are capable of conducting the current in one direction only (the forward direction), and will do so when the thyristor is forward-biased and

Design, Control, and Application of Modular Multilevel Converters for HVDC Transmission Systems, First Edition.
Kamran Sharifabadi, Lennart Harnefors, Hans-Peter Nee, Staffan Norrga, and Remus Teodorescu.
© 2016 John Wiley & Sons, Ltd. Published 2016 by John Wiley & Sons, Ltd.
Companion Website URL: www.wiley.com/go/Sharifabadi/ModularConverters

switched on by the gate signal. LCCs always consume reactive power in all operating modes (i.e. rectifier or inverter).

The converter typically consists of two six-pulse thyristor bridges, called *Graetz bridges*, which are series-connected on the dc side, and parallel-connected on the ac side. By using Y-Y transformer winding for one bridge and Y-D transformer winding for the other bridge, the six-pulse harmonics are theoretically cancelled, both on the ac and on the dc side, which reduces the filtering requirements. AC filters are required to limit the impact of the remaining low-order harmonics on the ac network and on the dc side. LCC-HVDC systems require connection to a strong ac network in order to ensure the commutation and to avoid operation instability of the converters.

Commutation is in essence the transfer of current from one phase to another between the converter's thyristor valves. The *commutation time* refers to the lapse of time during which a thyristor valve in one phase becomes reversed-biased and switches off, while a thyristor valve in the other phase is triggered to switch on. A *commutation failure* refers to the situation in which the commutation voltage, the driving force behind this process, reverses before the current is commutated from one phase to another. As a result, the valve that was previously conducting the current continues to conduct. This creates a short-circuit at the dc side, which prevents power transmission during a commutation fault. Commutation failures are more likely to happen in inverter mode and are normally caused by ac voltage sags originated from ac network fault conditions. The strength of the ac network is measured by the short-circuit ratio (SCR), which is the ratio of the short-circuit power of the ac network to the rated power of the converter. If this value is less than 2–3, the ac network is considered weak. As LCCs require a relatively strong synchronous voltage source in order to commutate, the LCC-HVDC terminals in weak ac networks must be supplemented by the installation of reactive power compensation equipment, such as a synchronous compensator, static synchronous compensator (STATCOM), or static var compensator (SVC).

These converters absorb reactive power, as the current is always lagging behind the voltage. The reactive power (VAr) requirement for the LCC terminals is in the order of 60% of the active power rating and depends on the power flow level. Alternative LCC designs utilize series capacitors between the transformer and the valves. This topology reduces the need for reactive power compensation equipment and reduces the reactive power flow through the converter transformer, hence reducing the converter transformer rating. This converter topology has some advantages with respect to the lower reactive power requirement, and a lower risk of commutation failure when compared with classic LCC-HVDC technology. The topology is known as a capacitor commutated converter (CCC).

In weak ac systems the capacitances of the ac harmonic filter banks tend to lower the eigenfrequency of the ac system, which could contribute to a resonance at the lower-order harmonics (e.g. second or third). The ac filters have to be designed carefully to avoid resonances and magnification of existing/background harmonics.

The dc bus current direction for an LCC-HVDC scheme is always the same, because the thyristor valves can conduct current only in one direction. Reversal of the power direction is therefore achieved by changing the dc-side voltage polarity. In addition, commutation failures cause a temporary collapse of the dc bus voltage. These limitations are a barrier to the development of large-scale multi-terminal LCC-HVDC schemes. The polarity reversal during the change of power direction also means that so far it has not been possible to use high-voltage extruded cross-bound polyethylene (XLPE) cables for LCC-HVDC schemes.

The LCC-HVDC transmission is the preferred transmission technology for bulk power transmission with overhead power lines over long distances. The LCC-HVDC transmission technology cannot be utilized in weak ac networks, such as islanded offshore WPPs.

10.3   Two-Level VSC-HVDC Transmission Technology

The evolution of voltage-source-converter high-voltage direct current (VSC-HVDC) transmission started with the development of the two-level VSC technology. This new converter technology with controllable semiconductors opened up a new era for HVDC transmission.

The first commercial VSC-HVDC scheme was the 50 MW Gotland scheme, which was commissioned in 1999. The VSC-HVDC technology utilizes voltage source semiconductors such as insulated-gate bipolar transistors (IGBTs). The VSC-HVDC is a self-commutated converter using semiconductors, which can be switched on and off in response to a gate signal. Because the VSC converters are self-commutated, commutation failures will not occur in the event of ac network disturbances, as is the case with LCC technology. VSC technology is capable of the independent control of active and reactive power. It can be connected to weak ac networks and can supply passive networks and energize a dead network (black-start). The development of VSC-HVDC triggered a fast development of offshore HVDC applications for the oil and gas industry and offshore WPPs. BorWin alpha was the world's first offshore wind farm connected via a VSC-HVDC scheme to the onshore network in Germany. The HVDC scheme was commissioned in 2009. The converter power rating is 400 MW at ±150 kV.

The basic equipment in a typical VSC converter station includes:

* VSC consisting of IGBTs and large dc capacitors;
* converter reactors;
* DC smoothing reactor;
* AC filters and RF filters;
* interface transformer;
* VSC control and protection systems;
* auxiliary systems (e.g. cooling systems, auxiliary power, fire protection system).

VSC-HVDC technology overcomes some of the shortcomings of the LCC-HVDC systems, however at an increased converter cost and higher converter losses. Converter losses for LCC technology is about 0.6–0.8% per converter, while the losses for the VSC technology is about 1.5–2% per converter. VSC-HVDC terminals can be integrated to weak ac networks independent of the grid's short-circuit capacity [1].

VSC-HVDC technology is capable of operating as an independent voltage source that can supply or absorb large amounts of active and/or reactive power. Hence the VSC-HVDC terminals can be connected islanded offshore ac networks such as offshore WPPs. The VSC-HVDC technology is able to supply the passive offshore network (black-start capability) and supply the required auxiliary power for the start-up of wind turbines.

10.3.1   Comparison of VSC-HVDC vs. LCC-HVDC Technology

VSC-HVDC transmission technology has a 40–60% smaller footprint compared with the LCC-HVDC technology. The VSC-HVDC terminals can support the ac networks with a wide range of ancillary services and are able to meet the most demanding grid-code requirements. The two-level VSC disadvantages are increased HVDC terminal costs and losses. As power reversal does not require voltage reversal of the HVDC poles, XLPE cables with polyethylene cross-linked insulation can be utilized to link the VSC-HVDC terminals. XLPE cables are cheaper, lighter, and have a smaller environmental footprint compared to the laminated or mass-impregnated cable technologies mainly utilized for LCC-HVDC schemes. XLPE cables can operate at higher temperatures compared with mass-impregnated cable technology, which results in a higher transmissible power per cable pair using the same conductor cross-section of the mass-impregnated cables. Cable fabrication (onshore and offshore) and installation costs of XLPE cables are much lower compared with the mass-impregnated cable technologies.

Two- and three-level VSC-HVDC schemes have higher losses compared with the LCC technology. The pulse-width modulation (PWM) high switching frequency (1–2 kHz) causes substantial converter losses. The converter losses, calculated according to IEC 61803, amount to approximately 1.6% of the rated HVDC transmission capacity (per station) at rated load. The no-load (standby) losses are approximately 0.2%. The main contributors to these losses are the IGBT valves ($\approx$1.1%), the converter transformers ($\approx$0.21%), and the converter reactors ($\approx$0.12%). The rest originates from the ac filters, the station service power, and the dc capacitor losses.

The high switching frequency of the IGBTs create high voltage derivatives, which stress the phase reactor connected in series with the converter terminal and additionally causes electromagnetic interference (EMI).

The two- and three-level VSC-HVDC technologies enable switching between two or three different voltage levels to the ac terminals, respectively. The main drawbacks of these technologies include high switching losses and the large filter requirements due to converter generated harmonics.

10.4   Modular Multilevel HVDC Transmission Technology

The need to improve the converter ac bus voltage waveform, reducing the converter switching losses, and eliminating the filter requirements has led to the development of MMCs. The MMC synthesizes a high-quality sinusoidal voltage waveform by incrementally switching a high number of series-connected voltage sources [2]. Furthermore, the switching frequency of the individual IGBTs is decreased from 1–2 kHz for the two- and three-level converters to typically 100–150 Hz in the MMC technology, hence the MMC losses are reduced to about 1% at each converter station; however, the losses are still higher compared with the LCC-HVDC technology but lower compared with the two and three-level VSC technology. The half-bridge MMC-HVDC transmission technology has superior advantages compared with alternative dc transmission technologies, which makes it a promising technology for export of offshore wind energy to onshore. MMCs with full-bridge topology are capable of blocking the dc pole faults. However, this comes at an increased cost in regard to the converter losses, owing to the increased number of semiconductors.

MMC-HVDC schemes planned or in operation are point-to-point installations. But the first multi-terminal projects are under development. There are several proposals on how to interconnect the existing point-to-point HVDC schemes and how to pave the way toward the development of large dc multi-terminal networks. Significant R & D activities are focussed on how to develop the future European super grid based on VSC-HVDC transmission technology.

It is foreseen that some of the point-to-point HVDC schemes will be further developed and interlinked on the dc side to create small multi-terminal HVDC networks. The interconnected HVDC schemes will be further interlinked in order to create regional MTDC networks. These small regional dc grids can be even further developed across larger geographical areas to create a large meshed MTDC network. However, such a move will require a strong economic support and political willingness to harmonize the regulatory frameworks across neighboring countries.

10.4.1   Monopolar Asymmetric MMC-HVDC Scheme Configuration

In its most simple form, a long-distance HVDC system can consist of two converters each connected between earth and a high voltage conductor linking the two converters. Current flows between the two terminals through the high voltage conductor and returns through metallic return conductor or the earth or sea electrodes, as shown in Figure 10.1b and 10.1c. This arrangement is known as *asymmetric monopole* with earth or sea electrode return. In recent years, environmental concerns that are largely due to risk perception, have made it difficult, and in some countries even impossible, to obtain permission for the construction of schemes using an earth or sea return current path. With submarine cables the return path conductor can be designed with a smaller cross-section conductor and low-voltage isolation, compared with the high-voltage negative pole cable (the return path cable does not require the full dc potential/voltage insulation). The return path cable isolation must be able to withstand the voltage drop along the conductor and the voltage rise during abnormal (fault) operating conditions. This cable design reduces the cable investment costs. Asymmetric monopole configuration with metallic return has higher transmission losses compared to the system with earth return. This is due to the higher resistance of the metallic return path compared to the earth return path. In asymmetric monopolar HVDC with overhead lines, normally the negative pole is the high-voltage pole in respect to the ground. The high-voltage negative pole reduces the corona effects in overhead lines. However, with the asymmetric monopolar configuration as the dc side of the HVDC transmission system is grounded and requires special transformer designs capable of withstanding the additional dc component stress [3].

10.4.2   Symmetrical Monopole MMC-HVDC Scheme Configuration

In this configuration, the dc side of converters is interlinked with two high-voltage conductors with opposite polarity, as shown in Figure 10.1a.

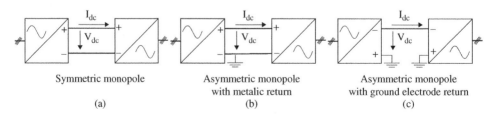

Symmetric monopole	Asymmetric monopole with metalic return	Asymmetric monopole with ground electrode return
(a)	(b)	(c)

Figure 10.1   Symmetric and asymmetric HVDC scheme configurations, (a) symmetric monopole, (b) asymmetric monopole with metallic return, (c) asymmetric monopole with ground or sea electrode return

The earth reference is provided with various methods such as direct grounding, resistive grounding, or high-impedance grounding. During normal operation conditions of the scheme, no current will flow through the earth. The converter grounding strategy has a major impact on the dc short-circuit fault current characteristics. Although in this configuration two conductors of opposite polarity carry power, they cannot operate independently, i.e. if one conductor is out of service, the healthy conductor cannot continue transmitting power utilizing earth as the return path. Standard transformers can be used to interface the converter to ac networks, as there are no dc components flowing through the transformers as is the case with asymmetric monopolar configuration. The majority of VSC-based HVDC systems to date utilize the symmetrical monopole configuration.

10.4.3   Bipolar HVDC Scheme Configuration

The bipolar system consists of four HVDC converters, as shown in Figure 10.2. A bipolar arrangement may greatly improve system reliability by using two converters at each HVDC terminal, one connected between ground and the positive pole and the other one between ground and the negative pole.

If continuous operation with earth return is allowed, only 50% of the transmission capacity is lost in the event of converter or cable failure, with the healthy pole remaining in operation in a monopolar configuration. If earth return is not allowed, the HVDC system has to be completely shut down in the event of a cable failure, unless any midpoints of the converters are linked with a low-voltage cable. In the event of a converter failure, the scheme can continue its operation using the other HVDC cable as a metallic return, provided that appropriate dc switchgear has been provided to allow such reconfiguration. However, it should be highlighted that for routine maintenance the downtime of the monopolar scheme is potentially low, provided that the auxiliary systems are designed with sufficient redundancy.

The bipolar configuration is more expensive than a monopolar configuration. Hence, a bipolar configuration is usually employed when the transmitted power is higher than the capacity of a single pole in a monopolar configuration. The bipolar configuration requires two HVDC cables carrying the dc-side current in opposite directions. During normal operation, the currents in each cable have the same amplitude, and there is no current in the return path.

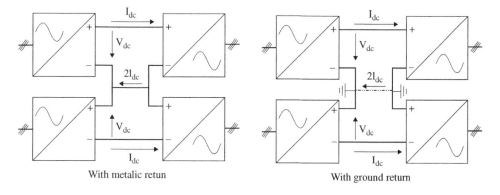

Figure 10.2   Bipolar configuration with ground electrode and return conductor.

In the event of imbalances, the current can usually flow through the ground if there are no environmental restrictions or through the metallic return cable.

10.4.4 Homopolar HVDC Scheme Configuration

In a homopolar configuration, two HVDC poles have the same polarity and the current return path is through earth, as shown in Figure 10.3. As in the asymmetrical monopolar configuration, usually negative polarity is chosen for the high-voltage overhead power line, owing to lower corona losses, as well as less radio interference.

Since the polarities of both pole conductors are identical, this HVDC transmission system configuration has the advantage of reduced insulation costs. In addition, it does not suffer from the main drawback of monopolar configurations, i.e. in the event of a single converter fault, the whole transmission system will not become offline. Instead, the converter that is not affected by the fault can be reverted to feed the remaining pole. The main disadvantage with the homopolar arrangement is the constant need of a return path and the high current associated with it, which can equal twice the nominal value in rated conditions, as the current from each pole shares the same return path.

10.4.5 Back-to-Back HVDC Scheme Configuration

Back-to-back configurations are usually employed when a connection between two asynchronous ac systems with equal or different ac line frequencies must be established but when there is no need for transmission lines or cable connections, as shown in Figure 10.4. In back-to-back configurations, the converter valve halls, control equipment, cooling devices, and transformers of both converter stations are located in the same area. As there is no need for dc cables, the currents are usually kept high, while voltages are kept low at approximately 150 kV or less. As the configuration does not require long-distance cables or overhead lines, the losses are only limited to the converter losses. The cost of HVDC components such as valves and dc reactors is voltage dependent. Additionally, there is no need for a telecommunication link between the two HVDC converter stations, which simplifies the control equipment.

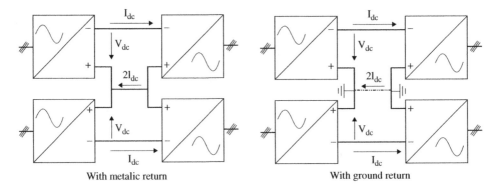

Figure 10.3   Homopolar HVDC scheme configuration.

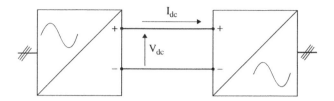

Figure 10.4 Back-to-back HVDC scheme.

10.5 The European HVDC Projects and MTDC Network Perspectives

In recent years, the development and integration of offshore wind to the European onshore grid has gained momentum, reflecting the commitment of European countries to reduce greenhouse gas emissions. The EU countries have made a commitment through a climate and energy package. This package consists of binding legislation which aims to ensure that EU countries meet their ambitious climate and energy targets for 2020. These targets, known as the *20-20-20 targets*, set three key objectives for 2020:

- a 20% reduction in EU greenhouse gas emissions compared to 1990 levels;
- raising the share of EU energy consumption produced by renewable generation to 20%;
- a 20% improvement in the EU's energy efficiency.

Under the Renewable Energy Directive, member states have taken on binding national targets to raise the share of renewable energy in their mixed-power generation by 2020. These targets reflect the member states' different starting points and potential for increasing renewable power generation.

10.5.1   The North Sea Countries Offshore Grid Initiative (NSCOGI)

In order to achieve the EU's renewable targets for 2020, the North Sea Countries Offshore Grid Initiative (NSCOGI) was formed as the responsible body to evaluate and facilitate the coordinated development of a possible offshore grid that maximizes the efficient and economic use of renewable-energy sources and the infrastructure investments. NSCOGI intends to connect approximately 40 GW of offshore wind power between several countries in northwest Europe by 2030. A Memorandum of Understanding was signed on December 3, 2010 by the neighboring countries around the North Sea. A significant share of renewable energy will be generated by offshore WPPs. The goal is to interconnect the offshore wind farms located in the North Sea and create a large offshore grid. The offshore grid will be integrated in the transmission networks of the European countries in order to ultimately create an interconnected offshore and onshore network. This large network will convey the offshore wind energy and the onshore generation sites to the load centers across the European countries.

HVDC interconnectors will connect European power markets and the electricity can be exchanged between all involved countries. This will also enable the optimization of renewable-energy generation, and make the overall interconnected system less susceptible to intermittency, thus making the system more robust. The idea is to use large hydroelectric power plants located in Scandinavia and in the Alps to support the interconnected system with the required balancing reserves during the peak load periods or during low wind conditions. There are many European HVDC interconnector projects under development (e.g. Norway/Germany and Norway/UK) that support the North Sea grid vision. The ultimate North Sea grid development will support a new European electricity market allowing for a more efficient combination of supply and demand across all interconnected countries. HVDC schemes across the Europe will play a significant role in the development of NSCOGI.

10.5.2   Large Integration of Offshore Wind Farms and Creation of the Offshore DC Grid

All offshore WPP connections to the mainland network are planned, designed, and developed based on individual radial ac or dc links. This design is seen as the most economical approach for the export of offshore wind energy to mainland networks. As an alternative to the radial connection of wind farms to the onshore network, the concept of offshore clusters has been proposed. Several offshore WPPs will create an offshore cluster of WPPs. The energy is collected and transmitted via one large offshore HVDC hub platform to the mainland network, as shown in Figure 10.5. This approach can be used to enable the optimized massive integration of offshore wind farms to the mainland network. The cluster approach has the advantage of sharing the development costs of the transmission infrastructure between several wind farm project developers. At the same time it has fewer environmental impacts as only one transmission link will connect several wind farms to the mainland instead of the dedicated radial export links from each wind farm to the shore.

Germany is pioneering the development of offshore HVDC collector hubs. Germany has planned around 58 offshore wind parks with approximately 4500 offshore wind turbine generators. Most of these WPPs have a distance of 120–250 km to the nearest mainland ac network. These clusters of WPPs will have dedicated point of common coupling to

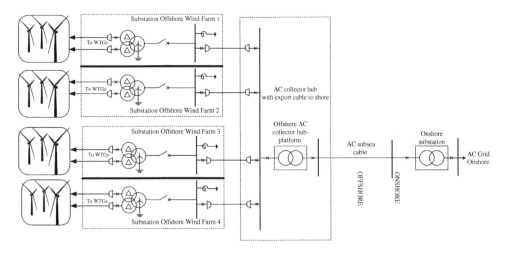

Figure 10.5 Cluster connection of offshore wind farms to a collector hub and export link to shore.

HVDC platforms offshore (e.g. DolWin, BorWin, SylWin, and HelWin HVDC platforms). The German transmission operator is responsible for the development and provision of the offshore grid access for the wind farm project developers. However, this development approach requires a well-managed offshore grid development whose design is coordinated between the various parties and projects.

The development of clustered offshore WPPs connected to HVDC hubs requires the committed involvement of wind farm project developers, the offshore grid operator (e.g. the offshore transmission system operator, or TSO), and regulatory authorities. The design and development of the offshore HVDC collector hubs can be challenging, as it should incorporate several wind farms of different operators and wind turbine manufactures with different development schedules. Planning and development of such clustered solutions is an extremely complex task that requires involvement and commitment from all stakeholders.

It is envisaged that in the later stages of the cluster development approach there will be initiatives to interconnect the offshore HVDC terminals and create the larger offshore network. Interconnecting the offshore HVDC converters on the dc busbars will be a decisive step toward the creation of MTDC networks. The development and realization of such large MTDC networks is likely to happen gradually. It has to incorporate a wide range of equipment from different manufacturers, some of which are already in operation. The interoperability of the HVDC equipment will be crucial for the development of MTDC networks. One of the most obvious obstacles will be the dc grid voltage for the interconnected system. Once a dc-side voltage is selected, it will define the dc transmission voltage for the entire interconnected system. Analogously to ac power systems, several voltage levels could be technically possible, but converting dc transmission voltages with dc-to-dc converters is more complex and has a higher cost and electrical losses compared to the ac transmission technology utilizing transformers to convert between different transmission voltages.

10.6 Multi-Terminal HVDC Configurations

LCC and VSC-HVDC transmission technologies that can be used to implement an MTDC grid have different features and limitations [4]. Among the available HVDC technologies, the MMC is apparently the best option for dc grids given that this technology presents low switching losses as well as a better fault performance.

The VSC-HVDC transmission systems that have been planned, designed, and in operation so far are based on point-to-point HVDC schemes. Recently, a few multi-terminal MMC-HVDC projects have been commissioned. In China a three-terminal MMC-HVDC with ratings of $\pm160\,kV/200\,MW/100\,MW/50\,MW$ connects local WPPs to Nan'ao island and the Guangdong power grid. In east China's Hangzhou Bay, a five-terminal MMC-HVDC scheme with ratings of $\pm200\,kV/400\,MW/300\,MW/100\,MW/100\,MW/100\,MW$ was put in service in 2014 [5].

A multi-terminal HVDC system has more than two converter stations, some of them operating as rectifiers and others as inverters. The simplest way of building an MTDC system from an existing two-terminal system is to introduce a tapping along the HVDC link with a new HVDC terminal. However, there are several different designs (e.g. series, parallel, or meshed configurations). The simplest MTDC system is a three-terminal serial or parallel configuration and the most complex configuration is a meshed MTDC network. Many intermediate topologies can be designed in between [6].

10.6.1 Series-Connected MTDC Network

In a series-connected MTDC network, all converter stations share the same dc transmission current, while their voltages will vary according to the power to be extracted or delivered to the dc network. The series-connected MTDC cannot be seen as a realistic future development and extension scenario of the existing VSC-HVDC schemes.

10.6.2 Parallel-Connected MTDC Network

In a parallel-connected MTDC network, all the converter terminals share the common dc system voltage. The simplest way of developing an MTDC system from an existing two-terminal system is to introduce parallel-connections along the point-to-point scheme, as shown in Figure 10.6.

As an example, it can be assumed that HVDC tapping converters can be implemented between an HVDC interconnector connecting two mainland ac systems. The tapping converters can be HVDC terminals that import offshore wind power to the dc link, and some other HVDC terminals that export power to offshore consumers, such as oil and gas installations [7].

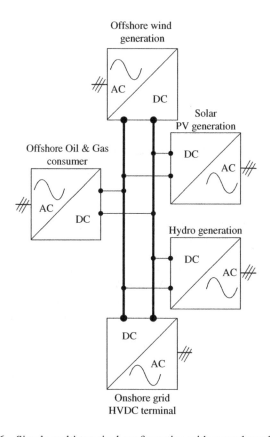

Figure 10.6   Simple multi-terminal configuration with several parallel tapping.

In MTDC networks, converter stations must contribute to the dc-side voltage control based on defined characteristics. Radial MTDC networks with parallel-connections control the dc power flow by only changing the dc bus voltage at the terminals. In this configuration the endpoint terminals will share the dc bus voltage-balancing task.

10.6.3   Meshed MTDC Networks

In a meshed dc grid, at least two converter stations have more than one connecting path or dc link connections. These links may establish a parallel path for the power flow between the terminals, as shown in Figure 10.7. The parallel links introduce some redundancy to the transmission links and will increase the reliability and availability of the system. The main purpose of a transmission system is to transmit the electricity from generation sites to load centers, cost-efficiently, and with minimum loss. Without any additional measures the current through a dc link will then be determined by the dc bus voltages of the converter stations, as well as the resistances of the parallel-connections. Depending on the topology of the MTDC grid and the operation strategy, dc grid controllers may not be needed.

Large meshed dc grids will require some form of centralized or decentralized dc grid voltage controller. In ac networks, a flexible alternating current transmission system (FACTS) is utilized to control the ac load flow by varying the transmission line's characteristic impedance, or the voltage phase angle. Reactive current, reactive power, frequency, and phase angle are all properties in ac grids that can be controlled, but they do not exist in dc grids. The dc link impedances are purely resistive. Consequently, the power flow control in dc networks can be made by controlling the dc-side voltage and current amplitudes. In an MTDC network with parallel dc links, the dc line power flow controllers (PFCs) can insert a certain dc voltage in series with the line, in order to control its current flow. However, a dc line PFC does not exist yet, although it has been proposed in various academic research papers. Research institutes and academia are currently discussing the requirements for the control algorithms of a dc-line PFC.

DC grid controllers in large networks should be equipped with communication links to the control and dispatch entity (i.e. the TSO), which is responsible for monitoring and controlling the dc and ac load flows based on the established market rules and requirements. The communication link provides the required interface to the system operator which can initiate the manual

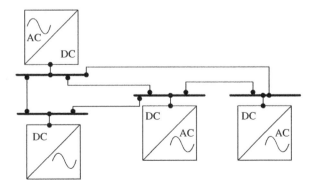

Figure 10.7   Example of meshed dc grid with a parallel path between at least two converters.

control intervention. The manual control includes coordination and control tasks related to, for example, the start-up or shutdown of converters or the execution of new set-points to the HVDC converters.

The stable dynamic behavior of HVDC terminals in a meshed network is an important design and operation criterion. Satisfactory dynamic behavior of the interconnected ac and dc systems requires the detailed modeling, simulation, and design of dc network. State-of-the-art simulation tools (e.g. real-time digital simulators and electromagnetic transient, or EMT, simulation tools) and models can provide an insight into the dynamic behavior of the system (e.g. dynamic overvoltages and overcurrents occurring during different operation and switching modes). In addition, small signal and stability analysis of the dc grid components and subsystems during ac and dc fault conditions are of utmost importance for design of reliable meshed dc networks.

10.7   DC Load Flow Control in MTDC Networks

The objective of power flow control in a dc grid is to control the dc-side voltage in the system within its limits and fulfill the planned power exchange to the connected ac networks, without exceeding maximum currents through the HVDC converter and dc power lines or cables.

In point-to-point connections the desired load flow is achieved by controlling the importing terminals keeping their steady-state voltages higher (to inject active power to the dc grid) and exporting terminals keeping their steady-state voltages lower to export active power from the dc grid.

However, the dc power flow in the parallel branches may differ from each other. Therefore, besides the control of voltage and the current through each converter, the branch powers also need to be measured and controlled. The branch current can be controlled by inserting a series resistance to the overloaded branch or by introducing an active voltage source in series with a branch, which makes it possible to control the current through that branch. The power rating of such a device could be relatively low, as it only has to overcome the resistive voltage drop of a line, which is normally in the order of a small percentage of the rating of the line.

The dc voltage of the multi-terminal HVDC grids must be maintained within the defined design parameters and limits of the dc network converters. The dc voltage in MTDC networks has the same function as the frequency in ac grids. A surplus of energy in ac grids will increase the system frequency and a deficit of energy will decrease the grid frequency. A surplus of energy in dc grids will increase the dc grid voltage and a deficit of energy will decrease the dc grid voltage. Thus, maintaining the dc grid voltage within a defined bandwidth in a dc circuit corresponds to maintaining the frequency in an ac system.

If energy storage systems are introduced to an MTDC grid, they can be utilized to balance a temporary mismatch between power export and import. Additionally, the dc grid can be equipped with "dc braking devices," which are capable of absorbing the excessive power for a limited period.

It should be noted that the operating concept of a converter station always affects the connected ac system. If, for example, a certain converter station is dedicated to control the dc bus voltage to a fixed value within certain power limits, this station will control its power exchange with the ac system accordingly.

The control functions of a dc grid should be differentiated, with respect to their dynamics, into HVDC terminal controls and the dc grid controls. The HVDC terminal controls determine the operating point of its converter such as control of its ac parameters (e.g. ac bus voltage, frequency, active and reactive power). In addition, the HVDC terminal controls the local dc bus voltage and current. The DC grid controller is responsible for controlling the dc grid/transmission voltage and the dc power flow.

10.8 DC Grid Control Strategies

Control strategies for point-to-point HVDC schemes are based on specific project requirements, i.e. the inverter side operates in constant power mode, while the rectifier side controls the dc bus voltage as slack bus. The control system of the VSC-based HVDC is realized by using a fast inner current control loop and several outer control loops, depending on the application and design.

For point-to-point HVDC schemes connecting offshore wind farms to the onshore ac network, the typical control scheme is based on the offshore terminal in power control mode tracking the power output of the wind farm, and the onshore terminal operating in the dc bus voltage control mode. Consequently, the offshore converter station is set to maintain the frequency of the respective wind park network by varying the power through the converter and can't control the dc bus voltage at the same time.

A stable HVDC scheme operation cannot be achieved if both converters of the two-terminal VSC-HVDC are assigned to a constant dc bus voltage control mode. If there are two converters with constant dc bus voltage control, there will be a hunting phenomenon, analogous to frequency hunting in ac grids in the presence of multiple synchronous generators with fixed frequency control. Similarly, in a multi-terminal VSC-HVDC configuration at least one terminal should control the dc grid voltage. If possible, more than one converter should be designed to contribute to dc voltage control in order to divide the balancing power between several converters and ac networks connected to the system [8]. The dc grid control strategy must be robust enough to define a new setting point of operation after any contingency event (e.g. converter or dc link trip following a dc fault). This new operating point shall be inherent to the control strategy and shall not rely on a master control or telecommunication links. The goal is to avoid a collapse of the dc grid or connected ac networks.

The control system of an MTDC network will consist of several different layers designed for specific tasks, depending on the design of the HVDC converters and the dc grid control strategy. A schematic representation of one such MTDC control system hierarchy is shown in Figure 10.8.

The inner control and outer current control loops are designed to provide the very fast converter control actions at the converter's ac and dc buses (e.g. controlling the active power, reactive power, and ac- and dc-side voltage). The converter controllers must function even when there is a loss of the communication link between the HVDC terminals or the master grid controller. A loss of communication is always possible. Therefore, all converters must be able to survive and operate with only local parameters. This is similar to ac power systems, where primary frequency control of generators ensures a reliable system operation without the need for communication.

In an MTDC network the converter controllers should have an interface to a master dc grid controller that is responsible for the dc voltage balance of the interconnected dc network. The master dc grid controller optimizes the overall performance of the MTDC network. This controller is much slower compared to HVDC converter controllers. The voltage optimizer controller will monitor the dc node voltages. Owing to variations of the load flow, the

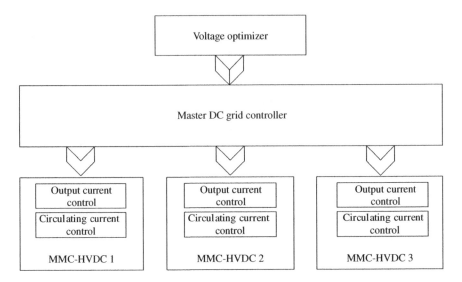

Figure 10.8   Schematic representation of MTDC control system hierarchy.

steady-state converter voltages may not be at their maximum designed dc bus voltage, resulting in higher dc transmission losses. Hence it will be suitable to reduce the losses by optimizing the dc terminal voltages. This controller may optimize the terminal voltages, when the steady-state load flow is achieved for a period of time.

These centralized dc grid controllers provide the individual converter station controllers with their control characteristics and reference values. Their responsibility is to keep the overall dc grid voltage under control and, if required, to optimize the connected HVDC converter voltages in order to reduce any transmission loss.

10.8.1   Dynamic Voltage Control and Power Balancing in MTDC Networks

A multi-terminal HVDC system has to operate with unplanned contingencies, such as power exchange variations, loss of a transmission link, or a terminal trip. Therefore, converter stations injecting active power into the dc grid should limit any possible overvoltage by reducing power import, and terminals exporting power should limit the undervoltage by reducing its power export. The chosen control strategy must be able to keep the interconnected system at a stable operating point (e.g. stay within the predefined voltage limits). Because of the long distances between substations and the risk of communication failure, a local control method seems to be a better solution for such a primary control as is the case with primary control in ac systems.

As already mentioned, the role of the dc bus voltage in an MTDC network is similar to the role of frequency in ac networks. In the event of an energy imbalance that flows in or out the MTDC grid, the voltage at the different nodes of the interconnected system will change. In the event of an energy deficit (e.g. an outage of a rectifier importing power to the grid), the dc grid voltage will decrease. In the opposite case (e.g. an outage of an inverter terminal exporting power to the ac system), the energy surplus will lead to a dc grid voltage increase. In an ac power system the frequency is the same over the synchronized ac system. The dc node voltage in a dc grid will not be the same across all dc buses, as there will be voltage differences due to the resistive voltage drops in the power lines or cables. However, it should be noted that the same dc node voltage across the entire dc network means that there is no current flowing in the dc grid. Therefore, dc node voltage cannot be seen as a true global parameter as it is the case with frequency in ac networks. As a result, it is more challenging to use the dc voltage as a reference for the grid balancing control. Despite these difficulties, in terms of control, the dc node voltage still appears to be the best indicator for the stable operation of MTDC networks. Therefore, the energy balance in the dc grid must be restored immediately in order to keep the voltage from falling or rising.

In point-to-point HVDC schemes the control is typically arranged by one terminal control-ling the dc-link voltage while the other terminal operates in the current or power control mode. This control philosophy of having only one converter controlling the dc grid voltage can also be applied to MTDC networks. This control strategy could be feasible for small radial MTDC networks with parallel tapping. However, with large MTDC networks, it will be difficult to assure the power balance if only one large HVDC terminal is responsible for the control of dc grid voltage.

An outage of the HVDC converter responsible for controlling the dc transmission voltage will have a significant negative impact on both the ac and the dc networks. Therefore, it is advisable to divide the dc grid voltage control task between several converters.

10.8.2   Power and Voltage Droop Control Strategy

The dc grid voltage can be controlled by two different control strategies: the current-voltage-based control or the voltage-power-based control [9]. These control strategies behave similarly in steady-state operation modes; however, the differences arise when there are large deviations in the dc grid voltages related to the reference value.

10.8.2.1   Current–Voltage Control

The current–voltage (I–V) relationship characteristic is used to control the dc-bus voltage in a current-based control strategy. The main advantage of a current-voltage control characteristic is that it reflects the linear control behavior in the sense that a voltage deviation will result in an equivalent current deviation. As the voltage is linked to charging or discharging the capacitors in the dc system, the control is linear and it is the same for all voltage deviations from any reference value [10].

10.8.2.2   Voltage–Power Control

In the power-based control strategy the dc voltage control is expressed in terms of active power. The power–voltage (P–V) relationship is assumed to be linear. The voltage droop control strategy for MTDC networks is similar to the voltage droop control traditionally implemented in ac power systems. In ac grids the load-dependent frequency variation is used as an input signal for the control system to adjust the generated power to meet the demand at all times. In ac power systems frequency variations are used as parameters to control the power injection from the generators. In a dc network, whenever a power imbalance occurs, the dc bus voltage at the different terminals will change. The dc-node voltage variation is related to the network power imbalance and results in a discharge or charge of the converter capacitors and cable capacitance.

The dc droop control strategy regulates the direct voltage within the system by adjusting the converter's current so that a power balance is achieved. If the dc grid voltage increases, it means that the interconnected system has power surplus; as a consequence, the regulating stations should increase the power export at the inverter terminals. When the dc grid voltage decreases, there is a lack of in-feed power and the regulating stations should increase the power import from the rectifiers. Consequently, the droop regulator controls the inner-current control of the HVDC converters, and will regulate the VSC-HVDC terminals' current.

The voltage–power-based droop control is a preferred alternative to droop control strategies, owing to the easy integration with the existing vector control schemes of HVDC converters. However, it is important to understand that the power-based droop control shows a nonlinear (hyperbolic) control behavior contrary to the current-based droop control. With power droop control the actual control of the dc voltage becomes nonlinear [11].

The current–voltage and voltage–power control strategies shown in Figure 10.9 are building blocks for dc voltage control strategies in MTDC networks. These control strategies can be expanded to many sophisticated control strategies such as voltage margin control, voltage droop control, ratio control, dead-band droop control, and priority control.

10.8.3 Voltage Margin Control Method

Voltage margin control is a combination of constant current/power control and constant voltage control. A converter with voltage margin control normally operates in constant current/power control mode while the converter is operating in its normal operation band. If the voltage deviation reaches the limit of the voltage margin band, the converter controller switches to constant voltage control, clamping the voltage at the margin limit to prevent further voltage deviation, as shown in Figure 10.10.

With the voltage margins control method, when the slack bus converter reaches its voltage control limits, and remains at a constant steady-state current/power operation, the converter transfers its voltage control task to another converter in the grid. As a result the converter taking over the control function becomes the slack bus converter and controls the dc grid voltage. Each terminal is given a marginal offset direct voltage reference, called a *margin*. The voltage margin is defined as the direct voltage reference difference between the two VSC-HVDC terminals. By implementing this method, the slack bus task can be passed forward and back between converters based on their availability and load. For smaller dc systems, this control might be an acceptable control system. But this control may influence the stability of connected ac grids. One of the major drawbacks of this control strategy is that the HVDC converter ratings and the system rating must be in the same range.

The voltage margin method (VMM) was first proposed by Tokiwa et al. in 1993 [12]. Some years later Nakajima and Irokawa implemented this control strategy for a three-terminal back-to-back MVDC scheme in the Shin-Shinano substation in Japan [13].

10.8.4 Dead-Band Droop Control

This control system is based on a droop control method with an additional dead-band compared with a direct voltage control strategy. At normal operation all converters except one terminal

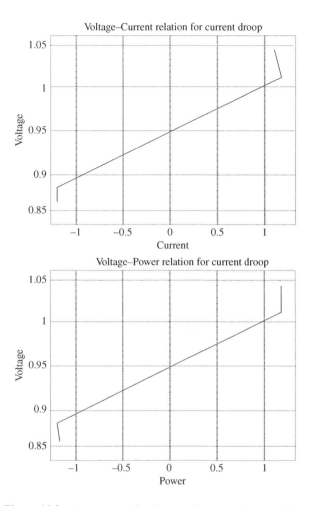

Figure 10.9 Droop control voltage and current characteristics.

are within their dead-band. Therefore, the control behavior is identical to direct voltage control. Droop control does not declare a single converter to be the slack bus, but several converters take part in the voltage regulation. Just like in the case of voltage margin control, the converter operates in constant current/power control mode as long as the voltage is within the normal operation band. If the voltage deviates to the limit of the dead-band, the controller switches to the droop control. If the power rating of the slack bus (or the chosen regulation band) is exceeded, the voltage control task is not passed on to a new slack bus (as in voltage margin control) but to all converters with a droop approach. It is not necessary that all converters take part in controlling the dc bus voltage. Droop-controlled converters and power-controlled converters can easily be integrated into the same dc grid. This is similar in ac grids, where some power stations take part in primary frequency control, while others do not.

This control method has been developed with the intention of combining the advantages of droop control and voltage margin control. At normal operation all converters except one

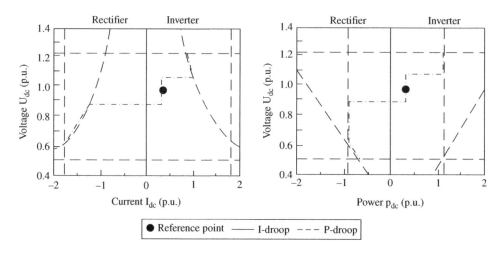

Figure 10.10   Current–voltage curve and power–voltage curve for voltage margin control.

can keep constant power and the system state is well defined, as in direct voltage control and voltage margin control. At larger disturbances, the consequences are divided among all converters, leading to a new equilibrant operating point, as in droop control. Even though this method is an improvement compared to voltage margin control, it still faces some problems in large MTDC systems, because operation outside the dead-band becomes more likely with an increased dc system size. This may impose stability challenges for the ac grid connected to the dc converter(s) acting as a slack bus.

10.8.5   Centralized and Distributed Voltage Control Strategies

In a centralized control strategy one converter controls the dc bus voltage to a constant value of the converter's dc bus voltage, and acts as a dc slack bus, while the other converters control their current/power. For smaller dc systems, this control can be a good choice, as long as the converter ratings match the system rating. The control concept is similar to the normal operation of a point-to-point VSC-HVDC system. This strategy cannot be implemented for large dc networks, since only one converter has to balance the entire dc network for all disturbances within the grid. This control strategy would also have a significant negative stability influence on the ac system connected to a converter acting as the slack bus terminal. The most critical problem, however, is how to handle severe disturbances, such as an imbalance exceeding the capabilities of the slack bus or when the dc slack bus terminal is out of operation [14].

With a decentralized voltage control strategy, it is not necessary that all converters take part in the control of the dc bus voltage similar to ac grids, where some power stations take part in the primary frequency control, while others do not. In a distributed control strategy for large MTDC networks, each dc bus voltage-controlling HVDC terminal is assigned to a specific direct voltage set-point control. By implementing the distributed control strategy, no single converter is left alone with the responsibility of balancing the power inside the dc network, i.e. the control of the dc system voltage is distributed between several nodes inside the MTDC network. Distributed control is a common practice in ac networks, where no single generator is left alone with the task of system balancing.

Within the dc network only the changes in the voltage droops on the lines determine the changes in the operation points of the droop-controlled converters, leading to small distributed control contributions.

10.9   DC Fault Detection and Protection in MTDC Networks

Most of VSC-HVDC projects are designed as point-to-point HVDC schemes. However, from a system perspective (e.g. reliability and availability) it is expected that, eventually, HVDC schemes are further developed and integrated in order to create a meshed MTDC network. The realization of large MTDC networks will depend on overcoming the challenges related to fast and reliable dc fault-detection and protection technologies. There is presently little experience with fault handling in multi-terminal systems. In general, fast fault clearing and power transmission recovery in the event of dc faults is important to maintain the power system's integrity and stability [15, 16]. The availability of dc breaker technologies is not enough to guarantee high reliability and availability of the dc grid under fault conditions. A suitable fault-detection and protection strategy needs to be available that can accurately detect and rapidly isolate the fault.

If dc faults are not detected and isolated quickly, it will propagate through the MTDC network and cause a collapse of the dc grid voltage and loss of dc power transmission. DC networks are almost inertia-less with low impedance; therefore, dc faults propagate rapidly across the network.

The dc fault-handling methodology in MTDC networks is different compared with the techniques employed in ac networks. In the event of a dc fault, interrupting the dc fault current represents a principal technical challenge in contrast to the ac current, as there is no natural zero crossing in dc current.

For ac systems communication-based protection solutions are the dominant technology. As an example, differential protection relays provide full selectivity, but require communication of measured voltage and current between the protection relays. Protection relays that require communication links are generally not suitable for use in dc networks, owing to the fact that fault transient waves propagate faster in dc systems than in ac systems, and are faster than the presently available communication speed (with fiber optics). Differential protection equipment with fiber optic communication between the terminals with a distance of 200 km will have approximately 1 ms propagation delay, while the fault detection and trip command to the breakers must be initiated in a timeframe of 2–3 ms after the dc fault occurrence. The need for extremely fast fault detection and discrimination implies that the comparative fault-detection techniques with telecommunication links aren't suitable for large-scale MTDC networks. The rapid rate of rise of the dc fault currents requires that the fault current be interrupted on the initial rising edge well before the peak value is reached. This implies that the protection should operate in the range of a few milliseconds, whereas protection systems for ac networks operate in the 25–100 ms range. As a result, conventional protection equipment for ac networks is not suitable for HVDC applications.

In the event of a dc fault, transients are dominated by traveling waves in the millisecond range. These waves propagate through the network with the velocity of $v = 1 \sqrt{LC}$, where L and C represent the inductance and capacitance of the transmission line. The fault clearing must happen in the same time range as the fault propagation time throughout the dc network. The propagation delay is an important parameter for the choice of the protection scheme and the fault-detection method. Thus, the dc network protection techniques require the development of new fault-detection solutions that are able to operate many times faster compared to ac protection equipment.

One important issue to be considered is that ac systems supplied by synchronous generators have high overload transient characteristics, while the half-bridge MMC-HVDC stations have a very limited overcurrent capability and the converter control and protection systems will block the converters in the event of dc faults. Blocking the HVDC terminals leads to the temporary loss of the whole dc network until the fault has been isolated or cleared. The maximum allowed loss of infeed power to ac networks is limited by the maximum available primary reserve on the ac system. This is related to the maximum loss of infeed (maximum active power loss criterion) that the ac system can ride through and survive without compromising its stability. Existing European grid codes have defined the maximum infeed loss of power generation units, including HVDC terminals feeding ac networks (3 GW in continental Europe, 1600 MW in the UK, and 1200 MW in Scandinavia).

In the event of a dc short-circuit fault, the fault will propagate through the dc circuit and will appear at the individual converter stations as a sudden dc-bus voltage drop. Without reliable, selective, and extremely fast dc fault-detection and protection technologies, the consequence of a fault occurring anywhere in the dc grid will be a general collapse of the whole dc grid. This will consequently impact the operation and stability of the ac networks connected to the dc system. Foreseen impacts can be, for example, major stability issues including frequency deviation, voltage collapse, and in the worst case blackout of the connected ac networks.

The following minimum requirements are essential for the design of dc fault-detection and protection technologies [15]:

- Selectivity: The protection system should only operate after a fault (not during normal operation), and only if the fault is in its protection zone.
- Speed: The protection system should be fast enough to interrupt faults before they reach unacceptably high values that may cause permanent damage to the dc grid components.
- Reliability: A good protection system is reliable and has a backup system in case the primary protection system fails.
- Robustness: The protection system should have the ability to detect faults in normal mode as in degraded mode, and to discriminate faults from any other operation occurring (e.g. set-point changes, normal switching operations).
- Seamless: After the fault clearance, the remaining part of the system should continue operating in a secure state.

DC fault-detection equipment must be robust with high selectivity and speed in order to detect and localize the fault and consequently initiate the breaker trip signals to isolate the fault. Fault-detection algorithms must be selective and able to differentiate between the transients originated from normal switching events (e.g. energizing a dc cable) and transients originating from fault conditions. This means that the protection system must not trip during an event resulting from a normal switching operation [17].

The required fast action will constrain the response time of the protection equipment, hence the fault-detection algorithms have to be built as much as possible on local measurements with a minimum need for telecommunication links between and across the fault-detection equipment.

Once the fault is detected, the location of the fault must be identified and the fault current must be interrupted, and finally the faulty equipment or line must be isolated.

In a dc system this can be achieved in three distinct ways:

- utilizing dc circuit breakers, interrupting the dc fault current, and isolating the faulty section;
- fast converter control action (e.g. converters with inherent fault-blocking capabilities);
- activating the ac breakers of all HVDC converters and isolating the dc grid from the connecting ac networks.

Different solid state and hybrid dc circuit breakers are under development. These breakers are large, heavy, and very expensive. The dc system circuit breaker should be able to isolate and clear the fault current very rapidly and much faster than an ac system frequency cycle. The dc fault current rises rapidly and consequently the energy which must be dissipated through the breaker increases.

Converter topologies with inherent fault-blocking capabilities are capable of blocking the dc fault current and allow clearing of the dc fault with existing ac breaker equipment under zero voltage/current. However, these converter technologies have higher steady-state losses and significantly increase the cost of converters compared with the half-bridge MMC topology. As MTDC networks will evolve gradually from existing point-to-point HVDC schemes, it is not realistic to assume that all converters will be upgraded to converters with fault-blocking capability.

A simple solution to remove the fault current is to isolate all HVDC converters from the connecting ac buses, but this could have a significant impact on the reliability and availability of dc grids and connected ac systems.

During a pole-to-pole or pole-to-ground dc fault, fault currents will reach destructive levels for the anti-parallel diodes conducting the fault currents. Figure 10.11 shows the development of the dc fault current for +/–320 kV.

The protection design must be robust with backup protection to handle a possible malfunction of the main protection system or a breaker failure.

10.10   Fault-Detection Methods in MTDC

DC fault-detection and protection techniques can be classified into two main categories of direct measurement techniques and signal processing based methods, as shown in Figure 10.12. The majority of these methods have been successfully applied in point-to-point HVDC schemes or ac power systems.

10.10.1   Overcurrent and Voltage Detection Methods

In this fault-detection method the magnitude of the current or voltage is measured and sampled. The sampled data is compared with a threshold value and used as the indicator that a dc fault has occurred. The protection relays issue the trip signals when the measured current

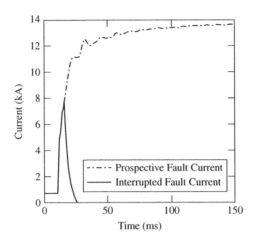

Figure 10.11   DC fault current development under pole-to-pole short-circuit [15].

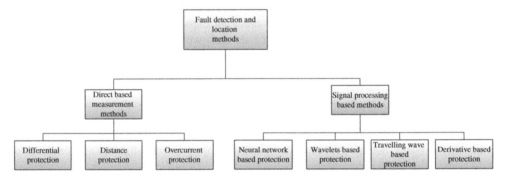

Figure 10.12   Fault-detection technologies.

or voltage exceeds the predefined threshold. However, a margin between the highest load current and the lowest fault current is needed to ensure that the protection system operates only during faults and not in converter overload conditions. The fault current magnitude becomes lower with distant faults from the converter terminals. In addition, the fault current magnitude also depends on the short-circuit capacity of the ac network connected to the HVDC and the impedance of the short-circuit fault. An overcurrent protection scheme is the simplest but least-selective solution. An overcurrent protection scheme can be used as a backup protection scheme for MTDC networks.

DC bus voltage level protection responds to dc voltage drops over a large time interval to detect faulty transmission lines or busbars. Thresholds and delays must be chosen in such a way that voltage transients caused by normal operation and switching do not trigger the protection system to operate. A common setup for the dc undervoltage protection scheme is the cascade configuration. The protection scheme is implemented using multiple levels with their respective threshold and time setting, so that a deeper voltage drop will have a much shorter response time. This cascade configuration provides a more optimized fault-detection scheme. DC undervoltage protection is commonly used as a backup protection system.

10.10.2   Distance Relay Protection

Distance protection is widely applied to ac power systems. This method estimates the fault distance using the impedance between the fault location and the protection device. The impedance is obtained using the local sampling data of the voltage and current. The measured and estimated impedance is compared with the total impedance of the protected zone in the R–X plane. If the estimated impedance is lower than the total impedance then the protection relay issues the trip signal. The fault distance is obtained easily since the line parameters are fairly constant. This method relies on the accurate measurement of the current and voltage. However, the line parameters vary significantly with the frequency. Moreover, dc transmission links contain harmonics sourced from the transient voltages and currents. The impact of these harmonics is normally neglected, hence the influence of the frequency results in the inaccurate fault distance estimations.

For dc grids, distance protection may have some limitations:

- The calculation of the impedance may be distorted by the converter action or other phenomena inherent to dc transmission.
- The influence of the frequency gives inaccurate fault distance estimations.

10.10.3   Differential Line Protection

Differential protection measures and compares the currents flowing in and out of the dc line or dc busbars. The difference between the two measured currents is known as the *differential*. The differential is calculated and compared with a threshold. If the threshold is exceeded, the protection relay initiates the trip command. The differential protection relays require communication links (e.g. fiber optics between the protection relays that cause time delays due to the propagation time of the telecommunication signals).

Differential current protection is generally used as a backup protection. This protection scheme is seen as suitable for the protection of multi-terminal configurations with limited distance between the HVDC terminals.

Differential line protection is a simple method which provides good and reliable protection coverage. However, for dc grids, it may have some limitations, owing to the signal propagation delay of the communication link that influences the response time of differential protection equipment.

10.10.4   Voltage Derivative Detection

With this method, the dc bus voltages and currents are measured and their derivatives are obtained using either analog or digital signal processing. The sign of the derivative of the current gives information about the location of the fault, i.e. a positive sign may indicate that the fault is located in the dc line and a negative sign may indicate that it is located in the dc yard of the converter. A weight sum of the derivatives is calculated and then compared with a threshold, as shown in Figure 10.13. If the threshold is exceeded, the breaker trip signal will be initiated.

Voltage derivative protection is capable of detecting faults within 2–3 ms from the fault occurrence.

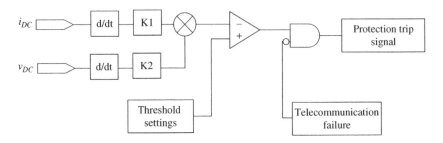

Figure 10.13 Block diagram of a voltage derivative protection scheme.

The disadvantages with the voltage derivative protection are:

- The derivative of the voltage is highly dependent on the fault loop impedance. High-impedance faults are difficult to detect using this method.
- The parameterization of the protection relays requires a detailed network study to ensure the protection only operates for dc faults and is not trigged by normal switching operations in the grid.

10.10.5 Traveling Wave Based Detection

This method is based on the traveling wave theory. This theory is based on the fact that any disturbance on a transmission line produces traveling waves along the transmission line. These traveling waves are the result of a charge and discharge of the line capacitance and line inductance of the transmission line. Each wave is composed of frequencies, from a few kilohertz to several megahertz, which have a propagation speed close to the speed of the light. Traveling wave based protection techniques are well suited for HVDC lines because a dc line has a simple structure and the converters reflect waves instead of refracting waves [18]. The traveling waves propagate along the transmission line generating high-frequency oscillations. These waves attenuate, owing to transmission losses and the corona effect. The frequency and damping of these oscillations are related to the transmission line parameters. The propagation of traveling waves when a fault occurs can be shown by a lattice diagram, as shown in Figure 10.14.

Traveling wave based detection methods require measurements of time-stamped arrival of fault generated surges at the terminals to determine the distance to the fault [19]. In dc transmission lines, the conventional current and voltage transformers can provide simple and cost-effective measurements. Global positing systems can provide an accurate measurement of the wave arrival times to determine the distance to the fault.

This method has the following limitations:

- The key to traveling wave fault location is the detection of the wave front. This method will fail if the wave front is not detected. Traveling wave signals become weak when the line is grounded through a large resistor or by a fault caused by a gradual change in the transition resistance. The detection method is vulnerable regarding weak signals that may not be detected. In addition, this method is very vulnerable to interference signals.

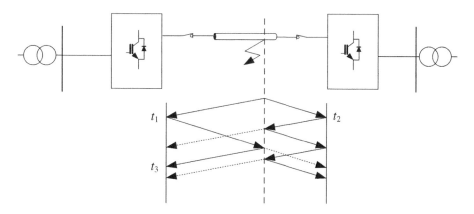

Figure 10.14   Lattice diagram to illustrate the traveling wave phenomena.

- The accuracy of the method depends greatly on the parameters of the line and measurement techniques, since the fault distance is the product of the wave front arrival time and the wave speed.
- The method depends on a high sampling frequency. Since the wave travels almost at the speed of light, a very high sampling frequency is needed. The fault locator requires a high sampling frequency of a minimum of 1 MHz. The higher the sampling frequency of the input signal, the more accurate the result.

10.10.6   Frequency Domain Based Detection

High-frequency components of the voltage and current contain information about a fault. The frequency spectrum is used to detect faults and calculate the fault distance. When a dc fault occurs, the generated wave travels from the fault point to the source end and is reflected back from the source end to the fault point. The frequency spectrum is composed of the natural frequency harmonics. This protection scheme is based on spectrum analysis to extract the natural frequency and its wave speed.

   This method has the following limitations:

- The response time of this method depends on the capacity of the signal-processing hardware.
- This detection method is practically impossible with long dc transmission cables, owing to the attenuation of traveling waves sourced from the large stray cable capacitances.

10.10.7   Wavelet Based Fault Detection

The wavelet transformation (WT) theory is based on signal analysis using varying scales in the time and frequency domain. Fourier transformation provides information in the frequency domain, but it cannot provide any information of the spectrum changes with respect to time. WT allows the representation of a signal in both the frequency and time domain through a time-window function. The window length determines a constant time and a frequency resolution. Wavelet analysis is a method well suited to detect abrupt changes in a signal (e.g. transient phenomena). Continuous WT is the convolution of the signal multiplied by the scaled and shifted version of the original wavelet. One disadvantage of the continuous process is that the representation of the signal is often redundant, i.e. many coefficients are produced. As a consequence, calculation requires powerful processing equipment to reduce the processing time.

In MTDC, when a WT of a current signal in one transmission line is compared with another transform in another line, it is important that both transformations are compared without any shift in time.

Wavelet-based protection schemes may use the voltage wavelet coefficients, the current wavelet coefficient, or both. In this method, based on the wavelet coefficient in each line, the fault is detected when a coefficient overpasses the specified threshold value.

There are some limitations for dc grid applications:

- This method does not take into account the distributed parameters of the transmission line. If the effects of the distributed parameters are ignored, however, the protection devices may not operate correctly.
- Rapidly detecting a fault is difficult, because a considerable number of calculations need to be performed.

Academia and the industry have proposed several fault-detection and protection methods. However, most of the proposals have only been verified by simulation tools or by small-scale dc grid lab demonstrations. At the time of writing, these methods are still under discussion and there are no MTDC fault-detection and protection products available on the market.

10.11   DC Circuit Breaker Technologies

The reliable operation of dc grids will depend on technologies for fast dc fault detection and protection. There are major differences between the requirements applied to ac and dc circuit breakers [20]. One of the main differences arises because of the absence of a natural current or voltage zero crossing in dc systems, as is the case with ac current and voltage.

VSC-HVDC transmission technology constructed with the half-bridge topology is vulnerable to dc pole fault conditions. Under dc fault conditions, the diodes in the circuit constitute an uncontrolled rectifier that feeds the short-circuit current from the ac side into the fault on the dc side even if the converter is blocked.

MMC-HVDC terminals constructed with half-bridge submodule technology cannot control and limit the dc fault current, as is the case with full-bridge converters [21]. Under dc fault conditions, the half-bridge converters are blocked and the ac-side circuit breakers have to be opened in order to clear the dc fault. After the converter ac breakers are tripped and the converters are isolated from the connecting ac networks, the magnetic energy stored in the dc systems will lead to a long tail current, which continues to flow in the circuit for several hundred milliseconds before decaying to zero.

The control and protection of MMC-HVDC schemes equipped with dc breakers requires special attention, as the converter overcurrent protection may block the converter if the fault current through the dc breaker exceeds the overcurrent limit of the converter. With the blocked converter, the dc bus voltage cannot be controlled and the power transmission service will be interrupted. It is important to design the rate of the fault current rise through the dc breaker with the maximum overcurrent limit of the converter, in order to avoid converter blocking due to overcurrent protection of the converter under operation of the dc breaker. One solution to this challenge is to employ fault-current limiter (FCL) reactors to the circuit. The disadvantage with this method is that it will introduce a delay in the fault interrupting time and will increase the dc energy that must be absorbed by the dc breaker. The other disadvantage is that the FCL increases the steady state losses of the scheme.

A dc circuit breaker has to fulfill the following requirements:

- The dc breaker must function and interrupt the fault current in a few milliseconds.
- The breaker must be able to absorb and dissipate large amounts of energy stored in the dc system.

An obvious counter-measure, which can interrupt the flow of uncontrolled currents into the dc grid, is to connect an electronic switch in the form of an IGBT stack in series with the dc pole of the converter, as illustrated in Figure 10.15. The switch may be unipolar if its only purpose is to limit the converter current during dc faults. In principle, the converter may perform a controlled exchange of reactive power with the connected ac system once the switch has been opened. The penalty for introducing the switch is added loss and increased space requirements. Additional losses caused by the electronic switch depending on the design are in the 0.1–0.2% range of the rated power.

An alternative solution to control the dc fault current is to employ converter topologies with an inherent dc fault current blocking capability. Full-bridge converters can oppose the ac terminal voltage, driving a fault current into the short-circuit in the dc circuit, thereby stopping the fault current. With full-bridge converters, offload ac circuit breakers can be employed to isolate the faulty branch under zero current conditions. These switch gears are fast-acting (40–60 ms) devices that allow the rapid reconfiguration of the dc grid and restart of power transmission.

The physics of breaking the direct current in a dc grid is quite different from the physics used in ac networks. AC circuit breakers take advantage of the zero-crossings that occur naturally in the line current. The direct current in dc networks can only be extinguished by inserting a counter-emf (electromotive force) that exceeds the driving emf, which is typically the dc system voltage. The dc breaker may insert this voltage either in series or in parallel, as shown in Figure 10.14. The fault current will decrease when the inserted emf exceeds the driving voltage source, U_{drive}. The ratio between the inserted voltage and the rated voltage will approximately fall in the range of 1.3–1.6 times larger than the rated voltage.

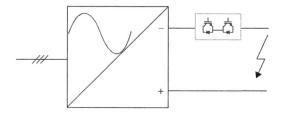

Figure 10.15 Electronic switch in series with an MMC.

As an example, assume the dc voltage is 500 kV and the current that shall be extinguished is 10 kA. In order to extinguish the dc fault current, the breaker has to insert a counter emf that is 1.5 times the nominal voltage, i.e. 750 kV. The instantaneous power then reaches 7500 MW. The energy accumulation in the breaker could accordingly become 7.5 MJ per millisecond, indicating that breaking must be very fast in order to keep the accumulated energy at a reasonable level. A metal oxide varistor (MOV) can be utilized to provide the counter-emf [22].

Academia and the industry have developed and demonstrated several different dc circuit breaker technologies. These breakers can be divided into three main categories:

- resonant dc breakers with serial or parallel resonance circuits;
- solid state dc breakers;
- hybrid dc breakers.

10.11.1 DC Circuit Breaker with MOVs in Series with the DC Line

Most dc breakers provide the current interruption by commutating the fault current into a high-voltage arrester consisting of MOVs. The nonlinear characteristic of MOVs is utilized to create the counter-emf and required transient interruption voltage. In addition, the MOVs are capable of absorbing large amounts of energy and dissipate it as heat [23].

When the MOV is located in series with the dc line, as in Figure 10.16(a), the MOV must be bypassed during normal operation. At the occurrence of a fault, however, it should be inserted within a few milliseconds. A straightforward method is to connect an electronic switch with turn-off capability (e.g. an IGBT) in parallel with the MOV keeping it in its ON-state during normal operation and block it when breaking capability is required. The switch must be designed to withstand the maximum MOV voltage, which is, as already mentioned, typically in the range of 1.3–1.6 times the nominal dc voltage for a symmetric monopole MMC-HVDC. Normally, a bidirectional switch will be required, which means that two IGBT valves connected in anti-series must be employed. The corresponding losses with this breaker is estimated to 0.1–0.3% of the maximum transferred power. Such high losses impose a severe penalty on the dc breaker with series-connected MOV. For this reason further refinement has been suggested which reduces the losses in the bypass switch. According to these solutions, the current during normal operation is conducted through a mechanical switch, in series with a power electronic switch with limited voltage handling capability. At zero-current conditions the mechanical switch can open very quickly and provide full voltage withstand capability in a few milliseconds. The power electronic switch typically consists

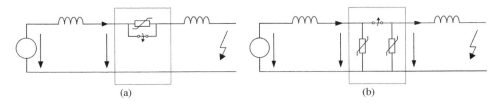

Figure 10.16 Basic principles of HVDC breakers: (a) energy absorbers inserted in series, (b) energy absorbers inserted in shunt.

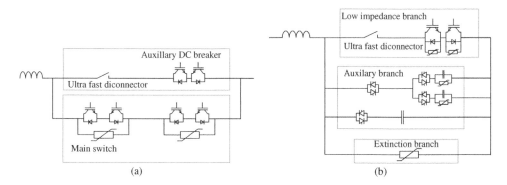

Figure 10.17 Hybrid HVDC breakers inserting MOVs in series: (a) hybrid HVDC breaker, (b) thyristor-commutated HVDC breaker.

of only one or a few IGBTs in series. Therefore its voltage drop is fairly small and the loss in normal operating conditions is acceptable. When a fault occurs, the line current is forced to commutate into a parallel branch with a low voltage drop. This eliminates the current through the mechanical switch, which can then be opened. Once the mechanical switch has established sufficient voltage withstand capability, the current will be forced to commutate again, this time into a further parallel branch containing MOVs. The latter inserts the high counter-emf that brings the current down to zero [22, 24].

Figure 10.17(a) illustrates the hybrid HVDC breaker. The MOVs are equipped with bypassing IGBT valves. Prior to the fault, the line current passes through the ultra-fast disconnector and the auxiliary dc breaker, causing only small losses. The breaking operation will be executed in a two-stage process:

1. At a breaking command the electronic main switch is switched on and the auxiliary dc breaker is blocked, which causes the line current to commutate into the electronic main switch.
2. The fast disconnector can then be opened, and when, after a few milliseconds, it has gained its full voltage withstand capability, the electronic main switch can be blocked, forcing the line current to flow through the MOV.

It is clear that the first step in this process may be executed already when the line current starts to increase, so that breaking can be executed without any delay once the current reaches the trip level. If it turns out that the trip level is never reached, the normal conduction path may be re-established.

Figure 10.17(b) illustrates the thyristor-commutated dc breaker. In normal operation the current flows through the ultra-fast disconnector and the small power electronic switch. A breaking operation is performed in five steps:

1. The power electronic switch in the low-impedance branch is turned off so that the current must flow through the (low-voltage) varistor.
2. The first thyristor switch in the auxiliary branch is triggered and the current in the low-impedance branch commutates into the uncharged capacitor, thereby eliminating the

current through the ultra-fast disconnector. The mechanical switch opens and establishes the required voltage withstand capability.

3. The current is commutated through a number of similar branches with varistors with increasing voltage. When a new branch is turned on, the preceding one turns off as the current initially commutates into the uncharged newly inserted capacitor.
4. The last capacitor in the auxiliary branch conducts the current until the voltage reaches the level when the metal-oxide extinguishing branch starts to conduct.
5. The MOV inserts a counter-voltage that makes the current drop to zero. The thyristor turns off when the voltage across the varistor drops.

10.11.2 DC Breakers with MOVs in Parallel with the DC Line

The counter-emf can alternatively be inserted in shunt which is schematically shown in Figure 10.18. In normal operation the current is carried by vacuum switches with mechanical contacts with negligible losses. The contacts are bridged by diodes. Furthermore, a controlled current pulse generator is provided in shunt, which consists of a diode in series with a capacitor charged to the full dc-side system voltage. A thyristor-controlled inductor is connected in parallel with the capacitor. When the thyristor is triggered, the capacitor voltage reverses through the established resonant circuit. The circuit is designed to perform the capacitor voltage reversal in less than a millisecond with sufficiently high current amplitude (tens of kilo-amperes).

The breaking operation in this case is executed in two stages. At the break command the opening mechanisms for the vacuum-switches are released. When the contacts in the vacuum switches start to separate, the current in one of the switches will commutate to its parallel diode, while the current in the other one will continue to flow through a plasma in the vacuum switch.

1. After a short delay (0.5 ms) the thyristor in the current pulse generator is triggered and a current pulse occurs through the pulse generator branch and the two external MOVs, which limit the negative voltage between the dc lines. The current in the vacuum switch that is still flowing (through the plasma) then reverses and the current commutates to its parallel diode at the created zero-crossing. Both vacuum switches are consequently switched off.

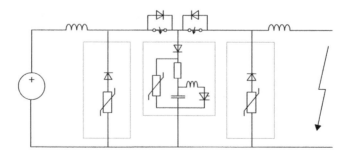

Figure 10.18 Breaker based on vacuum switches and thyristor-triggered resonant circuit.

2. The voltage in the capacitor in the pulse-generator branch starts to increase until the current going outward from the dc breaker is commutated to its external shunt MOV, and the current directed toward the dc breaker is commutated into the MOV in the pulse-generator branch. The MOVs will then quench the fault currents.

The vacuum switches can be realized by a series-connection of vacuum switches rated for some tens of kilovolts (30–50 kV). To achieve a proper functionality of this breaking method it is required that the vacuum switches can be operated in a fast and synchronous manner.

10.12   Fault-Current Limiters

The main function of a fault-current limiter (FCL) is to limit the fault current and its rate of rise to a more manageable level during a fault, without any significant negative influence on the electrical system. An FCL is a passive or active device which increases the resistive and/or impedance properties between normal conducting mode and current limiting mode to limit the prospective peak and/or RMS fault current to or below a desired value. The change in the resistive and/or impedance is due to the change in electrical conductivity or the magnetic permeability of the device or a combination of both, either self-triggered or externally triggered. An ideal FCL has the following properties:

- It is invisible to the grid during normal condition, with zero voltage drop and no losses.
- It is activated instantly when a fault current is detected and imposes a significant impedance during the fault.
- It steps out of operation instantly after the fault is cleared.
- It recovers immediately with no need for maintenance after each event, and is able of withstanding any number of repeated operations.

FCLs can be categorized as fault current limiting reactors, solid-state FCLs, or superconducting FCLs. Each type will be described separately in the following three subsections.

10.12.1   Fault Current Limiting Reactors

Fault current limiting reactors are passive elements that are permanently placed in the circuit and do not require an external signal to be activated. The fault current is reduced by the impedance of the reactor and limits the fault current due to the voltage drop across the terminals. However, it should be kept in mind that current limiting reactors also exhibit a voltage drop during normal operation and, therefore, present a continuous contribution to the overall system loss.

Some MMC-HVDC schemes are equipped with fault current limiting reactors in series between the terminal poles [25]. These reactors are normally built as dry-type air-core reactors as described in Section 2.4.

Since the impedance must remain constant during fault conditions only air-cored designs are suitable for this application. Design and sizing of the current limiting reactors is a trade-off between steady-state losses and the desired current limiting factors.

10.12.2   Solid-State Fault-Current Limiters

Solid-state fault-current limiters (SS-FCLs) utilize semiconductors as electronic switches to change the configuration of the circuit when a fault appears. SS-FCLs can be classified in three different types:

- impedance-insertion solid-state FCLs;
- resonant solid-state FCLs;
- bridge solid-state FCLs.

Each of these three types are treated briefly below.

10.12.2.1   Impedance-Insertion Solid-State Fault-Current Limiters

Impedance-insertion SS-FCLs make use of anti-parallel power semiconductor devices as switching elements. The semiconductors are normally connected in series with the line they protect, and are connected in parallel with a current limiting passive circuit that usually consists of an MOV in parallel with a passive snubber. In Figure 10.19(a), a schematic diagram of such a circuit is shown.

10.12.2.2   Resonant Solid-State Fault-Current Limiters

Resonant SS-FCLs are composed of an LC resonant circuit which is tuned to the grid frequency and can present different configurations by means of connecting the elements in series

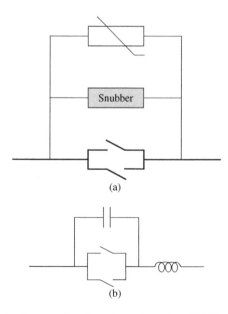

(a)

(b)

Figure 10.19   (a) Schematic diagram of an impedance-insertion SS-FCL, (b) schematic diagram of a resonant SS-FCL.

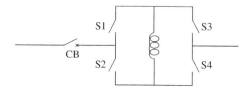

Figure 10.20   Schematic diagram of a bridge SS-FCL.

or in parallel. These limiters are usually simple but require a detecting circuit to operate. In Figure 10.19(b) a schematic diagram of a resonant SS-FCL is shown.

10.12.2.3   Bridge Solid-State Fault-Current Limiters

The bridge SS-FCL is composed of a semiconductor full-bridge and an inductor, as shown in Figure 10.20. Under normal conditions, the semiconductor switches are activated, and the line voltage magnetizes the DC limiting reactor. When a short-circuit occurs, the fault current reaches the steady state flowing through the reactor current immediately and is restricted by the reactor. Then, S1 and S2 are switched off, and for this reason the current in the reactor can flow out through S3 and S4. Finally, the circuit breaker (CB) connected in series with the FCL can be opened up and break the fault current.

10.12.3   *Superconducting Fault-Current Limiters*

Superconducting FCLs are made of low- or high-temperature superconductor materials. The superconductors show no electrical resistance when the current is below the critical current level. If the current exceeds the critical level of the superconductor, the material changes their property and undergoes a transition from the superconducting state to a resistive state [26]. This transition is known as *quenching* in superconducting materials. When the current through the FCL is reduced, the generated heat ($I^2.R$ loss) during the resistive stage must be dissipated with adequate cooling and the temperature of the superconductor material must be restored within its critical low temperature range.

Superconducting FCLs during the normal operation as long as the temperature, current, and magnetic field are lower than critical values, the superconductor impedance is almost zero. The cooling of the superconducting FCLs is usually provided by cryogenic systems.

10.13   The Influence of Grounding Strategy on Fault Currents

VSC-HVDC scheme configurations adapt different grounding strategies. HVDC schemes and converters can be designed with solid grounding, resistive grounding, or high- or low-impedance grounding. The scheme configuration and the grounding strategy influence the dc fault dynamics and behavior [15]. At the same time the grounding strategy impacts the HVDC scheme isolation coordination requirements and its protection strategy. Finally, the grounding strategy influences overall project and equipment costs.

A high impedance grounding results in significant voltage stresses during ac and dc fault conditions. In the event of low impedance grounding, the system will experience significant current stresses. During earth faults in a high-impedance-grounded HVDC scheme, the fault currents are effectively limited. However, the converters will be subjected to large overvoltages, up to twice the normal operating voltage.

For a symmetrical bi-pole configuration with an isolated grounding, very high short-circuit currents will flow through the system during a solid pole-to-pole fault. In the event of a single pole-to-ground fault, the system will experience lower short-circuit currents determined by the grounding impedance. Normally, symmetrical MMC-HVDC configurations are designed with high-resistive-grounding configurations. The single pole-to-ground faults will result in a dc voltage imbalance between the poles, and cause the un-faulted pole to be charged by up to 2 p.u. of the normal operating voltage. In the event of asymmetric configurations, pole-to-pole and pole-to-ground faults will cause high short-circuit currents to flow through the circuit.

Asymmetric configurations correspond to low impedance grounding. In normal operation, the metallic return conductor of an asymmetric monopole is operated at near-zero voltage and the positive pole voltage is equal to the nominal voltage of the converter. In order to limit the voltage rise on the metallic return, the asymmetric monopole needs to have low impedance grounding or have solid grounding. In the event of a pole-to-ground fault on the positive pole in a solidly grounded system, the steady-state post-fault voltage should not exceed the nominal voltage. The voltage on the metallic return remains near zero. In this case, the fault current shows a steep di/dt and will have a high steady-state value. Increasing the grounding resistance can influence the steady-state fault current and decrease it. However, this approach will increase the steady-state voltage on the metallic return and influence the steady-state loses. For the asymmetric configurations, fast-acting fault clearance is needed because of high fault currents.

10.14   DC Supergrids of the Future

Supergrids can be seen as the ultimate development stages of dc grids. The supergrid is seen as the solution that will allow the massive integration of renewable-energy sources into the European power system. It connects different remote renewable-energy sources to the existing grid while offering additional control. The supergrid offers balancing through geographic spread and allows a more diversified energy portfolio and increases the security of supply. The extreme transmission capacity of supergrids could reduce congestion and price differences in the system. The security of supply can increase. However, there are discussions in academia and the industry regarding the technical and economic feasibility of such grids. There are several challenges which should be met prior to the realization of such a complex dc grid over vast geographical and geopolitical areas. However, despite this, intense activity is taking place in the research and development of components and technologies that are required for the development of supergrids.

It is envisaged that the development of supergrids will occur in stages. As discussed in previous sections of this chapter, technical issues with dc-side voltage control in MTDC, dc load flow, and dc fault detection and protection must be addressed before complex MTDC grids such as supergrids can be constructed. Most importantly, the interaction between the existing ac system and the new supergrid needs to be investigated, focussing on reliability and security of supply. From an operational point of view, the main issues center on who will operate the grid and how this will be done in the current multi-zone environment that includes such a large number of stakeholders. So far there are no standards and dc grid codes to govern the development and operation of supergrids.

10.15 Summary

This chapter presents application of MMC technology for HVDC transmission. Different HVDC scheme configurations are presented and compared. Development strategies and various grid control strategies of the MTDC networks are presented and compared. DC fault-detection and protection technologies and methods are presented and compared. Finally, dc breaker technologies with passive components and solid-state hybrid breaker technologies are presented.

References

[1] CIGRE Technical Brochure 269, Working Group B4.37, "VSC transmission."
[2] M. Saeedifard and R. Iravani, "Dynamic performance of a modular multilevel back-to-back HVDC system," *IEEE Transactions on Power Delivery*, vol. 25, no. 4.
[3] CIGRE Technical Brochure 492, "Voltage source converter (VSC) HVDC for power transmission: Economic aspects and comparison with other AC and DC technologies."
[4] CIGRE Technical Brochure 364, "Systems with multiple DC infeed."
[5] C. Li, X. Hu, J. Guo, and J. Liang, "The DC grid reliability and cost evaluation with Zhoushan five-terminal HVDC case study," *Power Engineering Conference 50th International Universities*, pp. 1–6, 2015.
[6] CIGRE Technical Brochure 553, Working Group B4.47, "HVDC grid feasibility study."
[7] Friends of the Supergrid, "Roadmap to the supergrid technologies," http://mainstream-downloads.opendebate.co .uk/downloads/WG2_Roadmap_to_the_Supergrid_Technologies_2013_Final_v2.pdf, accessed Mar. 30, 2016, 2013.
[8] CENELEC TC8X/Sec0097/DC, "Technical guidelines for first HVDC grids."
[9] J. Descloux, P. Rault, S. Nguefeu, et al., "*HVDC meshed grid: Control and protection of a multi-terminal HVDC system,*" CIGRE symposium, Paris 2012.
[10] K. Rouzbehi, A. Miranian, A. Luna, and P. Rodriguez, "DC voltage control and power sharing in multiterminal dc grids based on optimal dc power flow and voltage-droop strategy," *IEEE Journal of Emerging and Selected Topics in Power Electronics*, vol. 2, no. 4.
[11] T. K. Vranaa, J. Beertenb, R. Belmansb, and O. Bjarte Fosso, "A classification of DC node voltage control methods for HVDC grids," *Electric Power Systems Research*, vol. 103, Oct. 2013.
[12] Y. Tokiwa, F. Ichikawa, K. Suzuki, et al., "Novel control strategies for HVDC system with self-contained converter," *Electrical Engineering in Japan*, vol. 113, no. 5, pp. 1–13, 1993.
[13] T. Nakajima and S. Irokawa, "A control system for HVDC transmission by voltage sourced converters," *Power Engineering Society Summer Meeting*, vol. 2, pp. 1113–1119, 1999.
[14] J. Beerten, D. Van Hertem, and R. Belmans, "VSC MTDC systems with a distributed dc voltage control: A power flow approach," *PowerTech, IEEE Trondheim*, 2011.
[15] D. Van Hertem, O. Gomis-Bellmunt, and J. Liang, "*HVDC Grids: For offshore and supergrid of the future*, IEEE Press Series on Power Engineering, Apr. 2016.
[16] M. Ghandhari and M. Delimar, "Technical limitations towards a SuperGrid: A European prospective," *Energy Conference and Exhibition, IEEE International*, pp. 302–309. 2010.
[17] B. Geebeln, W. Leterme, and D. Van Hertem, "Analysis of DC breaker requirements for different HVDC grid protection schemes," *AC and DC Power Transmission, 11th IET International Conference*, Edgbaston, Birmingham, Feb. 10–12, 2015.
[18] K. D. Kerf, K. Srivastava, M. Reza, et al., "Wavelet-based protection strategy for DC faults in multi-terminal VSC HVDC systems," *IET Generation, Transmission & Distribution*, vol. 5, no. 4, pp. 496–503, 2011.
[19] A. Burek, J. Rezmer, and T. Sikorski, "Evolutionary algorithm for detection and localization of faults in HVDC systems," *15th International Conference on Environment and Electrical Engineering (EEEIC)*, pp. 1317–1322.

[20] C. M. Franck, "HVDC circuit breakers: A review identifying future research," *IEEE Transactions on Power Delivery*, vol. 26, no. 2, 2011.

[21] Y. Wang and R. Marquardt, "Future HVDC-grids employing modular multilevel converters and hybrid dc-breakers," *15th European Conference on Power Electronics and Applications*, 2013.

[22] S. Norrga, L. Ängquist, and K. Sharifabadi, "Power electronics for HVDC grids: An overview," *CIGRE colloquium*, Lund, Sweden 2015.

[23] K. Tahata, S. Ka, S. Tokoyoda, et al., "HVDC Circuit Breakers for HVDC Grid Applications," *AC and DC Power Transmission, 11th IET International Conference*, Feb. 10–12 2015, Edgbaston, Birmingham.

[24] T. Eriksson, M. Backman, and S. Halén, "*A low loss mechanical HVDC breaker for HVDC grid applications,*" CIGRE Symposium, Paris 2014.

[25] E. Kontos, S. Rodrigues, R. T. Pinto, and P. Bauer, "Optimization of limiting reactors design for DC fault protection of multi-terminal HVDC networks," pp. 5347–5354, 2014.

[26] J. G. Lee, U. Khan, H. Y. Lee, and B. W. Lee, "Impact of SFCL on the four types of HVDC circuit breakers by simulation", *IEEE Transactions on Applied Superconductivity*, vol. PP, no. 99, 2016.

Index

Design, Control, and Application of Modular Multilevel Converters for HVDC Transmission Systems, First Edition.
Kamran Sharifabadi, Lennart Harnefors, Hans-Peter Nee, Staffan Norrga, and Remus Teodorescu.
© 2016 John Wiley & Sons, Ltd. Published 2016 by John Wiley & Sons, Ltd.
Companion Website URL: www.wiley.com/go/Sharifabadi/ModularConverters

Printed and bound by CPI Group (UK) Ltd, Croydon, CR0 4YY

05/02/2025

14638882-0001